26

Oil in Asia

Oil in Asia
Markets, Trading, Refining and Deregulation

PAUL HORSNELL

Published by the Oxford University Press
for the Oxford Institute for Energy Studies
1997

Oxford University Press, Walton Street, Oxford OX2 6DP

Oxford New York
Athens Auckland Bangkok Bogota Bombay
Buenos Aires Calcutta Cape Town Dar es Salaam
Delhi Florence Hong Kong Istanbul Karachi
Kuala Lumpur Madras Madrid Melbourne
Mexico City Nairobi Paris Singapore
Taipei Tokyo Toronto
and associated companies in Berlin Ibadan

Oxford is a trade mark of Oxford University Press

British Library Cataloguing in Publication Data
A catalogue record for this book is available from the British Library

ISBN 0-19-730018-9

Cover design by Holbrook Design Ltd, Oxford
Typeset by Philip Armstrong, Sheffield
Printed by Bookcraft, Somerset

PREFACE

This book is the result of a two-year project at the Oxford Institute for Energy Studies on the oil markets of the Asia-Pacific region. The project was made possible by the sponsorship of a number of companies and organizations, who we wish to thank. The sponsors were the following.

Arab Petroleum Investments Corporation
Bahrain National Oil Company
Exxon
Kuwait Petroleum Corporation
Mobil
Morgan Stanley
Petróleos Mexicanos (Pemex)
Petron
Qatar General Petroleum Company
Saudi Arabian Oil Company (Aramco)
Statoil.

In addition, we wish to thank a number of the benefactor companies of the Oxford Institute for Energy Studies who provided points of contact and commented on the study. These companies were the following.

BP
Cosmo Oil
Elf
Enterprise Oil
Mitsui
Mitsubishi
Neste
Repsol
Shell
Texaco
Tonen.

A panel was constituted of nominated representatives of the above sponsors and from the benefactors of the Oxford Institute for Energy Studies. This panel read through drafts of the study, and provided comments and the opportunities for interviews. Their expertise represented a major asset to draw on, and we are very grateful for their generosity with their time, and for their patience with what was a long study. The panel of representatives was comprised of the following;

Dominique Badel (Exxon), Robert Bever (Bahrain National Oil Company), Chris Brooks (BP), Iain Everingham (Shell), D.J. Forbes (Arab Petroleum Investments Corporation), Atsushi Fukunouye (Petrodiamond), Liz Gall (Enterprise), Amalio Graiño (Repsol), Steven Hewlett (Statoil), Dennis Houston (Exxon), David Jackson (Saudi Arabian Oil Company), Tor Kartevold (Statoil), Talal Khalil (Arab Petroleum Investments Corporation), Marcel Kramer (Statoil), Atsushi Kume (Mitsui), Isao Kusakabe (Cosmo Oil), John Lloyd-Davies (Texaco), Robert Maguire (Morgan Stanley), Alfredo Marquez (Pemex), Keizo Morikawa (Cosmo Oil), Yasuhide Nagamatsu (Tonen), Gary Nunnally (Conoco), Streuan Robinson (BP), Mark Sato (Mitsui), Hassan A Al Saud (Ministry of Energy and Industry, State of Qatar), Sharaf Majid A. Salamah (Petron), Wee Tai Seck (Neste), Nader Sultan (Kuwait Petroleum Corporation) and Paul Ying (Elf).

The involvement of the panel and the companies and organizations they represent in no way is an endorsement of the content of this study, and no endorsement of any fact or view contained in the study should be inferred.

Two long research trips were made as a part of this project, in July and August 1994 to Singapore, and in March and April 1995 to Thailand, Malaysia, Singapore, Hong Kong, the Philippines, Korea, Japan and the USA. Work towards the project was also carried out on a series of shorter visits to the region in the course of 1995 and 1996. Beyond the panel, many others helped by being interviewed, be it formally or informally, reading drafts, or providing help with data or research material. We would like to thank them all, and the above disclaimers equally apply to them. Their company affiliations are those that applied at the time they helped the project. When this has happened in different incarnations, both are noted.

Adrian Binks (Petroleum Argus), Beh Soo Hee (BP), Nick Black (Petroleum Argus), Daniel Carr (NYMEX), Richard Child (Petroleum Argus), Georgie Chong (SIMEX), Stephen Chow (Mobil), Jim Curtis (Exxon), Christopher Donville (Bloomberg), Tilak Doshi (Louis Dreyfus and then ARCO), Tony Dukes (Intercapital Commodity Swaps), Bob Ellison (Oil Trade Associates and FEOP), Han Chwee Juan (Neste), Onno Handels (Oiltanking), Dale Hardcastle (Petroleum Intelligence Weekly and then Bloomberg), Hironobu Ishikawa (Mitsui), Martin Jones (Statoil), T. Ken Kitahara (Neste), Karen Klitzman (NYMEX), Koo

Woon Kian (Texaco), Lim Choo Beng (SIMEX), Lim Khoo Eng (Singapore Trade Development Board), Andy March (BP), Paul Mason (BP), Mohamed Merican (Van Ommeren), Tadasu Mitsui (Tonen), A. Mizuta (Mitsubishi Corporation), Tetsuji Murai (Petrodiamond), Ng Weng Hoong (Singapore Oil Report), Uichu Okina (Tonen), Francisco Oñate IV (Petron), Roger Osborne (Seapac Services and APPI), Ian Perks (British Gas), Wilma Rodriguez (Petron), Kazuhiro Takahashi (Cosmo Oil), C.S. Tay (Texaco), Piers Tonge (Petroleum Intelligence Weekly), Xavier Trabia (Elf), Al Troner (Petroleum Intelligence Weekly and Asia Pacific Energy Consulting), Yoshitaka Tsuchiseto (Mitsubishi Corporation), Barbara Vogelsang (Platt's), Michael Walker (Intercapital Commodity Swaps), Karen Yip (Singapore Trade Development Board), John Young (Gotco) and Yashushi Yoshikai (Mitsui).

Without the help of all the above, this project would have been impossible. In several cases the degree of help and hospitality went far beyond the call of any duty. The author is particularly indebted to Alfredo and Maria Marquez for their hospitality when he was in Tokyo, and to Al and Barbara Troner in Singapore and also Kuala Lumpur. The project would have been far harder and considerably less pleasant without them. Special thanks also to Julia Horsnell for her patience and support, and to Kate Horsnell for providing distractions.

CONTENTS

x

Appendices

TABLES

FIGURES

ABBREVIATIONS

ADNOC	Abu Dhabi National Oil Company
ANS	Alaskan North Slope
ANWR	Arctic National Wildlife Refuge
AOC	Arabian Oil Company
AOT	Approved Oil Trader
APICORP	Arab Petroleum Investments Corporation
APPI	Asian Petroleum Price Index
Aramco	Saudi Arabian Oil Company
ASEAN	Association of South East Asian Nations
ASTM	American Society for Testing and Materials
b/d	Barrels per day
bb	Billion barrels
BHP	Broken Hill Propriety
BJP	Bharatiya Janata Party (India)
BNOC	British National Oil Corporation
c+f	Cargo and freight
CFD	Contract for Differences
CME	Chicago Mercantile Exchange
CNOOC	China National Offshore Oil Corporation
CNPC	China National Petroleum Corporation
CPC	Ceylon Petroleum Corporation (Sri Lanka)
CPC	Chinese Petroleum Corporation (Taiwan)
CRINE	Cost Reduction Initiative for the New Era (UK)
cst	Centistokes
DPK	Dual Purpose Kerosene
EFP	Exchange of Futures for Physical
ENI	Ente Nazionale Idrocarburi (Italy)
EOR	Enhanced Oil Recovery
ERB	Energy Regulatory Board (Philippines)
FAO	Food and Agriculture Organization
FCC	Fluid Catalytic Cracker
FEOP	Far East Oil Price
fob	Free on board
FSU	Former Soviet Union
GATT	General Agreement on Tariffs and Trade
GDP	Gross Domestic Product
GNP	Gross National Product
GPW	Gross Product Worth

GTCs	General Terms and Conditions
HKFE	Hong Kong Futures Exchange
HP	High Pour
HSD	High Speed Diesel
HSFO	High Sulphur Fuel Oil
ICP	Indonesian Crude Price
IEA	International Energy Agency
IMF	International Monetary Fund
IOC	Indian Oil Corporation
IPE	International Petroleum Exchange of London
IPG	International Petroleum Group
JDA	Joint Development Area
kgoe	Kilograms of oil equivalent
KPC	Kuwait Petroleum Corporation
LP	Low Pour
LNG	Liquefied Natural Gas
LPG	Liquid Petroleum Gas
LSWR	Low Sulphur Waxy Residue
mb/d	Million barrels per day
MG	Metalgesellschaft
MGRM	Metalgesellschaft Refining and Marketing
MITI	Ministry of International Trade and Industry (Japan)
MMBtu	Million British Thermal Units
MOPPAS	Mean of Platt's and Argus Singapore
MOPS	Mean of Platt's Singapore
MPM	Ministry of Petroleum and Minerals (Oman)
MTBE	Methyl Tertiary Butyl Ether
mtoe	Million tonnes of oil equivalent
NAFTA	North American Free Trade Agreement
NGLs	Natural Gas Liquids
NYMEX	New York Mercantile Exchange
NIOC	National Iranian Oil Company
OAPEC	Organization of Arab Petroleum Exporting Countries
OECD	Organization for Economic Cooperation and Development
ONGC	Oil and Natural Gas Commission (India)
OPEC	Organization of the Petroleum Exporting Countries
OPSF	Oil Price Stabilisation Fund (Philippines)
OSP	Official Selling Price
OTC	Over the counter
PCS	Petrochemical Corporation of Singapore
PdVSA	Petróleos de Venezuela
Pemex	Petróleos Mexicanos

Pertamina	Perusahaan Pertambangan Minyak Dan Gas Bumi Negara (Indonesia)
Petromin	General Petroleum and Mineral Organization (Saudi Arabia)
Petronas	Petroliam Nasional Berhad (Malaysia)
PIL	Petroleum Industry Law (Japan)
PML	Provisional Measures Law for Selected Petroleum Product Imports (Japan)
PNOC	Philippines National Oil Corporation
PPP	Purchasing Power Parity
PRC	Pacific Resources Corporation
PRT	Petroleum Revenue Tax (UK)
PSA	Port of Singapore Authority
PTT	Petroleum Authority of Thailand
PUB	Public Utilities Board (Singapore)
RFCC	Residue Fluid Catalytic Cracker
RON	Research Octane Number
S$	Singapore Dollar
Samarec	Saudi Arabian Marketing and Refining Company
SEZ	Special Enterprise Zone (China)
SFE	Sydney Futures Exchange
SIMEX	Singapore Monetary Exchange
Sinochem	China National Chemicals Import and Export Corporation
Sinopec	China Petrochemical Corporation
SKO	Superior Kerosene Oil
SPC	Singapore Petroleum Company
SPEX	Shanghai Petroleum Exchange
SR	Straight Run
SRC	Singapore Refining Company
TAPS	Trans-Alaska Pipeline System
TOCOM	Tokyo Commodity Exchange
TWO	Transworld Oil
UAE	United Arab Emirates
ULCC	Ultra Large Crude Carrier
VLCC	Very Large Crude Carrier
VGO	Vacuum Gasoil
WTI	West Texas Intermediate
YPF	Yacimentos Petroliferos Fiscales (Argentina)

PART I

Oil in Asia

CHAPTER 1

INTRODUCTION

The twenty-first century has been dubbed, even before its advent, as the Pacific century. While perhaps still premature, the epithet is representative of a general perception that the centre of the world economy is increasingly moving towards the Asia-Pacific region, and the distribution of economic power is steadily moving away from Europe and the USA. This study is concerned with one aspect of this transition, the impact on the energy industries and in particular oil.

Most of the attention on Asian energy markets has focused on the impact of high rates of energy demand increases derived from consistent economic growth. Economic growth is a gradual and, in the recent Asian context, seemingly inexorable process. The impact of increased energy demand is one which brings about long-term structural change. The motivation for this study is, not in any way to dismiss the importance of this growth, but to regard it as a constant background factor, while the short-term dynamics (which may affect and be affected by energy demand patterns) carry force. In particular, our premise is that structural change in the industrial organization of the oil sector, often created by changes in energy market regulation and through liberalization, is fundamental to an understanding of the short- and medium-term operation of those markets in Asia.

An obvious example where changes in government policy have swamped the underlying impact of economic growth is China. In the early 1990s, China was seen as perhaps the major reason for bullishness in the Asian market. Surging economic growth and oil demand seemed to usher in an era of unbroken opportunity for the oil industry, where China was set to be the main positive driving force. Yet, at the time of writing in 1997, for three years China has been a depressing factor in the Asian oil market. The imposition of greater regulation and import controls has meant that China's positive impact has fallen well below prior expectations. Concentration on the beacon of fast growth distracted the market from considering the far greater immediate impact of changes in the regulatory regime. For reasons that are discussed further in a later chapter, it was not the economic growth rate that should have attracted all immediate attention, but the inflation rate, the factor which proved crucial to the degree of regulation. While economic growth is still there as a consistent positive factor for oil markets, the short-run dynamics of regulatory impact have proved far stronger.

Other examples abound. While growth provides the impetus for capital to flow into the Asian oil industry, particularly into refining, it is government policy timetables and the induced changes in the structure of domestic industries that will determine precisely where and when that capital will enter. Likewise, the impact of deregulation (and in some cases extra regulation), ranks higher in the current concerns of many Asian energy ministries and companies than the longer-term impact of economic growth.

This book seeks to fill two main gaps. We would contend that Asian energy markets are under-researched. Within this research deficit, the bulk of attention has been paid to the impact of growth on energy balances. The normal conclusions, i.e. that the region faces an increasing resource deficit and will become the world's largest energy market, tell us little of the transition to that state or of the evolution of markets. This is the first gap, and in response to it we wish to give industrial structure and government policy a more central role. A general feature of liberalization throughout Asia is the homage paid in statements of policy to the desirability of exposing the functioning of domestic oil markets to the discipline of international markets and prices. Given this emphasis, there is perhaps surprisingly little documented on the nature, structure and operation of those markets in Asia that deal in oil products and crude oil. This is the second gap that this study seeks to fill.

The development of oil trading markets in Asia has until recently been stalled, primarily due to the lack of involvement of most Asian national oil companies and because of the nature of regulatory structures. With little exposure to the market through floating prices, there had been little need for many state or private oil operations to manage risk. Changes in relative prices could simply be pushed on to consumers, or, more commonly, simply absorbed by government or through the dissipation of some other rent within the system. Most countries have operated under highly uncompetitive systems with distorted prices and constraints on private capital, producing oil industries that have been inefficient, and companies that have faced little pressure to optimize operations. In short, there has been considerable slack in the industry. Changes in policy have brought changes in the operating environment, and taken away some of the insulation from risk and competitive pressures that government had previously conferred.

It is reasonable to start with a definition of what comprises Asia in the context of this book. We have considered Asia to be made up of those countries east of and including Pakistan, running around the Pacific Rim to northern China and Japan. Our definition excludes two main areas. First, we have not considered Afghanistan and the ex-USSR republics of Central Asia.[1] As yet, these countries, all minor energy consumers, are also not strongly linked in any role of energy producer to the rest of Asia. We have considered them to constitute a separate system with separate energy sector policy issues. Secondly, we have also not included the Russian Far East economy centred on Vladivostok. This area does have the (as yet unrealized) potential to be a major supplier of energy into Asia,[2] and as a consumer its trade flows, while currently minor, also link into this system. Its exclusion largely arises in practical terms from the data problems inherent in stripping this area away from Russia.

Any work on Asia tends to be awash with terms indicating a finer geographical distinction. We have tried to minimize their use, given that there is no broad consistency between authors on their exact definition. Cross-referencing between sources becomes an exasperating endeavour given the widespread employment of radically different geographical groupings all under the same label. To minimize the exasperation of others, it is as well to be absolutely explicit at the start about our own groupings. For finer geographical subdivisions of Asia we have adopted the following taxonomy. Southeast Asia comprises the seven member countries of the Association of South East Asia Nations, namely Brunei, Indonesia, Malaysia, the Philippines, Singapore, Thailand and Vietnam, with the addition of Burma, Cambodia and Laos. East Asia consists of China, Japan, Mongolia, Taiwan and North and South Korea, while South Asia comprises the Indian sub-continent. Where used, the term Far East is substitutable with Asia as a whole.[3]

To provide a context for the magnitudes involved in the regional market, Table 1.1 gives an overview of the global oil economy. This shows a snapshot of oil consumption, production and the net balance as at 1995, with Asian economies disaggregated and listed in descending order of oil consumption.[4] Two fundamental characteristics of the Asian market emerge from the statics of the situation as shown.[5] First, the level of oil demand at 16.97 million barrels per day (mb/d) is large on a global basis, representing 25 per cent of world oil consumption. Within the region, more than 70 per cent of consumption is accounted for by Japan, China, South Korea and India. On a global basis these are large consumers. In 1995 there were six economies that consumed more than 2 mb/d, three of which are Asian.[6] Japan ranked second in the world in terms of oil consumption, with China third and South Korea sixth.

The second key characteristic is that these are overwhelming oil import dependant economies, within a region whose overall oil deficit is already greater than that of either Europe or North America. Only four Asian countries are net oil exporters, namely Indonesia, Malaysia, Vietnam and Brunei. Among the net importers, only China and India have crude oil production that is significant relative to their consumption level, with the rest of the countries being almost exclusively reliant on imports. The

Table 1.1: Oil Balances 1995. Million Barrels Per Day.

	Oil Demand	Crude Oil Production	Oil Balance
Japan	5.78	0.02	-5.76
China	3.31	2.99	-0.32
South Korea	2.01	0.00	-2.01
India	1.51	0.79	-0.72
Indonesia	0.81	1.58	+0.77
Taiwan	0.73	<0.01	-0.73
Thailand	0.69	0.05	-0.64
Singapore	0.51	0.00	-0.51
Malaysia	0.43	0.74	+0.31
Philippines	0.35	<0.01	-0.35
Pakistan	0.31	0.06	-0.25
Hong Kong	0.21	0.00	-0.21
Vietnam	0.08	0.15	+0.07
Bangladesh	0.05	<0.01	-0.05
Brunei	0.01	0.18	+0.17
Others	0.15	<0.01	-0.15
ASIA	16.94	6.56	-10.41
Australia	0.78	0.58	-0.20
New Zealand	0.12	0.03	-0.09
Papua New Guinea	<0.01	0.10	+0.10
Australasia	0.90	0.71	-0.19
North America	20.25	13.75	-6.50
Europe	15.33	6.62	-8.71
Former Soviet Union	4.31	7.23	+2.92
South America	4.14	5.65	+1.51
Middle East	3.88	20.06	+16.18
Africa	2.17	6.97	+4.80
WORLD	67.93	67.54	

Sources: *BP Statistical Review of World Energy 1996* and *International Petroleum Encyclopaedia 1996* and various industry sources.

overall net import bill for the region amounted to over $60 billion in 1995, with Japan alone importing crude oil and oil products to the value of about $35 billion. The size of the import bill and the degree of dependency on other regions serve as a reminder that in the Asian context oil can not be divorced from its macroeconomic and strategic implications.

A large part of this book is concerned with a detailing of the structure and operation of the Singapore oil market. This plays the pivotal role of the price setting market for oil products in Asia, and sources of data and other research about it merit separate mention. Given its key role in

Asian energy markets, the Singapore oil industry has attracted surprisingly little research. Further, occasionally we found that rather basic data were difficult or impossible to access. Ng Weng Hoong summed up both these aspects very neatly in the *Singapore Oil Report*.

> As Asia's leading oil centre, Singapore's lack of institutional research into energy issues stands out as probably its most notable failing. While the industry has built a world-class refining and trading centre here over the last 35 years, how this was accomplished is little understood as publicly available research, data and information are lacking and difficult to access.[7]

Difficulty of access to data is of course the perennial complaint of a researcher. However, in this case many data problems, for instance in the area of domestic Singapore energy demand data, are such that they do impinge on the functioning of the industry and on its planning processes. Flows of oil industry information in Asia as a whole are relatively poor. Much is contained in commissioned reports that move in small circles. However, the public domain research vacuum is not total. In particular, there is the excellent ground-breaking work by Tilak Doshi[8] and also Shankar Sharma.[9]

Despite these bright spots, overall there has been little continuity in research. It is probably not unfair to say that this has been reflected in planning failures within the Asian oil industry, or at least in a lack of diversity in views both within and between the key participants. Events have all too often been taken as surprises, suggesting a failure to ask the 'what if' questions as counterfactuals to implicit assumptions. All too often, it has been the more seemingly absurd 'what if' questions that have come to fruition, and these have caused the higher costs resulting from planning failures. The distinction between 'tell me what will happen' and 'tell me how to interpret existing structures and processes so as to understand what can happen' is a major one. The burden of most Asian energy market research seems to have fallen on the former, designed for quick consumption and frictionless planning, rather than on the latter, harder to consume but ultimately with far more value added. Put simply, it is ultimately more useful to be sometimes wrong for the right reasons, than to be right occasionally through what amounts to pure divining.

Point projections of the future are best left to those with a ready supply of the appropriate body parts of amphibians, a cauldron, and the left-over costumes from a performance of Shakespeare's Scottish play. As most of the fauna of the Oxford area is protected, we have abstained from making such point projections. However, in parts of the study we have provided ranges for some key variables. In all such cases the projections should be taken as being illustrative of the logic of the underlying dynamic processes, not as an indication of the author's assumption of clairvoyance. While point projections are extremely common in the oil industry (and often come at a considerable cost to the user), we believe that they are useless without a sound process based logic.

This is particularly true of linear interpolation of growth rates, which is, sadly, far too prevalent in oil research. Thus, the forward looking analyses where they occur in this study should be taken as being examples of the potential impact of some of the underlying processes and as a way of looking at the world, rather than any attempt to be an oracle. If those processes occur with different timing or different relative force, we hope we have provided the reader with a logic that can adjust the implications, at least in a qualitative direction.

In relation to the two forms of research question posed above, the emphasis we have tried to put is not on knowing the future but on providing an explanation of the dynamics that create that future. To take but one example, there is a conventional wisdom that sees oil production in Asia heading through slow growth and stagnation into permanent decline. While noting that is the implication of the current statics of the situation, we have tried to explain the processes whereby the inevitability of the conclusion is overturned. The failure of the conventional wisdom in the 1980s that North Sea oil production would fall consistently throughout the 1990s should be taken as a salutary warning about the danger of extending the present forward on the supply side of a market, rather than thinking in terms of underlying factors and dynamics.

The approach taken in this study is in common with the ethos of the work, particularly that which has been concerned with the operation of markets, undertaken at the Oxford Institute for Energy Studies since its foundation in 1982. That approach can in one aspect be expressed thus. A full understanding of markets rests upon the analysis of their operation and their historical and institutional development. Effective forecasting and the testing of theory are only possible with an appreciation of how the structure of markets operates. Our motive is thus not to influence policy or planning, but rather to provide a description of processes and dynamics in such a way that policy and planning are facilitated.

Taking this approach has several consequences. In the context of this study it means that the material covers a very wide range of issues. Where we believe that historical developments are important in understanding structures, for instance in Japanese regulation or the development of the Singapore oil industry, we have outlined those developments even when this takes us back to the nineteenth century. Likewise, a view that energy regulation in China can not easily be understood except in the context of the economic reform process, and indeed that reform has caused increases in energy regulation, necessitates a discussion of general reform and the economic boom-bust cycle.

Despite the predictable researcher's complaint on official data made above, data availability was not in any way a constraint on this study. A variety of data sources have been utilized, and here we note some of the major elements. As in previous studies, the author is immensely grateful to Adrian Binks, the publisher of Petroleum Argus, for the use of two large databases. These are central to the work contained in Parts III and IV, and indeed the study could not have been undertaken without them. The Petroleum Argus Crude Oil Deals and Products databases, built up through Petroleum Argus' operations, provide an exceptionally detailed view of the markets, unavailable in any other form. Much of the value added in this study has come from their use, and we also wish to acknowledge the many Petroleum Argus staff who have expended considerable time in the compilation and maintenance of these databases.

In the absence of a large corpus of published research, with the two main exceptions noted above, we have often relied on oil trade journals. Both *Petroleum Intelligence Weekly* and *Weekly Petroleum Argus* (together with their respective sister publications) have provided excellent coverage of the Singapore market and Asian markets in general. I have also benefited greatly from discussions with their reporters. Among the trade journals the coverage of *Energy Compass* and *Platt's Oilgram* has also been closely followed. The one Singaporean journal whose coverage intersects with this study's is the *Singapore Oil Report*, whose themed issues have provided considerable interest.

The general Singaporean press, in particular the *Straits Times* and the *Business Times* have also proved at points to be invaluable sources. In particular, on several occasions we have found that data which are unavailable as a corpus can be built up from their constituent parts through newspaper back issues. When such data are presented, the advantage of this method is that the compilation of parts produces a whole that is truly public domain, even if that whole is not generally available.

The important aspect of this is that at no point have we used data which are privileged or sensitive. In cases where the compilation of public domain material produces something that might be considered sensitive, we apologise to those affected but of course can make no apology for using public sources in that compilation. We have also found *Asian Energy News*, which provides a compilation of press reports from across Asia and beyond on Asian energy issues, an extremely valuable, and we suspect generally underutilized, research tool.

This book has five parts. The first consists of this introduction, and an overview of economic growth and energy demand in the region to serve as a background to the rest of the study. Part II consists of four chapters concerned with the changes in the structure of the oil industry in a series of countries. The two main oil consuming nations are covered in separate chapters, with the Chinese market detailed in Chapter 3, and the Japanese market in Chapter 4. Five further economies, namely India, Korea, the Philippines, Thailand and Taiwan are covered more briefly in Chapter 5.

Liberalization of oil markets leads on to two further strands which are addressed in Parts III and IV. Oil-importing countries of course face the full macroeconomic impact of changes in world market prices. However, at the microeconomic level of the operations of the companies involved in the domestic oil industry, government price regulation has often tended to reduce or remove the sensitivity to international price changes. The removal of that cushion then means a more widespread exposure for many regional oil companies to market prices, and an associated impetus towards entering regional markets and also towards at least the consideration of risk management. This raises the issues of how those market prices are created, the structure of the markets that generate them, and the development of risk management instruments within the region. The response to all of these questions comes from an analysis of the key price setting Singapore oil products market, which is the focus of Part III. The four chapters in this part consider in turn the growth of the Singapore oil centre and its current status, the physical trading markets, the informal paper markets, and oil futures trading.

Some of the technicalities of Part III have generated the need for the study to have associated appendices. The first of these provides an explanation of price assessment in the informal Singapore oil markets and thus the origin of the price information used at various points. To reduce the need to break the main text with explanations that many readers may find superfluous, a second appendix provides a brief and simplified description of the refinery technologies encountered in the text. In addition, a third appendix provides a brief listing of the refinery structure in the countries not covered in Part II.

Part IV focuses on crude oil markets and trade flows. In the first of three chapters, crude oil production indigenous to the region is considered together with its associated markets. In the second, the Middle East oil market is examined in relation to Asia, with an analysis of the formulae now used in OPEC pricing and of the key, albeit fragile, Dubai market. The final chapter in this part attempts to take a more global view of the evolution of the world crude oil system and its implications for Asia. In particular, the implications of the dominant share of non-OPEC producing countries in meeting recent increases in world oil demand are examined. Part V contains a single chapter on refinery margin dynamics, and Part VI provides some conclusions from the study.

Notes
1. i.e. Kazakhstan, Kyrgyzstan, Tajikistan, Turkmenistan and Uzbekistan.
2. For a detailed analysis of the energy supply potential of the Russian Far East, see Keun-Wook Paik (1995), *Gas and Oil in Northeast Asia, Policies, Projects and Prospects*, The Royal Institute of International Affairs, London.
3. Dictionary definitions of Far East, reveal that in common usage it may refer to the whole of Asia, Asia minus the Indian sub-continent, or East Asia alone.
4. In Table 1.1, global oil demand does not equal global crude oil production primarily due to inventory level changes, refining volume losses and processing gains.
5. The development over time of oil demand is covered in Chapter 2, and that of oil production in Chapter 10.
6. The others were the USA (16.94 mb/d), Russia (2.94 mb/d) and Germany (2.88 mb/d).
7. Ng Weng Hoong, editorial, *Singapore Oil Report*, February 1996.
8. Tilak Doshi (1989), *Houston of Asia: The Singapore Petroleum Industry*, Institute of Southeast Asian Studies, Singapore.
9. Shankar Sharma (1989), *Role of the Petroleum Industry in Singapore's Economy*, Institute of Southeast Asian Studies, Singapore.

CHAPTER 2

THE ASIAN ECONOMIC BOOM AND OIL DEMAND

1. Introduction

Economic growth provides the backdrop to, and often motive force behind, a process of structural change in the Asian oil industry and of market development. This chapter seeks to provide an overview of that growth, and its consequent impact on oil demand. The *dramatis personae* of development success in Asia has been ever expanding. After the takeoff of the Japanese economy in the 1950s and 1960s came the success of the 'Four Little Dragons' or 'Four Tigers', namely Singapore, Taiwan, Hong Kong and Korea. In the last two decades, the focus was first expanded to the newly industrializing economies of Malaysia and Thailand. More recently, and in terms of absolute GDP increments more significantly, the giant economies of China and India have shown their economic potential. This chapter includes this introduction followed by three further sections. In Section 2 we provide a brief overview of economic growth in Asia, noting that most of the continent is still at a low income level, with the process of growth in a relatively early stage. The third section considers the evolution of primary energy demand, and briefly considers the markets for energy other than oil. Section 4 concentrates on oil demand, and also considers the state of conventional wisdom on the future path of Asian oil demand.

2. The Asian Economic Boom

Two main schools of thought exist on the factors behind Asian growth. A broadly neoclassical economic view sees it as the result of getting markets right, with direct government interventions normally not only unhelpful, but also dangerous, as they leave open the possibility of policy failure and rigidities. The second school sees a more positive function for state intervention and provision of incentives, beyond the conscious policy of improving market performance, and sees a role for the longer-term historical context of mercantilism within a country.[1] The World Bank has termed its approach, which is broadly consistent with the latter school, as the 'Market Friendly View'.[2] This sees growth as the result of a set of central policy prescriptions; macroeconomic stability, trade and capital market openness, an emphasis on education and training, and a policy environment that encourages private sector development and capital growth. Within this a role is left for selective intervention, often in response to perceived market failures, as long as no ratchet effects operate, i.e. if unsuccessful interventions can be rapidly withdrawn.

The maintenance of sustained growth in the high performance Asian economies has certainly relied on a general promotion of competition, private sector development and a relatively open trade orientation. However, we would note that in these economies, with the significant exception of Singapore, the oil industry has historically operated outside of the maxims of the market friendly view. The application of these precepts to the industry has often been during a late stage in overall economic development. The roots of tight regulation of the oil industry have often been in ideology, based on economic, strategic, foreign policy or nationalistic reasons. In terms of macroeconomic management the oil industry represents one of the commanding heights of the economy, making it an obvious candidate for government control, if not full nationalization, under an interventionist economic philosophy. If that philosophy swings to one of closer market orientation, the oil industry also tends to be among the last in the queue for liberalization.

Oil has a strategic function, which has sometimes combined with the imperative to demand some control for military and defensive reasons to provide a motivation for government intervention. The macroeconomic implications of oil prices create a motivation in oil import dependent economies to protect against risks to the general economy, just as it does in energy export dependent economies. Resource nationalism may create reasons for control in exporting countries, just as nationalism can often motivate the desire for a national force to either compete with or to supplant foreign capital.

A summary of the growth performance of Asian economies is given in Table 2.1, where the countries are ranked in terms of GNP per capita. A first stark feature is the extent to which the performance of Japan, the four dragons, and Malaysia and Thailand has taken them far ahead of the other countries. Despite widespread growth, most Asians are still in low income economies, with considerable disparities in economic performance across the continent. As is shown in Table 2.1, only seven economies had a GNP per capita in 1994 of above $1000. These seven contain less than 7 per cent of the population of Asia, with the remaining 2.9 billion living in low income economies, 2.1 billion in China and India alone. In 1994, Japan's total GNP was twice that of the rest of the Asian countries combined. The implication of the above is that for Asia as a whole the development process is still in an early stage, with most of the potential for economic growth still unrealized.

Table 2.1: Summary Economic Statistics for Asian Countries. Real GDP Growth Rates, Total GNP, GNP Per Capita and Population.

| | Population Millions | Growth Rates of Real GDP Per cent per annum | | | | | Total GNP $ billion | GNP per capita |
	1995	81–90	91	92	93	94	95	1994	1994$
Japan	125.1	4.3	4.3	1.4	0.1	0.6	2.0	4638.9	37,120
Singapore	3.0	6.3	7.0	6.2	10.4	10.2	8.9	65.8	23,360
Hong Kong	6.2	6.9	5.1	6.3	6.1	5.4	4.6	126.3	21,650
Taiwan	21.2	7.8	7.6	6.8	6.3	6.5	6.1	251.1	11,930
South Korea	44.9	10.7	9.1	5.1	5.8	8.6	9.0	366.5	8220
Malaysia	20.1	5.2	8.6	7.8	8.3	9.2	9.6	68.7	3520
Thailand	59.4	7.9	8.5	8.1	8.3	8.7	8.7	129.9	2210
Philippines	70.3	1.0	-0.6	0.3	2.1	4.4	4.8	63.3	960
Indonesia	195.3	5.5	8.9	7.2	7.3	7.5	8.1	167.6	880
Sri Lanka	18.0	4.2	4.8	4.4	6.9	5.6	5.6	11.6	640
China	1204.9	10.4	9.3	14.2	13.5	11.8	10.2	630.2	530
Bhutan	1.7	7.5	3.5	4.1	6.3	6.5	5.5	0.3	400
Pakistan	129.8	6.2	5.6	7.7	2.3	4.5	4.4	55.6	440
Mongolia	2.3	5.8	-9.2	-9.5	-3.0	2.3	6.3	0.8	340
Laos	4.7	-	3.4	7.0	5.2	8.0	7.0	1.5	320
India	916.5	5.7	0.8	5.1	5.0	6.3	3.3	278.7	310
Cambodia	10.2	-	7.6	7.0	4.1	4.0	7.0	2.4	240
Bangladesh	116.9	4.1	3.4	4.2	4.5	4.2	4.1	26.6	230
Burma	44.7	-0.1	-0.6	9.7	5.9	6.8	7.7	9.0	200
Nepal	20.6	4.9	6.4	4.6	3.1	7.1	2.1	4.2	200
Vietnam	74.0	7.1	6.0	8.6	8.1	8.8	9.5	13.8	190

Sources: Asian Development Bank, *Key Indicators of Developing Asian and Pacific Countries*; International Monetary Fund, *International Financial Statistics* and various.

Economic growth is a long-term phenomenon, which, as shown in Table 2.1, has been maintained at high rates in the majority of Asian countries throughout the 1980s and into the 1990s. Its key feature is that there is no internal dynamic that leads on to a conclusion that Asian growth is in any way slowing down. Even among the four dragons, there is still a large margin until full catch up with OECD economies is achieved, and, as Singapore's performance in the 1990s has shown, these economies are still capable of periods of extremely fast growth. Among the lower income economies, starting later from a lower base means that many years of very strong growth need to be sustained to even reach a position comparable to the current level of development of the dragons.

A second important feature of Asian growth is that it has not been, at least recently, dependent on Japan. As shown in Table 2.1, the slowdown in the Japanese economy after 1991 did not cause the booms elsewhere to be reined in. Of course, we do not have the counterfactual of how much faster other economies could have grown if there had been no Japanese slowdown, but the data do seem to imply that the old idea that Japan served as the major engine of growth for the Asian economies as a whole is no longer relevant. While not totally decoupled from the Japanese economy, the rest of Asia does not seem to require Japan to play the role of the principal regional economic driver. Growth in internal markets, the development of greater intra-Asian trade, together with further trade penetration into OECD economies (even when the latter are in recession), have all created the dynamic for booms that have proved resilient to downturns in the industrialized world. Again, the experience is enough to reveal that it would be premature to suggest that any slowdown is visible or predictable, save in circumstances of very major deteriorations in foreign relations or the internal structural cohesion of political and economic systems. Armageddon scenarios may make interesting reading, but to predicate a base case view of Asian economic growth on them would appear to be less than wise.

The GNP per capita figures shown in Table 2.1 are converted into a US dollar basis using actual exchange rates rather than purchasing power parity (PPP) exchange rates. The difference between the two measures can be profound, for example the low level of prices in China produces estimates for a PPP figure for GNP per capita between two and three times that shown. On the basis of Table 2.1, the total size of the Hong Kong economy is 20 per cent of the size of the Chinese economy, a rather large economic addition to incorporate into China in July 1997. Using PPP, the relative size of the Hong Kong economy, while still very significant, becomes considerably smaller. While Table 2.1 uses actual exchange rates, we would note that for measures of many issues other than international economic power or purchasing power on world markets, correctly evaluated PPP adjustments can often be appropriate. The problem is of course correctly evaluating a consistent set of adjustments.[3]

Among the economic successes there are also the relative failures, which provide their own pointers as to the key to developmental success. Some, such as Cambodia and Laos, have rather unique and catastrophic historical circumstances. Mongolia has had the characteristics of a transition economy with the associated painful restructuring, and indeed its relatively fast turnaround in the 1990s is something of an economic triumph. The Philippines managed only 1 per cent per annum growth over the 1980s, and its improvement in the 1990s has been slow in comparison with other economies. The Philippines has been seen as a case of getting politics and markets wrong, with distorted capital and labour markets, and failures of the political system.[4] State intervention had been heavy, and for a long period the system had been hostile to foreign, and in particular non-US, capital. The distortion resulted in the need for the drastic reform and liberalization we describe in Chapter 5 in the context of the oil sector, and the desire to get capital markets to work in a more frictionless and efficient manner.

A similar failure is evident in Burma where, after a decade of negative growth, progress began to be made from 1992. This growth should be seen in the context that Burmese GNP per capita had fallen to one-tenth of that in neighbouring Thailand, and where the economy had been so distorted as to resist the strong pull factor of its neighbour for so long. Burma has one of the very poorest economies in Asia, with an economic record that has been inferior even to Laos. Developmental backwardness has followed from a combination of political failure and distorted capital markets, leaving an economy with an enormous but unfulfilled potential. A further potentially important Asian economy is not shown in Table 2.1, namely the Russian Far East. This area plays only a small role in this study, however we would note that its economy is now fully a part of the Asian system, both in terms of general macroeconomics, and also in the role of a potentially significant energy consumer and energy supplier.[5]

We now turn to the implications of economic growth for energy demand. This mapping is reinforced by economic growth induced structural changes that have important resource implications. We would identify four main such shifts. First, there is the move towards urbanization, manifest in the rapid expansion of the Asian 'megacities', with their implications for power and transport infrastructures and demand. Secondly, there is the changing balance between the economic importance of internal and external markets, with the growth in the former leading to further implications for the transport sector. Thirdly, there is the change in private economic aspirations backed by increasing purchasing power, leading to the realization of pent-up demands for private transport and the use of commercial rather than traditional non-commercial energy sources in the domestic sector. Finally, particularly for countries at an earlier stage of development, such as India and China, there is the sectoral balance shift away from traditional agriculture and towards energy intensive manufacturing and heavy industry. As we note in the following sections, during some stages of development these structural shifts are liable to cause dislocations and make the relationship between GDP growth and oil demand highly volatile.

3.   Primary Energy Demand

The pattern of primary commercial energy demand across fuels in Asia (with the major consuming countries also shown separately), and primary energy consumption per capita is shown in Table 2.2, with the USA and the UK included for comparison.[6] A major feature is the generally low level of energy use in Asia compared to developed countries. In 1995, the USA with about one-twelfth of the population of Asia, consumed almost exactly the same amount of primary energy. Likewise, while India has a population about sixteen times greater than that of the UK, it consumed only marginally more energy.[7]

Table 2.2:   Primary Energy Demand in Asia, USA and UK, and Share by Energy Source. 1995. Million Tonnes, Kilograms of Oil Equivalent and Per Cent.

| | Total Energy Use mtoe | Energy per capita kgoe | Oil | Shares of Total Primary Energy | | | |
				Natural Gas	Coal	Nuclear	Hydro-electric
Bangladesh	9.2	76	25.0	71.7	2.2	0.0	1.1
China	833.1	682	18.9	1.9	76.9	0.4	1.9
India	227.3	244	31.9	7.5	56.4	0.9	3.3
Indonesia	69.9	361	55.2	37.6	6.0	0.0	1.1
Japan	490.2	3915	54.5	11.2	17.5	15.2	1.6
Malaysia	35.1	1746	57.3	37.3	4.0	0.0	1.4
Pakistan	31.6	243	48.4	38.3	7.0	0.3	6.0
Philippines	18.8	275	89.4	0.0	8.0	0.0	2.7
Singapore	16.1	5384	91.9	8.1	0.0	0.0	0.0
South Korea	149	3326	63.6	6.2	18.3	11.6	0.3
Taiwan	66.1	3103	53.6	5.9	25.7	13.8	1.2
Thailand	49.2	817	67.7	16.9	14.2	0.0	1.2
ASIA	2081.3	667	38.6	8.2	46.0	5.1	2.0
USA	2069.4	7868	39.0	27.0	23.9	8.8	1.2
UK	218.7	3738	37.4	30.1	21.9	10.5	0.2

Source: Own calculations from *BP Statistical Review of World Energy* and various.

Growth in primary energy demand in most countries is then coming from a low base in per capita terms, most markedly in the countries of the Indian sub-continent. Within this generally low level of energy demand, both South Korea and Taiwan have a usage that is now approaching that of OECD countries, with Singapore even higher than the UK. Considering the shares by energy source in total commercial primary energy, the fuel mix used in the region shows a similar percentage reliance on oil to the USA and UK. Natural gas is relatively less utilized, and the reliance on coal is considerably greater. The aggregate figures are however heavily influenced by the size of coal burn in one country. China is the largest coal user in the world accounting in 1995 for about 29 per cent of global consumption. The disaggregation of the fuel mix by country reveals a split between the predominantly coal based energy economies of China and India, the bias to gas in Bangladesh (within an extremely low per capita use of energy), and the rest of Asia which is primarily reliant on oil.

We would note that the rapid pace of energy use expansion in China and India, and their reliance on indigenous coal as a base load source, is enough to render discussions of stabilization of global carbon dioxide emissions as largely academic. Emissions stabilization in those countries is not on the political agenda, and there can be no pretence that the environmental effects of Chinese and Indian growth will be benign.

Beyond oil and coal, other fuels primarily impinge at the margin, in the market for power generation. Hydroelectric power is relatively unimportant and untapped, although it is still significant in terms of electricity supply in China, India, the Philippines and Sri Lanka. However, while large projects are in construction or planning, particularly in China, the pace and scale of development suggests that hydroelectric provision is increasing only slightly faster than overall energy demand. The implication is then that it will not move significantly higher than the level of below 2 per cent of all primary energy in Asia shown in Table 2.2. Nuclear energy is important in Japan, Korea and Taiwan, and in addition further significant capacity is planned or being constructed in a series of other countries, most notably China, India and Indonesia. However, like hydroelectric power, the pace of development does not imply an increase in provision faster than overall energy growth, even before considerations of the problems of public attitudes to nuclear power, particularly in the more developed economies, are introduced.

Beyond the forms of primary commercial energy shown in Table 2.2, in many Asian countries, non-commercial energy sources are important, particularly biomass. Assessing non-commercial energy use carries the obvious data problems, however Table 2.3 provides an overview of the share of biomass in total energy consumption in selected countries. Most notably in the countries of the Indian subcontinent, biomass has a large share of total energy use, and is particularly dominant in the residential sector. The high share of biomass in the residential sector represents a source of considerable future demand growth for oil products, and in particular LPG. As we detail below, LPG demand in Asia has indeed grown at extremely fast rates over a sustained period. However, domestic market penetration is in many cases limited by infrastructure, and even in relatively high income economies biomass use continues to grow in rural areas. As an example, Table 2.4 shows the pattern of energy use in Thailand since 1982.

Biomass energy use in Thailand is certainly growing far less rapidly than conventional energy and its share has fallen sharply, however its

Table 2.3:   Biomass as a Share of Total and Residential Energy Use. Selected
Countries. Various Years. Per Cent.

	Year	Share of Biomass in Final Energy Consumption	Share of Biomass in Residential Sector
Bangladesh	1992	73	89
Bhutan	1991	82	n.a.
Burma	1991	74	n.a.
China	1992	10	25
India	1992	33	78
Indonesia	1992	39	73
Laos	1991	88	n.a.
Malaysia	1992	7	15
Nepal	1992/3	92	97
Pakistan	1992/3	47	83
Philippines	1992	44	66
Sri Lanka	1990	77	93
Thailand	1994	26	65
Vietnam	1991	50	n.a.

Source: FAO Regional Wood Energy Development Programme, reported in *Asian Energy
News*, May 1996.

Table 2.4:   Composition of Energy Use in Thailand. 1982–94. Thousand Tonnes of Oil
Equivalent.

	1982	1986	1990	1994
Total Final Energy Consumption	16,221	23,749	30,893	43,849
of which Conventional Energy	9371	16,023	21,684	32,372
Biomass	6850	7726	9209	11,477
of which Wood Energy	4903	5282	6835	8360
of which Fuelwood	2676	2958	3426	3902
Charcoal	2227	2324	3409	4458
Share of Biomass	42.2	32.5	29.8	26.2

Source: FAO Regional Wood Energy Development Programme, reported in *Asian Energy
News*, May 1996.

absolute level has still been rising appreciably. The use of charcoal has
doubled between 1982 and 1994. Relative to biomass, oil products are
luxury goods and are also less environmentally destructive. For the absolute
level of biomass use to fall, requires incomes to rise beyond a critical level
in rural areas.

An important factor in the mapping from economic growth onto primary energy demand and then onto oil demand, is the extent to which natural gas continues to increase its share in energy use. The Asian gas market splits into that for LNG (Liquefied Natural Gas) and pipeline gas. LNG is currently imported into Japan, Taiwan and Korea from a series of LNG production trains in Indonesia, Malaysia, Abu Dhabi, Australia, Brunei and Alaska, with a further two trains in Qatar coming on stream.[8] On the demand side there is potential for LNG imports into Thailand, India, the Philippines and China, and on the supply side there is a large raft of potential new trains.[9] We would note that estimates of the delivered cost into Asia for new projects are extremely closely clustered, particularly in the range of $3.50 to $4 per MMBtu, with the implication that the decision as to whether very few or very many new LNG projects are viable is very price sensitive. A full analysis of the potential for LNG expansion is beyond the scope of this study, and has already been detailed by the IEA. However, we would add a series of observations that follow from our general theme. Just as Asian oil markets are becoming more competitive and also price sensitive, so are power and gas markets in the face of liberalization. This sensitivity also carries over to the returns from investment. LNG projects are extremely long term, and consist of lumpy, indivisible technology with high upfront capital costs in both trains and reception terminals, where the largest proportion of variable cost lies in transportation. Further, the pace of technological advance and cost reduction in LNG has been exceedingly slow. Providing LNG at acceptable cost in these conditions, yet still leaving a netback that provides an acceptable return to the producer, becomes difficult in energy markets that remain predominantly soft, particularly while the price clauses in most LNG contracts are in some way linked to oil prices. Most proposed LNG projects are currently extremely close to the margin on those terms. Increasingly, market liberalization also adds regulatory risk, especially rate of return risk, as well as the risk of structural change in the domestic market into the equation, particularly relevant given the long-time span of a LNG project. In short, we would consider that any major boom in LNG, before, say, 2005, may not be justified on strictly risk and option value adjusted economic terms under current conditions.

There has also been significant growth in the use of pipelined gas, primarily in the member countries of ASEAN,[10] among whom Indonesia, Brunei, Malaysia, Vietnam and Thailand have significant gas reserves. Natural gas consumption is shown in Table 2.5. Japan's use of LNG makes it the largest gas consumer in the region, but note the rapid development of domestic gas use in Thailand, Malaysia, and Indonesia. Overall, gas use in Asia has grown strongly, up from 82 mtoe in 1985, to 122 mtoe in 1990 and 173 mtoe in 1995. This equates to a growth rate

Table 2.5: Natural Gas Consumption in Asia. 1985–95. Million Tonnes of Oil Equivalent.

	1985	1990	1991	1992	1993	1994	1995
Bangladesh	2.6	4.3	4.6	5.1	5.4	6.0	6.6
China	11.5	13.2	13.4	13.6	14.6	14.9	15.8
India	3.5	11.2	12.7	14.3	14.7	15.7	17.0
Indonesia	12.3	18.0	19.5	20.3	21.5	24.0	26.3
Japan	35.9	46.1	49.2	50.4	50.7	54.3	55.0
Malaysia	2.4	6.8	8.1	9.5	11.7	12.3	13.1
Pakistan	7.3	10.4	10.6	10.9	11.6	11.9	12.1
Philippines	-	-	-	-	-	0.1	0.1
Singapore	-	-	-	1.0	1.4	1.4	1.4
South Korea	-	3.0	3.5	4.6	5.7	7.6	9.2
Taiwan	1.0	1.7	2.7	2.8	2.7	3.6	3.9
Thailand	2.8	4.9	6.3	6.8	7.6	8.6	8.3
Other	2.7	2.4	2.3	2.4	2.8	2.9	3.0
ASIA	82.0	122.0	132.9	141.7	150.4	163.2	172.7

Source: *BP Statistical Review of World Energy.*

of 8.3 per cent per annum since 1985, and a slower 7.2 per cent per annum rate since 1990.

Before leaving the discussion of the patterns of overall energy use, the potential for energy efficiency merits some consideration. We noted above that primary energy use is very limited in China and India in per capita terms. However, when we consider this in GDP units, a radically different picture emerges. While China and India use very little energy given the size of their populations, they use a lot given the size of their economies. Table 2.6 shows the total primary energy used, expressed in kilograms of oil equivalent, per thousand dollars of GDP, for selected Asian countries together with the UK and USA. This measure is affected by a large number of factors. The first is the structure of the economy, particularly the balance between energy intensive sectors such as industry and less energy intensive sectors such as services and, at least in less developed countries, agriculture. The second factor is the technology of energy use, and within that the energy efficiency with which that technology is employed. The relative price of energy compared to labour and capital represents a third major factor. In addition the measure is sensitive to the choice of exchange rates, and a whole host of social, climatic and logistical factors.

In Table 2.6, China and India are shown to have extremely high levels of energy use per unit of GDP, especially when compared to the extremely energy efficient Japanese economy.[11] The complexity of the factors listed

Table 2.6:   Energy Requirements Per Unit of GDP. 1993. Kilograms of Oil Equivalent Per Thousand Dollars of GDP.

Japan	115
UK	229
Taiwan	280
Philippines	308
USA	320
Thailand	371
South Korea	395
Malaysia	483
Indonesia	491
India	785
China	1264

Source: Own calculations.

above as underlying the measure make interpretation difficult. However, note that India and China are still at an early stage of development where peasant agriculture is far more important than in most other Asian countries. *A priori*, considering only the level of development, they would not be expected to emerge at the extreme of the measure. The data are then strongly suggestive that either the relative price of energy has been set too low, or the production technology of energy use is inferior, or energy use is inefficient, or (and in reality) a combination of all of these. The implication is that the mapping of economic growth onto energy use for these countries is complicated by the presence of a considerable scope for efficiency savings.

4.   The Demand for Oil

By 1995, oil demand in Asia accounted for just over a quarter of the global total. At the margin, it represented the majority of global growth. As shown in Table 2.7, between 1990 and 1995 its level increased by about 4.1 mb/d. Over the same period, this compares to a 4.1 mb/d decrease in the Former Soviet Union, and a 2.3 mb/d increase in the rest of the world. Growth in oil demand in Asia over that period was 5.6 per cent per annum. Excluding the mature, and dominant in absolute size, Japanese economy, the rate of oil demand increase was 8.6 per cent per annum in the rest of Asia. In 1985, Japan accounted for 45 per cent of Asian demand, by 1995 this proportion had fallen by 34 per cent. Of the increase between 1990 and 1995, around one-half was provided by South Korea and China alone.

Table 2.7: Oil Demand in Asia and Growth Rates. 1985–95. Thousand b/d and Per Cent Per Annum.

	1985	1990	1991	1992	1993	1994	1995	Growth % p.a. 85–95	Growth % p.a. 90–95
Bangladesh	35	45	35	40	40	45	50	3.6	2.1
China	1810	2255	2410	2660	2915	3145	3310	6.2	8.0
Hong Kong	100	140	130	165	170	170	180	6.1	5.2
India	885	1200	1220	1285	1300	1400	1510	5.5	4.7
Indonesia	460	645	675	730	785	775	810	5.8	4.7
Japan	4435	5305	5410	5540	5455	5765	5780	2.7	1.7
Malaysia	195	270	290	295	330	370	430	8.2	9.8
Pakistan	155	220	230	250	270	295	305	7.0	6.8
Philippines	150	235	225	280	290	305	345	8.7	8.0
Singapore	225	370	380	400	425	495	510	8.5	6.6
South Korea	535	1040	1255	1520	1675	1840	2010	14.2	14.1
Taiwan	355	550	570	585	625	665	725	7.4	5.7
Thailand	235	410	445	490	555	615	690	11.4	11.0
Others	205	215	225	245	265	295	310	4.2	7.6
ASIA	9780	12,900	13,500	14,485	15,100	16,180	16,965	5.7	5.6
Asia minus Japan	4675	6795	7305	8155	8815	9545	10,285	8.2	8.6

Sources: *BP Statistical Review of World Energy*; IEA, *Annual Statistical Supplement to Monthly Oil Market Report.*

The primary determinant of the fast rate of oil demand growth is of course the underlying rate of economic growth, a rate which we argued above shows no internal dynamic to cause it to slow. The process of economic growth causes structural changes in domestic oil markets, which we illustrate with reference to Table 2.8, which shows the composition of oil demand over time in Japan, China, the rest of Asia and Asia as a whole.

The stages of economic growth have an asymmetric effect on the demand for oil products. In the domestic sector, growth first leads to a substitution away from traditional fuels towards oil products, and in particular towards Liquid Petroleum Gases (LPG), and most particularly butane. Later stages of development normally see the development of infrastructure enabling a further switch to occur towards natural gas or electricity in heating and cooking. Development, as shown in Table 2.8 for China and other Asia outside Japan, has been associated with a very rapid expansion of LPG demand. Over the 1985 to 1994 period, LPG demand increased by 13 per cent per annum in China, and 12.2 per cent in other Asia. As development proceeds, the expansion of LPG begins to slow, given that it is primarily driven by wholesale once and for all fuel

Table 2.8:   Oil Demand by Product. Japan, China, Other Asia and Total Asia. Selected Years. Million b/d and Growth Per Annum

	1975	1980	1985	1990	1994	Growth % p.a. 75–85	85–94
Japan							
LPG	0.40	0.47	0.57	0.60	0.62	3.6	0.9
Naphtha	0.57	0.49	0.42	0.54	0.70	-3.1	5.8
Gasoline	0.74	0.59	0.63	0.76	0.86	-1.6	3.5
Aviation Fuel	0.10	0.05	0.06	0.07	0.09	-4.8	4.1
Kerosene	0.39	0.43	0.47	0.54	0.56	1.8	2.0
Gasoil/Diesel	0.73	0.75	0.80	1.11	1.25	1.0	5.1
Residual Fuel Oil	2.23	1.64	0.95	0.91	0.89	-8.2	-0.7
Other	0.62	0.54	0.54	0.77	0.72	-1.3	3.2
Total	5.79	4.97	4.43	5.29	5.69	-2.6	2.8
China							
LPG	0.02	0.04	0.05	0.08	0.15	10.3	13.0
Naphtha	0.01	0.05	0.12	0.21	0.31	25.6	11.2
Gasoline	0.19	0.23	0.33	0.44	0.63	5.8	7.5
Aviation Fuel	n.a.	0.01	0.01	0.02	0.04	n.a.	14.3
Kerosene	0.06	0.07	0.07	0.06	0.05	1.3	-3.0
Gasoil/Diesel	0.29	0.34	0.39	0.55	0.75	3.3	7.3
Residual Fuel Oil	0.40	0.58	0.56	0.66	0.74	3.4	3.2
Other	0.35	0.43	0.33	0.30	0.39	-0.7	1.8
Total	1.32	1.75	1.86	2.32	3.06	3.3	5.3
Other Asia							
LPG	0.04	0.07	0.17	0.31	0.46	14.0	12.2
Naphtha	0.08	0.16	0.24	0.33	0.60	11.2	10.6
Gasoline	0.25	0.32	0.38	0.62	0.86	4.0	9.5
Aviation Fuel	0.13	0.15	0.18	0.28	0.35	3.5	7.6
Kerosene	0.21	0.31	0.33	0.45	0.55	4.4	5.8
Gasoil/Diesel	0.54	0.85	1.08	1.61	2.20	7.1	8.3
Residual Fuel Oil	0.88	1.29	1.14	1.68	2.14	2.6	7.3
Other	0.09	0.11	0.13	0.21	0.24	3.6	7.3
Total	2.23	3.26	3.63	5.51	7.39	4.8	8.0
Total Asia							
LPG	0.46	0.58	0.79	0.99	1.24	5.4	5.2
Naphtha	0.67	0.70	0.78	1.08	1.61	1.6	8.3
Gasoline	1.18	1.15	1.34	1.82	2.34	1.2	6.4
Aviation Fuel	0.23	0.21	0.25	0.38	0.48	1.0	7.3
Kerosene	0.67	0.81	0.87	1.05	1.16	2.7	3.3
Gasoil/Diesel	1.55	1.94	2.27	3.27	4.19	3.9	7.1
Residual Fuel Oil	3.51	3.51	2.65	3.26	3.77	-2.8	4.0
Other	1.06	1.08	1.00	1.29	1.35	-0.6	3.4
Total	9.34	9.98	9.93	13.12	16.14	0.6	5.5

Source: Own calculations from International Energy Agency data.

switching. Further, at higher stages of development, the switching away from LPG towards gas and electricity would be expected to occur. This is shown by the modest 0.9 per cent per annum growth since 1985 in Japan. Japan remains the major LPG market, but its share of 87 per cent of Asian LPG demand in 1975 had been eroded to exactly 50 per cent in 1994.

A second major structural change occurs when petrochemical industries are expanded in response to both domestic sector demand, and also petrochemical input demand from the industrial sector. Petrochemical development leads to rapid increases in the demand for naphtha, as shown in Table 2.8, as a feedstock. As in the case of LPG, demand from outside Japan has been the major source of growth, with Japan's importance slipping from 86 per cent of Asian demand in 1975, to 43 per cent in 1994.

The main structural change as growth progresses lies in the transportation sector. Incremental Asian demand has been, and will remain, heavily skewed towards transport fuels, particularly as these are the areas in which the scope for substitution by other fuels is the most limited. Of the total 6.8 mb/d total increase in oil demand between 1975 and 1994 shown in Table 2.8, gasoline, aviation fuel and diesel have contributed 4.1 mb/d. The earlier stages of development involve the expansion of the relative importance of diesel, due to the development of the internal domestic market and the associated industrial demand for transportation of goods. Most Asian economies are still moving through this stage, resulting in fast demand growth for diesel and expansion of its share within the demand barrel as shown in Table 2.8. This has been reinforced in several countries (most notably India and Thailand) by the tendency, through tax/subsidy systems, to keep the price of gasoline relative to diesel well above world market levels. Over time this has encouraged the development of a vehicle fleet heavily biased towards diesel powered vehicles. Later stages of development tend, dependent on government tax and automobile production and importation policies, to see a switch towards faster gasoline demand growth.

The growth of transportation fuels skews the demand barrel towards the lighter products, and in Asia particularly towards middle distillates. This tendency has been reinforced by changes in the pattern of residual fuel oil demand. In Asia as a whole, fuel oil accounted for 38 per cent of the demand barrel in 1975, but just 23 per cent in 1994. The primary market for fuel oil has been power generation, where the scope for substitution by other fuels is at its greatest.[12] As shown in Table 2.8, over time Japan has largely achieved this switch, with fuel oil demand falling by about 1.3 mb/d between 1975 and 1994. Even with its rapid electricity demand growth, China has managed to suppress increases in fuel oil demand and use more coal, resulting in a growth rate since 1985 which is modest compared to other fuels. For countries without indigenous coal resources or immediate access to imported gas or nuclear programmes, fuel oil has served as the major source of incremental power station feed, resulting in the high growth rate shown in Table 2.8 for other Asian countries. However, over time economic growth tends to bring infra-structural development, facilitating some switch to either LNG or pipeline gas, and in some cases towards the development of nuclear power.

The direct impact of economic growth is not however the only factor that has an impact on Asian demand. We would identify three other main determinants. The first is the relatively low real price of oil in the 1990s, compared to the two previous decades, reinforced in some countries by currency appreciations against the dollar. The second is the relatively low current share of commercial energy in total energy supply in many countries noted in the previous section. These two factors together combine to produce the expectation of a continuation of fast demand growth, an expectation that we quantify below in the context of a series of published demand forecasts. The third determinant can act as either a dampener or an accelerator to oil demand growth, i.e. government policy, and is considered further below.

To summarize the main features of the state of conventional wisdom about the future growth of demand in Asia, in Table 2.9 we show the incremental demand (relative to a 1993 base) from three sets of forecasts which we would consider as being broadly representative. Together, they suggest a range of expected demand increment by 2000 of 4.1 mb/d to 6.1 mb/d. This then implies an average annual increment of between

---

Table 2.9:   Incremental Asian Oil Demand Forecasts from 1993 Base. 2000, 2005 and 2010. Million b/d.

Forecast	2000	2005	2010
(a)	4.3	7.4	-
(b) (i)	6.1	-	14.7
(b) (ii)	4.1	-	9.4
(b) (iii)	4.9	-	11.6
(c) (i)	4.6	-	13.1
(c) (ii)	4.1	-	11.9

KEY:
(a)   Koyama (1995) [13]
(b)   Fesharaki et alia (1995) [14]
(i)   High case, (ii) Low case, (iii) Base case
(c)   IEA (1996) [15]
(i)   Capacity constraints (ii) Energy savings

about 600 thousand b/d and 900 thousand b/d. For 2010, the range for the total increment from 1993 is between 9.4 mb/d and 14.7 mb/d (these two extremes being the low and high case from the same forecast), implying average annual increments between 2000 and 2010 within the range of 500 thousand b/d and 900 thousand b/d. The IEA, as reported in forecast (c), in addition predicts global totals in 2010 of about 92 mb/d and 97 mb/d for its two cases, implying that over 40 per cent of incremental world demand will come from the Asia-Pacific region.

A more detailed view of these forecasts is shown in Table 2.10. They are strictly not fully comparable in absolute terms, given differences in definitions and coverage which result in the divergences shown in their base figure for 1993. Forecast (c) uses definitions broadly comparable to the data shown in Table 2.7. Forecast (b) uses a narrower definition of oil demand, and in particular omits the direct burning of crude oil in power stations resulting in the considerably lower figure for Japanese demand. Forecast (a) uses comparable definitions to (b), but employs a narrower country coverage. Figures for the main individual countries are shown, except in the case of (c) where the source makes no further disaggregation than the one presented.

On the basis of Tables 2.9 and 2.10 we would from a synthesis of conventional wisdom as represented in the forecasts, isolate three main features beyond the continuation of strong demand growth. The first is the assumption of a declining growth rate for oil demand (note the forecasts tend to have near constancy in the average *absolute* growth in demand). The second is the feature that growth in China is forecast to be the highest in the region, with Chinese oil demand in absolute terms reaching about 4 mb/d in 2000, and surpassing 6 mb/d in 2010. The strong forecast rate of growth in India (included under South Asia in (c)) represents a third main feature (but note the implied divergence in the magnitude of that growth).

Forecasts certainly serve a useful function in terms of planning. However, the basic problem with oil demand forecasting is the 'what if' questions referred to in Chapter 1, and in particular the fact that a major series of these arise from government policy. The solution taken in forecasting is to treat policy as largely exogenous, when in reality it is highly endogenous in respect of the nature and magnitude of oil demand growth. Forecasts can use a *ceteris paribus* assumption, however when the government is part of the *ceteris* it has a tendency not to be *paribus*. One could almost suggest, only slightly tongue in cheek, that the usefulness of the forecasts is then perhaps at a secondary stage, i.e. to act on what would be the implications if other agents, including government, acted on the basis of the forecasts themselves. On the one hand this becomes highly circuitous if all agents follow this course. On the other, in refinery

Table 2.10: Comparison of Forecasts of Oil Demand in Asia. 2000, 2005 and 2010. Million b/d.

Forecast		1993	2000	2005	2010
Koyama	Total	14.2	18.5	21.6	-
	of which				
	China	2.7	4.0	5.0	-
	Japan	4.8	5.1	5.2	-
	South Korea	1.6	2.2	2.7	-
	India	1.3	1.9	2.3	-
Fesheraki (i)	Total	14.9	21.0	-	29.6
(ii)	Total	14.9	19.0	-	24.3
(iii)	Total	14.9	19.8	-	26.6
	of which				
	China	2.7	3.9	-	6.3
	Japan	4.7	5.1	-	5.1
	South Korea	1.7	2.4	-	2.9
	India	1.3	1.9	-	2.8
IEA (i)	Total	15.6	20.2	-	28.7
	of which				
	China	2.9	4.2	-	6.7
	East Asia	5.0	6.7	-	9.8
	South Asia	1.7	2.6	-	4.7
	OECD Pacific	6.0	6.7	-	7.5
IEA (ii)	Total	15.6	19.7	-	27.5
	of which				
	China	2.9	4.2	-	6.7
	East Asia	5.0	6.4	-	9.0
	South Asia	1.7	2.5	-	4.6
	OECD Pacific	6.0	6.5	-	7.3

KEY: See Table 2.9

construction in Asia, for example, it would have produced a far more profitable strategy than that which most companies pursued on the basis of forecasts.

The forecasts give a central role to China and India. In oil product markets that role is in fact even more central, as in later chapters we note the tendency for oil refining surpluses to arise in other major consuming countries. This results in a growing number of countries looking to China and India to take the role of a 'demand sink' for surplus products. However, China and India are the two countries where the oil market is the most distorted by government actions, and where the government plays the central role in the evolution of that market. Government policy alone can produce a large swing in oil demand, particularly as both countries have features of quantity rationing in the oil market. In other words, while we might attempt to forecast the path of notional (i.e. unrationed) oil demand on the basis of pure macroeconomics, the paths of effective and realized demand are almost entirely dependent on government policy. Further considerations arise in the presence of actual or perceived foreign exchange constraints, where the projection of large oil deficits neglects a strong tendency for endogenous policy actions for demand suppression to constrain a deficit.

A further set of effects arise from fiscal policy, where taxes can not be treated as being exogenous to oil demand growth. The possibility of demand dampening tax policy is ever present, particularly when growth moves up the barrel to concentrate on gasoline in the more advanced stages of the development process. Gasoline taxes do not have the same impact on industrial growth as diesel taxes, and in Asia tend to be progressive in terms of income distribution. Where fiscal systems are relatively undeveloped, gasoline taxes also represent a relatively efficient way of widening the fiscal base. It should also be noted that in many countries the importation of automobiles attracts large import duties or *de facto* rationing, providing a further set of variables for government action.

In Asia generally, market deregulation and liberalization has also impacted on oil demand. First, these processes can have strong price effects. In some cases, such as Japan as discussed in Chapter 4, this is a positive impact on demand as domestic prices fall through greater competition. In other cases liberalization dampens oil demand, as it involves the lifting of either explicit or implicit oil product price subsidies. Some countries have had oil price stabilization funds, which, as we will see in Chapter 5, have tended to require net transfers from the government and thus have subsidized as well as stabilized. Secondly, liberalization affects the composition of demand between sectors and products, as pre-liberalization distortions are generally not equally distributed, competition effects are not neutral, and where subsidies exist they tend to be cross-subsidies. The mapping from economic growth through to oil demand may not then always be a straightforward one.

Notes

1. Perhaps the best example of this school of thought is the *tour de force* represented by W.G. Huff (1994), *The Economic Growth of Singapore: Trade and Development in the Twentieth Century*, Oxford University Press, Oxford. See also Sanjaya Lall (1996), *Learning from the Asian Tigers, Studies in Technology and Industrial Policy*, MacMillan Press, Basingstoke.
2. See World Bank (1993), *The East Asian Miracle: Economic Growth and Public Policy*, Oxford University Press, Oxford. The view is also described in World Bank (1991), *World Development Report 1991*, Oxford University Press, Oxford.

3. A famous (while tongue in cheek) measure for PPP is the 'Big Mac' index compiled by The Economist, i.e. the exchange rate vector that equalizes international prices of the eponymous delicacy. For the record, the Big Mac PPP for China in 1996 implied yuan undervaluation of the order of a dollar to yuan rate almost exactly twice the actual rate.

4. See in particular Rob Vos and Josef T. Yap (1996), *The Philippine Economy, East Asia's Stray Cat*, Institute of Social Studies, MacMillan Press, Basingstoke; Yoshihara Kunio (1994), *The Nation and Economic Growth: The Philippines and Thailand*, Oxford University Press, Oxford.

5. For a detailed description of energy supply projects in the Russian Far East and East Siberia see Keun-Wook Paik (1995), *Gas and Oil in Northeast Asia, Policies, Projects and Prospects*, The Royal Institute of International Affairs, London.

6. Singapore's domestic energy use is notable for the very high proportion of bunker fuel for ship refuelling. Out of domestic oil demand of 510 thousand b/d, 230 thousand b/d was for bunker fuel for non-Singaporean flagged ships. Including this amount greatly distorts the measure of energy use per capita. International bunkers have thus been excluded from Singapore's energy use, but note that energy per capita in Singapore is still by far the highest in Asia.

7. 1995 was the first year that Asia consumed more energy than the USA, and India consumed more than the UK. The gap had closed rapidly over the previous decade. In 1985 the USA consumed 35 per cent more than Asia, and the UK consumed 60 per cent more than India.

8. For details of an Asian LNG trade see International Energy Agency (1996), *Asia Gas Study*, International Energy Agency, Paris.

9. IEA (1996), *Asia Gas Study*.

10. The members of the Association of South East Asian Nations, as of the start of 1997, are Brunei, Indonesia, Malaysia, Philippines, Singapore, Thailand and Vietnam.

11. If PPP exchange rates were to be used, China would still be at the end of the scale, albeit less dramatically. However, the measure shown is one of resource cost at international prices, and therefore we would contend that a PPP adjustment would be inappropriate in this case.

12. Demand for fuel oil as shown in Table 2.8 also includes its use for refinery fuel, and as a transportation for ships, i.e. bunker fuel.

13. Ken Koyama (1995), 'Outlook for Oil Supply and Demand in Asia-Pacific Region and Role to be Played by Japan's Oil Industry', *Energy in Japan*, March 1995, no. 132.

14. Fereidun Fesharaki, Allen Clark and Duangjai Intarapravich (1995), *Pacific Energy Outlook: Strategies and Policy Imperatives to 2010*, East-West Centre, Honolulu, Hawaii.

15. International Energy Agency (1996), *World Energy Outlook*, International Energy Agency, Paris.

PART II

The Changing Structure of the Asian Oil Industry

CHAPTER 3

CHINA: REFORM AND THE MARKET

1. Introduction

China has become an important element in Asian, and indeed world, energy markets over the last decade. Its high rate of economic growth, and in particular the boom in the southern provinces, has swiftly turned it from a significant oil exporter into an importer. As we saw in the previous chapter, those who make long-term oil forecasts are increasingly focusing on China to provide a major boost in oil demand over the next few decades. The role of China in such forecasts appears central, with huge crude oil and oil product deficits projected. Yet experience tends to suggest that the quantification of magnitudes of change in the Chinese oil industry has often, and indeed generally, been subject to what is best described as extreme 'hype'. The Chinese oil sector has tended to underperform the quantitative expectations mapped out, both internally and externally. Past projections of its crude oil production and the number of giant oilfields it contains have not been fulfilled by a considerable breadth. While also true onshore, it is particularly true offshore, where finding rates, size of reserves and the path of production have all run far behind the initial claims made. Downstream, the pace of new refinery construction, particularly that involving foreign capital, has lagged far behind the original speculative timetables. The claims of new dawns in terms of liberalization have been made almost on a monthly basis for several years, but have to date proved false.

There seems then to be a pattern to projections about the Chinese oil industry. In particular, they either never come true, or they do so far slower than envisaged. Given this, the huge projected deficits bear some examination, particularly as they are often the result of straight line forecasts. In Chapter 1 we noted a common tendency to concentrate on the long-run effects of economic growth, and to underplay the endogeneity of government policy. China demonstrates this well, with swings in policy that in the short run often swamp the gradual impact of growing notional demand for oil.

The concept of the notional demand for oil is an important one. Notional demands are those that arise in an unrationed market from the interplay of free market forces. In contrast, the effective demand for oil is that which can be realized, the difference being attributable to rationing in the oil market itself or as the spillover from rationing in another market, normally as a result of direct government policy. We would contend that the projection of extremely large import deficits in the near future ignores the likely responses of the state, and the ability and willingness to sustain the foreign exchange implications. In short, even with changes in the nature and geographical balance of political and economic power in China, the panoply of devices for tightening oil demand and oil import suppression policies are still effective.

The Chinese import gap is then a variable that we believe will cause at least partially offsetting shifts in government policy. In addition, we would contend that there are two major factors that provide an understanding of oil sector liberalization. The first is its link with general economic liberalization. An overall reduction of state intervention in the economy has throughout Asia found reflection in policies towards the energy sector. However, due to deficiencies in the fiscal and monetary control systems, the link between general economic and oil policy in China is actually often a negative one. General economic liberalization has at points led to greater state control over oil. In the absence of other effective mechanisms, oil policy has often been a direct method of macroeconomic control, particularly over inflation. To a large extent oil policy is related (counter-cyclically) to the swings of general economic policy.

Whereas in other countries the climate of general economic liberalization can be taken as a given, and also as a motive for oil sector liberalization, those assumptions are not valid in China. We have therefore in Section 2 provided an overview of the general macroeconomic reform process, to provide a context for the later discussion of the impact of reform on the petroleum sector. Reform has been anything but a linear process, indeed it has shown major discontinuities. The economy has gone through several boom-bust cycles, with there being an inability to create any smooth sustainable growth pattern. Instead, the economy has moved through exaggerated cycles, with attendant changes in policy. We believe that the cycles in oil sector liberalization might appear to be an almost random process without consideration of the overall reform process.

The other major factor behind the process is dislocation, or perhaps more generally decentralization. The degree of policy and institutional homogeneity within the oil sector and in oil policy and trade has been greatly reduced. The balance between central and regional control has changed, and differential rates of economic growth have led to an increasing North–South divide in oil. The apparatus of the state oil industry has become increasingly heterogenous, with a level of competition between different elements of the state. These themes are explored in the other sections of this chapter. Section 3 considers reform in the oil industry, with frequent policy changes hiding an overall slow movement towards liberalization. Section 4 provides an overview of crude oil production in China. The sector has failed to achieve the expectations policy makers had, and now the exportable surplus has been removed by strong domestic demand increases. Section 5 considers oil refining, and finds a structurally dislocated sector, combining units of high complexity with an absence of some basic processes. Section 6 discusses the current and future impact of China on the general oil market, and the final section provides some conclusions.

2.   The Economic Reform Process

It is often tempting to see Chinese history as a series of unique incidents, punctuated by watershed events. The post-1949 history can be compartmentalized into the 'Great Leap Forward', the 'Cultural Revolution' and so on. Following such a view, the reform process began at the Third Plenary session of the Eleventh Party Central Committee in December 1978. This is often seen as providing the blueprint for reform following the rise to power of Deng Xiaoping, and the start of the period of market socialism. In fact there was no blueprint, and no sudden immersion of the economy in market forces. Change has been very gradualist, a step by step process often of trial and error rather than a sequence of watershed policy changes. Economic reform has certainly not progressed in any remorseless linear fashion. Cycles have emerged during which fast growth and decentralization have overshot, leading to inflationary dislocations and other pressures which have caused reverses in policy and economic slowdown, and then into another phase of liberalization and fast growth. However, the trend in policy has been towards reform, the cycles are more like a spiral, each movement up making a return to the start of the spiral more difficult.[1]

The dominant theme of the reforms has been to graft a market sector onto the central planned economy, in such a way as not to antagonize vested interests. Russia attempted to simply replace the state economy with a market economy, and found that the short-run dislocations carried a huge political cost. By contrast, China has attempted to launch a market economy that will shrink the importance of the state sector through its greater growth, rather than through the direct dismantlement of state institutions.

The components of the state sector were however given incentives to become more market orientated. In 1979 they were allowed to sell all output over the plan quota levels at market prices. Such a system not only helped expand the market economy, it also gave the state sector a direct motive for greater efficiency. It brought state organizations into the reform process, creating lobbying pressure from them for less central control and interference, i.e. to have lower quotas under the plan. To further keep them on the side of reform, loss-making activities were still subsidized, and the government showed an inclination to back off and grant special treatment if reform bit too hard into the enterprises.

A second major theme has been what Shirk has described as 'particularistic contracting'.[2] This essentially involves treating relationships between the centre and other agents as a series of bilateral tailor-made contracts, rather than attempting any uniformity of policy implementation. Again, there was no great overall blueprint, just a series of individual relationships formed by special circumstances or expediency. One of the most important facets of this policy was the setting up of four Special Economic Zones (SEZs) in 1979, increased to five in 1988 with the addition of Hainan Island. The SEZs were taken out of the planned economy, and allowed to make their own fiscal arrangements with foreign capital entering the zones. Fiscal authority was widely delegated more towards the regions, but the SEZs were given preferential treatment. Figure 3.1 shows the location of the zones. The placement was directly motivated by a desire to facilitate capital inflow from what were seen as the most likely entry points, namely Hong Kong, Macau and Taiwan.

From their foundation the SEZs became a focal point for the disagreements between reformers and non-reformers. Their socialist theoretical basis was questioned, and they were a key target in the campaign against 'spiritual pollution' launched by the anti-reformers in 1983. For the reformers, losing the battle over SEZs would have been considered a grave defeat for reform, and they tried to push ahead with the scheme, for instance allowing wholly foreign owned enterprises to be set up. There were forces operating in favour of reform. State enterprises saw the zones as a source of foreign exchange and wished to sign into joint ventures. Provincial governments with zones were obviously in favour, but other provinces wanted SEZs or something similar in their own jurisdictions.

The programme was accelerated in 1984 with the creation of fourteen open coastal cities, shown in Figure 3.1, in which special development areas were set up that could attract foreign capital. In addition, in 1985 three 'growth triangles' were formed, in the deltas of the Min, Pearl and Yangtze rivers, and in several waves, greater autonomy and incentives were granted to selected inland cities. The zones, and in particular the two SEZs in Guangdong and the Pearl River triangle, soon meshed into the economies of Macau, Hong Kong and Taiwan.[3] Guangdong province was a particular driving force behind growth, having three SEZs, two open coastal cities, and bordering onto the two key capital inflow points

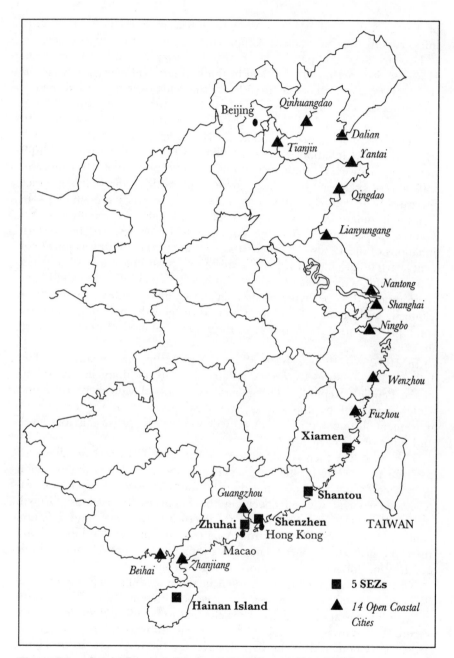

Figure 3.1:   Special Economic Zones and Open Coastal Cities in China.

of Hong Kong and Macau. Yet, as shown in Section 5, its energy infrastructure was insufficient compared to its economic importance.

Growth over the period of reform has averaged over 9 per cent. However, it is perhaps misleading to think in terms of overall averages as the total growth has been achieved very unevenly over the period. Figure 3.2 shows the rate of growth of GNP and the inflation rate from 1980 to 1994. The economy has shown signs of overheating at three points, with resultant sharp increases in inflation. The sustainable growth rate with medium to low inflation is certainly high, in the order of perhaps 8 to 9 per cent. However, whenever growth has gone above 10 per cent, inflationary pressures have been great, and, with a one-year lag, price increases have tended to move out of control. The strong growth of 1984–5 brought inflation above 8 per cent. It should be remembered that Chinese policy makers had no comparable experience to the western inflations of the 1970s. The highest yearly rate in the 1970s was just 2 per cent. Partial agricultural price liberalization had brought a level of 6 per cent in 1980, which was quickly reined in. The 1985 inflation would have then appeared to be hyperinflation in the Chinese context for the State Council, and again reform went into reverse with the reimposition of some price controls, and delays in the implementation of new reforming measures.

In 1987 growth was back above 10 per cent, and again in 1988. Inflationary pressures were now greater than at any point in the history of the People's Republic. The general economic slowdown of 1989 and 1990, deepened by the effects of the suppression of the democracy movement, brought the rate of growth well below the sustainable level, and inflation was back to 2 per cent.

The cycle was then repeated for a third time. Growth rose above 13 per cent in 1992 and 1993, and again inflation soared. On this occasion growth could not so easily be reduced, especially as much of the macro-economic control was effectively in the hands of the regions. Growth stayed above 10 per cent for a third consecutive year, and inflation continued to spiral out of control. A pattern to Chinese development has then emerged, and in particular it has swung between periods of economic overheating followed by attempts to regain control.

This pattern is even more marked in those leading edge regions with SEZs and coastal cities. These areas have benefited first from the upwards cycle of liberalization, and also been the first to be affected by slowdowns. Figure 3.3 shows the annual real growth in income in Guangdong province.[4] Compared to the national average the three booms of 84–85, 87–88 and 90 onwards have been greatly amplified, and the slowdowns are more marked.

The degree of economic takeoff that China has achieved is testimony

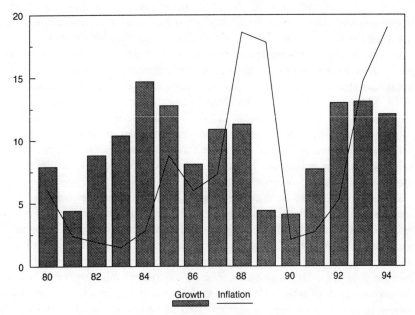

Figure 3.2: Chinese Growth and Inflation, 1980–94. Per Cent.

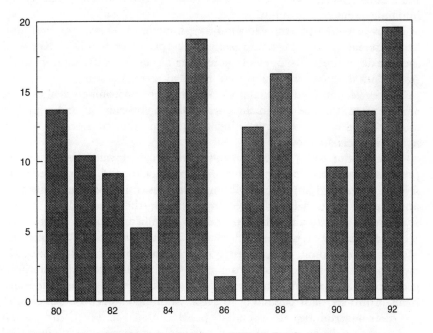

Figure 3.3: Growth in Guandong Province, 1980–92. Per Cent.

to the success of the reforms. However, there are some important internal stresses within the system that suggest the course of the economy will continue to be anything but smooth. The cycles in growth and inflation suggest that the centre is incapable of managing a steady sustained growth path. Inflationary pressures can be intense, fuelled by slack control over fiscal and monetary policy. Macroeconomic policy management has tended to be reactive, and too closely identified with internal policy differences. The necessary slowdowns after growth surpasses the natural rate, leave the reformers keen to accelerate growth as soon as possible, lest the slowdown be seen as a victory for the forces of anti-reform. When market forces are unleashed again, the desire to force the pace pushes the economy too fast, and the cycle repeats itself.

Inflationary peaks and troughs tend then to coincide with political initiatives. The 1980 inflation was used by the anti-reformer Hua Guofeng as a means of implementing reverses in policy. The slowdown in inflation was the signal for Hu Yaobang to launch his urban reform measures, and its reappearance led to him being purged. Likewise, the inflation following the reform measures of Zhao Ziyang increased the pressure for his dismissal, which eventually came after Tiananmen. The failure to maintain any sustainable rate has therefore had a direct political cost for the reformers, and led to the oscillations in policy. As shown in Figure 3.3, the changes in policy have had a particularly pronounced effect on the booming areas of the south.

There are other pressures beyond the problems of achieving a consistent accommodatory rather than reactive macroeconomic policy. Regional income disparities have opened up, causing large-scale internal migration. As a result disguised unemployment has grown overt, adding to the social problems of burgeoning city populations. The environmental pressures caused by reform are low in the current list of government priorities, but as time progresses and general living standards rise, pressure for policy action is likely to grow.

Three major issues remain on the agenda to be tackled by the reform process. First, state enterprises remain inefficient and overstaffed, but reform, particularly in the non-booming regions, raises obvious difficulties and incentives to delay change. In particular the potential for labour shedding is of a totally different scale to all other countries where state sector rationalization has been attempted. Secondly, eventually the question of how far economic reform can go before there is political reform will have to be addressed. Thirdly, to continue the movement to an enterprise economy, the financial sector needs to be modernized, as lack of development in this sector is beginning to act as an impediment to businesses.

Other uncertainties are essentially political in origin. The boom has led to a shift in perspectives and views in the regional bureaucracy compared to the centre. While still dominant, the relative power of the centre has certainly been eroded, with there being some strong regional pressures in operation.[5] There is a question of the ability of the centre to effectively exercise power in all circumstances, and thus of how the interaction and structure of power centres has altered policy formation.

The move to the era beyond Deng Xiaoping is perhaps less of an issue than the ongoing changes in the structure of power. The question of continuity in policy does not really arise, since, as we have argued above, there has been little continuity beyond the broadest brushstrokes during the Deng era. Retracement of the general thrust of economic policy is probably both practically and ideologically impossible. What is defined in the transition of power is the stance of the central authorities on some of the largely political issues noted above, and the size of the appetite to tackle some of the economic and emerging social problems.

3. Reform in the Oil Sector

We saw in the previous section that general economic reform has tended to evolve in spirals. The general movement has been towards liberalization, but the course of policy has been uneven, resulting in the boom-bust cycles shown in Figures 3.2 and 3.3. Reform of the oil sector has also followed a gradual path within spirals. As with the economy as a whole there has been no 'big bang' of market liberalization. There have been three main directions of change. First, control of the sector has gradually been moved away from the ministries to state enterprises, and these state enterprises have then been subject to competition and sometimes collaboration between them (creating a rapidly changing list of acronyms). In line with other state sectors there has been price reform, and finally there have been some opportunities for a role for foreign capital in selected parts of the industry. However, throughout this process there has been the understanding that the sector has a wider importance than most of state industry. Its prices have the greatest inflationary consequences, and lack of oil has served as a major brake on growth. In these regards reform in the sector has been perhaps more gradualist than in other areas, and it has always presented the temptation of having a use for macroeconomic control. As we detail later, this had the effect of creating a set of policy uncertainties.

In 1979 the control of the industry rested mainly with individual state ministries. The Ministry of Petroleum controlled oil production, and refining was split between that ministry and a series of others.[6] There was only one parastatal firm with a significant role, the China National Chemicals Import and Export Corporation (Sinochem). Sinochem was founded in 1950, with the original role of handling the marketing and procurement of oil outside China.

Much of the history of reform in oil has been concerned with the shifting of responsibilities within the state sector, with the structure of state interests being changed on a regular basis. The first of the new parastatals was the China National Offshore Oil Corporation (CNOOC), set up in 1982 to control the state's interests in the offshore oil provinces which had previously been under the aegis of the Ministry of Petroleum. CNOOC was, and remains, a relatively small company, but 1983 saw the creation of a giant state enterprise. In the refining industry, control of most major refineries, bar some units in the oilfields, was taken from the ministries and vested in a new parastatal, the China Petrochemical Corporation (Sinopec).

Together with the creation of CNOOC and Sinopec, there was also pricing reform. As we saw in Section 2, one of the first general economic reforms was to allow state firms to find their own market and price for above plan quota output. In oil a further distinction was made between low price and high price quota oil. Low price oil was a small fraction of world prices, high price closer to parity. Any production by a field beyond the combined low and high price quota could be placed on the free market. As with general policy, implementation was selective rather than generalized. The share of foreign offshore producers was considered to be all above quota, and CNOOC's production was all high price quota or above quota. Overall the majority of production was low price quota. In 1983 out of 2.12 million b/d production, 66 per cent was low price quota, 28 per cent high price quota, and a residual 6 per cent was above quota.[7] Ten years later, of crude oil produced within the sectoral plan, only 37 per cent was low price, and 53 per cent was high price.[8]

In 1989 the fourth major state enterprise was set up, taking over the onshore oil production and remaining refinery responsibilities of the Ministry of Petroleum. The ministry itself was abolished, its residual functions given to a new Ministry of Energy. The new company was the China National Petroleum Corporation (CNPC), responsible for onshore oil production, and the small oilfield refineries not under the control of Sinopec.

Throughout all these changes, the major constant had been the position of Sinochem, and its monopoly over international trading. But in the late 1980s it began to face pressure, not only from the new state enterprises, but also from the southern coastal provinces who saw the lack of oil as a constraint, and wished to be able to delegate Sinochem's functions. Faced with these pressures and a government policy that stressed the importance of competition within the state oil sector, the monopoly status was being severely jeopardized. Sinochem had been going through internal change and restructuring. In 1988 it became an international market player, and entered both physical and paper oil markets. In collaboration with Coastal, it took a stake in the US refining industry with the acquisition of the appropriately acronymed Pacific Resources Corporation (PRC) and its refinery in Hercules California, which was soon importing Chinese crude oil in quantity into the USA. Downstream integration has been followed by some major exporters, especially Venezuela, Kuwait, Libya and Saudi Arabia, primarily as a means of securing outlets. But the rationale for other countries hardly applies to China, where by 1988 security of outlets was not a major issue. The problem was how to cope with a fast diminishing exportable surplus. Internationalization was perhaps a goal in itself, the reaction of any domestic monopoly facing liberalization is normally to internationalize and diversify their activities. The joint venture was not a major success, and the refinery was idled in 1994.

Sinochem also stepped up its trading activities in 1988, and began taking more overtly speculative positions. They even entered the forward Brent market and became one of the six most active players of outright price deals in both 1989 and 1990. At least for a while, they started making some notable profits, which of course merely attracted the attention of their detractors in China who wished to be able to perform the same functions.

While Sinochem internationalized, the first cracks in its import monopoly appeared. In 1988 the four Special Economic Zones were allowed to import directly from foreign companies. Four major export refineries (Maoming, Guangzhou, Shanghai and Zhenhai) were then also allowed to sidestep Sinochem involvement in trade. The allowances to the SEZs and the four refiners eroded Sinochem's position, but the monopoly was not completely abolished until the start of 1993. Both Sinopec and CNPC had been actively lobbying for the right to trade on their own account, and, while this was still not allowed, the pressure brought about some reform. The government responded to the pressure by creating yet more companies. Chinaoil, a joint venture between CNPC and Sinochem, was created with the right to export crude oil. Unipec, a joint venture between Sinopec and Sinochem, was allowed to import crude oil, and both import and export oil products. From having just one state oil company, China now had six.

The distinctions between the state companies started to become less clear. While Sinopec had achieved, albeit by a proxy, a role in international trade, Sinochem retaliated by proposing to move into refining, joining two joint ventures for new refineries with foreign partners, in Dalian and Huangdao. To confuse matters further, the Hong Kong subsidiary of Sinopec also joined in the Dalian deal with Total. CNOOC also announced proposals for a putative refining project with Shell in

Huizhou.[9] When liberalization of energy industries is discussed in most countries, it normally involves the liberalization of state monopolies, with the next problem being how to open up the market for competition. The Chinese solution was different, allowing the delineation of activity between the existing state firms to become blurred, and then creating a few more for good measure. Further liberalization came in 1991 when China began to allow third party processing in its coastal refineries.

Having passed new powers and the scope for new responsibilities to the state companies, the next stage was to dissolve the Energy Ministry after its brief five-year history, which was done in March 1993. This has some curious echoes of the Thatcher revolution. The UK government had passed its energy responsibilities to the private sector and to regulatory bodies. This left a much diminished role for the Energy Department which was then abolished, with its remaining functions subsumed into another department. In China the role of the Energy Ministry had been passed to the state firms themselves, and the now essentially powerless ministry disappeared. As shown in Figure 3.4, the organizational structure of the oil sector had become somewhat less than parsimonious. The difference is however that the UK government knew that it would not

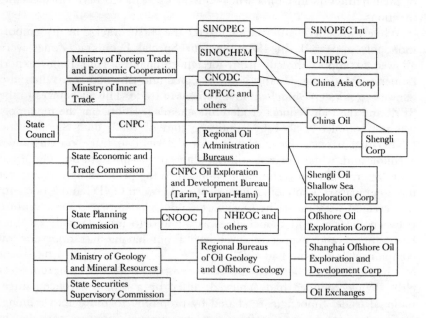

Source: Vioshiki Ogawa (1995) 'Present Situations and Problems of Overheating Chinese Oil Markets', *Energy in Japan*, no. 135.

Figure 3.4:   Organizational Structure of Chinese Oil Sector.

want to use the energy sector as a direct policy instrument again. Chinese reformers may have thought the same, but within a year the imperative had returned, but now within a framework that was perhaps less able to effectively channel the power of the state council.

Matters soon got worse for Sinochem. The devaluation of June 1993 caught it between buying at international dollar prices but selling in China in domestic currency. Further, the competition from its domestic rivals intensified. Sinopec (perhaps slightly disillusioned with Unipec) even opened its own trading office in Singapore, closely followed by CNOOC. All four of the major state enterprises were now trying to vertically integrate, with the two new joint-venture companies beginning to carve out a more independent role.

In late 1993 the official position had still been to go ahead with full deregulation of the oil sector. The aim was to abolish import licensing and deregulate prices by 1995. The move towards liberalization was in full cry. However, the uneven nature of growth has brought about inflationary cycles and short-term reverses in policy. As inflation began to get out of control, deregulation was swiftly reversed. The desire to limit growth in order to reduce inflationary pressures and any potential for social unrest, led to a resumption of state control.

In early March 1994 a temporary import ban was introduced to cover both crude oil and products, and fixed internal prices were introduced. Imports had risen sharply for two main reasons. First, the imposition of a value added tax, originally due for implementation in January, was to become effective at the start of April. As a result buyers were attempting to place as much as possible in storage before the deadline. Secondly, low world prices had resulted in imports of products being cheaper net of transport costs than domestic sourcing from the north. Imports rose, while unsold product filled up storage at state refineries. For the booming areas of the south, imports were not only cheaper than state supplies, they were of better quality, primarily due to contamination in the rail containers used to move product south.

The March ban left importing as the sole preserve of the state companies Sinochem, and Unipec, with Chinaoil being added in May. Refiners and users could no longer operate under municipal licences but had to conduct their foreign transactions through the state firms. With prices regulated and volumes rationed, competitive pressures were merely forced elsewhere, in this case an active market in import licences swiftly arose. Unable to take recourse to the market, the increasing use was made instead of *guanxi* (connections).

The ban has effectively continued into 1997, albeit with some changes in terminology. The apparent closure of the oil sector was a potential bar to Chinese entry into the World Trade Organization (WTO). Perhaps to facilitate WTO entry, in 1995 import quotas and licences were replaced with a system of registration.[10] This may be less overtly *dirigiste* in terms of language, but the net effect on operations is the same and the system preserves the monopoly of the three state companies in official exports.

We consider the impact of the controls on trade in Section 6, and comment on the root causes in the conclusions. However, there has still been a dynamic of change, primarily in the organization of state companies. Sinopec has continued a move towards what can best be described as a holding company, including the partial flotation of shares in some activities (most notably the Zhenhai and Shanghai plants). A similar form of organization has also been mooted for CNPC, aimed at leaving a state owned holding company with a number of subsidiaries in which it holds majority shares. It should however be noted that the wholesale reorganization of CNPC, with its size of payroll (the largest of any world oil organization) and diversification across many activities, raises the same political issues on state enterprise restructuring to which we drew attention at the end of the previous section.

4. Chinese Crude Oil Production

At the birth of the People's Republic in 1949, the new government inherited only a very minor oil industry, and three small producing oilfields. There was very little oil industry technology available, and very little expertise in geology and survey techniques. Indeed, the three pre-1949 fields were located by tar and gas seepages rather than by any active exploration.[11] In the first decade of the republic, the major advances in energy exploitation were confined to coal, an elevenfold increase in output by 1959. Coal totally dominated primary energy production, representing over 95 per cent of the total. However, in 1960 coal production reached a peak that was not reattained until 1972, and throughout the 1960s development of oil reserves was the major advance in the energy sector. Two fields dominated this process, Daqing and Shengli.

During the honeymoon in Sino-Soviet relations during the 1950s, Soviet technology and technicians had been involved in the Chinese industry, and some modest advances had been made. The real breakthrough came in 1959, with the discovery of the giant Daqing field in Heilongjiang province, the mainstay of China's oil production ever since. By Chinese oilfield standards it is unusual, being primarily one large contiguous field of relatively shallow depth. However, development of the field was always going to be demanding, given the extremely harsh weather of the far north of China and the lack of expertise due to the absence of any significant indigenous oil industry before Daqing.

The difficulty was increased by the abrupt split between China and the Soviet Union. In 1960, just as Daqing production was starting, the Soviet technicians withdrew, in the process not only cutting off the Chinese from expertise but also from the necessary heavy capital equipment. The solution was to rely on labour. Teams of men took the place of heavy machinery, working long hours in an impossible climate. The heroes of Daqing, such as the 'Iron Man' Wang Chin-hsi were used as an exhortation to other Chinese industrial workers. The slogan became 'Emulate Daqing'. Given this highly labour intensive mode of development, Daqing's output grew only slowly. By 1965 it had reached just 85 thousand barrels per day. Compared to the production profile of typical major oilfields outside China, Daqing is unusual for the very gentle slope of output increase. But, unlike most of Chinese industry, Daqing output still managed to grow throughout the Cultural Revolution. The state planning authorities succeeded in keeping a tight grip on the whole petroleum sector, even through the height of the revolution.[12] While the oil industry was kept out of the worst of chaos, industrial machinery output was severely impacted on, and hence Daqing's slow increase in output was still being constrained by extreme capital scarcity.

The second major field was Shengli, discovered soon after Daqing, and located in the slightly more favourable climate of Shandong province. Like Daqing, development had to be highly labour intensive, and output increases were again very slow. Shengli's development was made even slower by the priority given to Daqing which was (correctly) perceived as the larger accumulation, and its development was thought to be easier. Shengli is in fact not one oilfield but a large number of fields scattered over an area of several hundred square kilometres, all with a more complex geology than the relatively easier Daqing reserves. Given that Shengli oil is also far inferior in quality to Daqing, it is hardly surprising that what limited capital was available went north.

Predictions of the supply capacity of non-OPEC oil-producing regions have, in general, been prone to supply pessimism. In the North Sea and elsewhere production has always run ahead of projections, and even in declining areas the rate of decrease has often been more modest than forecasters have implied. Indeed, it has been a generic tendency for non-OPEC supplies to surprise on the up side. China however has always been different. In terms of both internal and external observers, production, both onshore and latterly offshore, has never lived up to its perceived potential. Despite many predictions to the contrary, growth in production in the 1980s, while large in absolute terms, was slow compared to projections. Policy makers believed that there must be many more fields such as Daqing to be found. After all, if Daqing had been found after relatively little exploration across China, and with the most basic of technology, with improved techniques and more extensive exploration surely many more would be found. The official talk in the late 1970s was of China producing 10 mb/d or more by the end of the century.

In the 1970s production certainly began to pick up sharply. Table 3.1 shows Chinese output by major fields from 1960 to 1995, with the location of the fields being shown in Figure 3.5.

Table 3.1:   Chinese Crude Oil Output by Field, 1960–95. Thousand b/d.

Field	1960	1965	1970	1975	1980	1985	1990	1993	1995
Daqing	16	85	353	845	1038	1106	1129	1101	1120
Shengli	-	15	90	300	401	554	679	656	607
Liaohe	-	-	3	81	90	181	276	267	314
Zhongyuan	-	-	-	-	-	103	128	159	84
Xinjiang	32	80	29	25	82	99	138	144	158
Huabei	-	-	-	-	226	207	109	92	96
Offshore	-	-	-	-	-	6	26	90	168
Others	52	51	125	239	283	224	313	410	439
TOTAL	100	230	600	1490	2120	2480	2798	2919	2986

Sources: Various.

Over the course of the 1970s, production increased by some 1.5 mb/d. However, the industry was still very centred on Daqing and Shengli which provided two-thirds of this increase. Based primarily on the performance of these two fields, China had gone beyond the goal of crude oil self-sufficiency, and exports began to grow. However, outside these fields production increases had been disappointing. To the largest part this was due to an over concentration on production from existing fields, and, as noted by Smil (1988),[13] still very little exploration or wildcat drilling.

These deficiencies were compounded by poor technology. The base of Chinese technology was still the legacy left by the Soviet advisors, and they were far behind the West in deep drilling and exploration techniques in particular. As a result, resource nationalism had to be abandoned as a matter of pragmatism when China wished to explore for offshore reserves, and foreign companies were allowed in.

It was clear to those in the Chinese upstream industry at the end of the 1970s that the onshore growth seen in the previous ten years could not be sustained. Indeed, from 1980 to 1985 production grew by a more modest 0.36 mb/d, with again Daqing and Shengli providing two-thirds of the increase. The perception grew that the great hope was offshore development. However, while Table 3.1 shows that offshore production

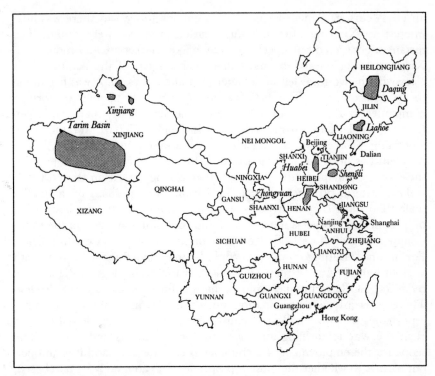

Figure 3.5: Provinces and Major Oil Fields of China.

represents a useful addition to output, the results have been extremely disappointing compared to the early perceptions. The growth of Vietnamese offshore production, while also affected by disappointments, illustrates the potential, but so far there have been no strikes of similar magnitude in Chinese acreage. Prospects still remain, mainly in waters subject to territorial dispute, but the idea that offshore production could be a quick fix to the slowing down of onshore increases has long been abandoned.

As economic reform progressed, the rate of growth of demand began to outstrip supply expansions, and the exportable surplus began to shrink. The pressure grew to increase production, no longer to provide a source of direct foreign exchange but rather to help limit China's oil import dependency in the late 1990s. With the failure of the offshore provinces to live up to expectations, attention turned back to onshore exploration. The focus shifted to west China, and in particular to Xinjiang province and the potential of the Tarim basin. The west had been the focus of the initial exploratory efforts in the 1950s, given the three known fields at the time were there, but the discovery of Daqing had left the area marginalized. A cursory glance at Figure 3.5 is enough to see why there was little interest in Xinjiang after the huge eastern reserves were found. The distances between oil prospects and any major market are enormous, with Xinjiang being one of the most inaccessible areas of the country. It is clear why the west was left until patience, and perhaps, hope was beginning to be lost offshore. The first Tarim acreage was opened to foreign companies in 1993, but to date interest has been fairly marginal, given the relative lack of prospectivity of the blocks offered, and the over-whelming problem of the lack of transportation infrastructure.[14] The problem has been encapsulated thus. While Tarim could be the jewel in the crown, 'unhappily the jewel is rather a long way from the crown'.[15]

Longer-term prospects may lie in Tarim, albeit perhaps on a smaller scale than some current hyperbole might suggest.[16] In the medium term, oil production prospects must lie with the continuing increase in the so far disappointing offshore fields, and in maintaining output in existing fields. For instance, Daqing production could be maintained for some time with an infusion of capital (still a constraint), better technology, enhanced oil recovery (EOR) techniques. and perhaps a little judicious extra drilling. There is clearly a determination not to let Daqing output slip, with, reportedly, the town hall and main railway station being demolished to clear the way for drilling and rigs sprouting in streets and parks.[17] The scope for the employment of technology is considerable, and it is an open question as to whether the most appropriate of it can merely be purchased, were there no capital constraint. A case could be made that the emphasis on confining foreign involvement purely to marginal exploration areas is misplaced, and the most productive, if politically sensitive, use of those potential resources lies in assisting with the redevelopment and optimiza-tion of existing fields.

It should be noted however that shortfalls in capital utilization and in the use of technology mean that Chinese output does fall short of what is feasible. There would still be considerable potential for increase, even without Tarim. However, at time of writing, there are few signs that the full potentiality will be realized quickly, or that the pattern of recent gradual increases in production will change in any dramatic fashion in the near future.

5. The Refining Sector

There are about 110 oil refineries in China. Many of these are extremely small topping units run by CNPC in or near oilfields, or by local authorities, which make little contribution to the overall total capacity. Utilization rates are generally low, with there being a large discrepancy between supposed nameplate capacity and what is actually feasible given the technical inefficiency of the majority of plants. Table 3.2 shows figures released in 1996, representing the official statistics on Chinese capacity.

Table 3.2: Chinese Refinery Capacity. 1996. Million b/d.

Operator	Nameplate Capacity	Actual Throughput	Utilization Rate Per Cent
Sinopec	3.36	2.29	68.4
CNPC	0.50	0.33	65.8
Other	0.36	0.20	56.2
TOTAL CHINA	4.22	2.83	67.0

Source: Sinopec reported in *Oil and Gas Journal*, 2 December 1996.

The gap between purported capacity and throughput is a function of three main variables. The first is oil import policy, and to an extent the refining system is facing input rationing. More important is the low level of efficiency, which suggests that the maximum sustainable throughput rate falls very far short of the 4.22 mb/d shown in Table 3.2. The third variable is policy towards the involvement of foreign capital in the refining sector. In particular, there may be a tendency to inflate refinery capacity figures to attempt to support the internal lobbying viewpoint that China has no need for grassroots refineries with foreign involvement, but just needs capital for debottlenecking operations. Table 3.3 shows the major refineries, together with their locations and provinces (the provinces were shown in Figure 3.5). Of the refineries shown in Table 3.3, all but seven are run by Sinopec.[18] Refinery location is biased to the north, and in particular to areas contiguous to the Daqing and Shengli fields. However, increases in oil demand have been biased towards the south. A major structural feature of the Chinese refining industry is its poor logistics. Put simply, it is in the wrong place. For the first three decades after the revolution, the industry was seen more in geopolitical terms than economic. There was a fear of invasion from the south, perhaps an incursion from Taiwan, and as a result little capacity was built in the area. There were also fears in the north, given the flammable nature of Sino-Soviet relations, and refineries were kept close to oilfields and away from consuming areas.

Such geographical dislocation between refining centres and consuming areas is common elsewhere, for instance the US East Coast has a large product deficit, as does the central area of Europe. The problem arises in the lack of sufficient transport infrastructure. The US East Coast has the Colonial and Plantation pipelines to move product in from the US Gulf

Table 3.3:   Major Chinese Refineries by Province. 1996. Capacity in Thousand b/d.

NORTH AND NORTH EAST			EAST		
Heilongjiang	Daqing	110	Anhui	Anqing	80
	Daqing CPF	20	Henan	Luoyang	100
	Harbin	30	Hubei	Jingmen	100
	Linyuan	30		Wuhan	60
Jilin	Jilin PC	85	Jiangsu	Jinling	140
	Qianguo	50		Yanzi	110
Liaoning	Anshan/	80	Shandong	Jinan	60
	Liaoyang			Qilu	170
	Dalian	140		Qingdao	50
	Dalian II	100	Shanghai	Shanghai	105
	Fushun	200	Zhejiang	Gaoqiao	150
	Jinxi	100		Zhenhai	140
	Jinzhou	110			
	Liaohe	50	**SOUTH**		
Beijing	Yanshan	150	Fujian	Meizhouwan	50
Tianjin	Tainjin	75	Guangdong	Guanzhou	105
Hebei	Cangzhou	30		Maoming	170
	Shijiazhuang	70	Hunan	Baling	100
			Jiangxi	Jiujiang	50
WEST					
Gansu	Lanzhou	100			
	Lanzhou	20			
	Yumen	30			
Xinjiang	Dushanzi	70			
	Kelamayi	30			
	Urumqi	100			

Sources: Various.[19]

Coast, and Europe has pipelines and the Rhine barge trade to move product out of the Rotterdam area. In China the pipelines from the north go no further than Jiangsu province, leaving internal transfers to be conducted inefficiently by rail. With bottlenecks and size constraints in the southern ports, internal coastal movement by tanker is also problematic.

The second structural feature is the imbalance between product demand and supply capabilities. Chinese refineries have in aggregate a considerable amount of fluid catalytic cracking (albeit of different vintages of technology). With this form of upgrading, together with the constraints on and cost of private car ownership, China once had no difficulty in satisfying gasoline demand other than the geographical dislocation noted above. However, gasoline demand increased by over 75 per cent between 1990 and 1994, moving China into a small net import position. The problem is more severe for middle distillates, and in particular gasoil. Gasoil demand nearly doubled between 1990 and 1994, moving above 1 mb/d. With upgrading based on catalytic cracking rather than hydrocracking, the capability to supply this quantity domestically has long since disappeared. Gasoil demand has been driven primarily by the demand for diesel fuel for lorry transport. Given the high elasticity of road haulage demand with respect to industrial growth, particularly as the internal market develops, demand growth remains very heavily biased towards gasoil.

Despite the complexity of the refinery system, the third structural feature is the deficiencies in some key units which cause an inflexibility in the system. Refineries have long been geared to a primary diet of Shengli and Daqing. While these are heavy crudes that have necessitated the need for cracking, they are also low in sulphur. As a result the amount of desulphurization capacity is very low. This would be no problem as long as China could bias its imports in favour of low sulphur Far East crudes, but in the longer term a greater proportion of its sourcing might have to be from the Middle East and of higher sulphur content.

The problem is not simply solved by adding desulphurization units, there would also have to be an upgrading of the quality of steel in pipes and distillation towers to cope with corrosion.[20] The second deficiency is in gasoline reforming, alkylation and isomerization units, making it difficult to produce high quality gasoline of sufficient octane. The difficulties with coping with sulphur are reflected in the sourcing of Chinese imports, as shown in Table 3.4. Daqing crude oil has a sulphur content of 0.1 per cent, and Shengli 0.9 per cent, and virtually all the imports flows are of similar quality. The major sources of imports have been Indonesia, particularly Minas crude with a sulphur content of 0.08 per cent (and which is, like Daqing, waxy), and Oman crude with a sulphur content of 0.9 per cent. To a large extent, the refinery system has been optimized on the two domestic crudes with limited flexibility available, and so the first call on imports has been to find close matches to the base load.

In total, the Chinese refining system is geographically dislocated, unable to meet the product demand slate, and also shows inflexibility in the range of qualities that can be either input or output from the system. Existing refineries need upgrading in the deficient units, and there is also a growing overall shortage of effective rather than nameplate refinery capacity.

Plans for new refineries are primarily joint ventures involving foreign capital, state oil companies alone or in combination, and state and local authorities.[21] Joint ventures have also been discussed for the expansion and upgrading of existing refineries. The first of these plans to come to fruition has been a second Dalian refinery with 100 thousand b/d of

Table 3.4:   Chinese Crude Oil Imports by Source. 1988–94. Thousand b/d.

	1988	1989	1990	1991	1992	1993	1994
Indonesia	2	26	25	55	95	77	94
Oman	12	22	17	53	61	81	67
Yemen	-	-	-	-	9	33	25
Papua New Guinea	-	-	-	-	10	11	16
Vietnam	-	-	-	-	6	6	12
Angola	-	-	-	-	4	24	7
Malaysia/Brunei	3	11	7	4	10	10	6
Pakistan and Other Asia		1	2	2	2	12	5
Saudi Arabia	-	-	-	2	4	4	3
Other Africa	-	-	-	-	6	21	3
Australia	-	-	1	-	11	8	1
Iran	0	5	6	1	2	1	1
UAE	-	-	-	1	5	11	1
TOTAL	17	65	58	119	226	312	246

Source: China Customs, reported in Y. Ogawa (1995), 'Present Situations and Problems of Overheating Chinese Oil Markets', op.cit.

crude distillation capacity. However, the history of this project does not augur well for the smooth or fast implementation of other plans. The Dalian refinery is a joint venture between CFP Total (who have a 20 per cent share), Sinochem, Sinopec, the state Chemical Industry Ministry and the local governments of Dalian and Daqing. The original commissioning was due at the end of 1994, but the start of full operations was delayed by two years. Cost overruns have been over 50 per cent, and the project has been marked by a continual struggle over terms and various other obstacles. The attractiveness of China is the potential of its domestic market, yet Dalian would currently be obliged to export a minimum of 70 per cent of output. In designing the ideal site for an export refinery, China would not be an immediate candidate.

Similar problems, particularly over the rules for access to the domestic market, are likely to cause frustration for other potential projects. The most important of these grassroots projects are shown in Table 3.5, together with major proposals for joint venture expansions of existing refineries.

The total expansion represented by these projects is 1.5 mb/d; however we express considerable reservations about their progress. Many of the above are at an early negotiation stage, and the difficult aspects of rates of return, management structure and domestic market share have not yet been made concrete. However, if there is no change in the current position on domestic market access, then most are unlikely to proceed.[22] Given the

Table 3.5: Potential Refinery Projects in China Involving Foreign Capital. Location, Participants and Capacity. Thousand b/d.

Location	Province	Capacity		Major Participants of Expansion
		Current	Expanded	
Beihai	Guangxi	-	100	Parkview, HK MG
Haikou	Hainan	-	120	Star Refinery and Petroleum Corporation
Huizhou	Guangdong	-	160	Shell, CNOOC, Sinopec
Ningbo	Zhejiang	-	100	Concord Oil
Qingdao	Shandong	-	200	Aramco, Ssangyong, Sinochem
Shenzhen	Guangdong	-	150	Sunkyong, Sinopec
Meizhouwan	Fujian	50	200	Amoco, Sinopec
Guangzhou	Guangdong	100	220	Exxon, Sinopec
Jinling	Jiangsu	140	210	Caltex, Sinopec
Maoming	Guangdong	170	340	Aramco, Sinopec
Zhenghai	Zhejiang	140	320	Arco, Sinopec

Sources: *China OGP*, 15 September 1995; *Energy Compass*, 8 March 1996, and various.

increase in the number of Asian countries with at least temporary excess refining capacity, all hoping that China will help absorb that excess, adding what would be primarily export refineries in China itself makes little economic sense. Even if these terms do soften, there is considerable competition for international capital flows into the refining industry.

The experience of the Dalian project may also lead to a reappraisal of some projects in the light of alternative uses of capital.[23] In particular, the large and discrete capital sums required for a new refinery rarely tend to be forthcoming when there is policy risk. In the presence of such risk, entry into China by foreign capital is more likely in areas where the size of the venture capital is comparably smaller, for instance storage and other infrastructure projects, lubricant sales and so on. It should also be noted that the view of foreign involvement in refining held by policy makers is not uniformly welcoming, both in central government and in the oil sector parastatals. Dalian also shows the potential for considerable slippages in timing. In all other than the Dalian project, we do not expect any other new grassroots joint-venture refineries to begin operations on any fast timetable, i.e. not before 2000, and dependent on policy, perhaps not before 2005 either.[24]

China's current refinery policy is to concentrate on debottlenecking and expansions at existing plants, while continuing to limit oil demand growth by both direct rationing and internal price movements. The possibility of significant expansions of foreign involvement is then to an extent a safety net in the case that effective refinery capacity does not advance as fast as the (highly constrained) level of oil demand. The underlying processes that will distinguish between the high or more modest capacity growth cases are primarily those that arise from policy and regulation. A sudden and consistent change provides circumstances amenable to the high capacity growth case. While this can not of course be ruled out, we would add that at time of writing we see no necessary dynamic at work that inevitably leads to such a change. Further, as noted throughout Sections 2 and 3, policy changes have tended not to be consistent, causing the spirals that would not normally be thought of as being conducive to capital accumulation in the form of such large and discrete amounts as oil refining necessitates.

6.   Chinese Trade and the Market

The evolution of Chinese oil trade between 1985 and 1995 is shown in Figure 3.6 for crude oil and Figure 3.7 for oil products. The impact of the 1994 restrictions is evident in the sharp discontinuity in 1994 and 1995 compared with the earlier part of the 1990s. Crude oil exports fell

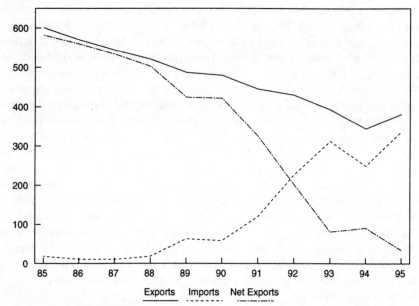

Source: A. Troner and S.J. Miller (1995), *Energy and the New China*, op.cit. and IEA Monthly Oil Report, various issues.

Figure 3.6:   Chinese Crude Oil Exports, Imports and Net Exports. 1985–95. Thousand b/d.

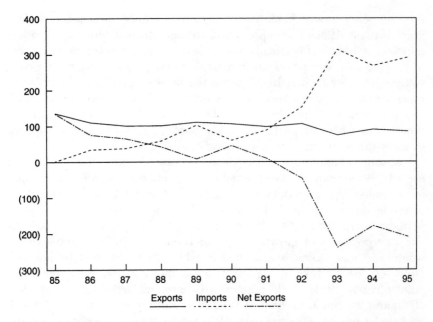

Source: A. Troner and S.J. Miller (1995), *Energy and the New China*, op.cit. and IEA Monthly
          Oil Report, various issues.

Figure 3.7:   Chinese Oil Product Exports, Imports and Net Exports. 1985–95.
                  Thousand b/d.

continually between 1985 and 1994, as production of Daqing and Shengli
was diverted into domestic refineries. A significant level of imports only
began to emerge in 1989, and then grew fast over the early 1990s, reaching
a level of over 300 thousand b/d in 1993. The impact of the north–south
imbalance is very evident in Figure 3.6. While exports have continued out
of the north, the rise in imports has been primarily due to demand from
the south (given the limited refining capacity in that region), as well as
from the Shanghai area. Imported crude has also flowed into the five
coastal refineries (Dalian, Maoming, Nanjing, Shanghai and Zhenhai)
where third party processing has been allowed. The southern refineries,
and in particular the large Maoming refinery in Guangdong, have moved
to an import biased crude input slate.

In total, as shown in Figure 3.6, the exportable surplus fell sharply, led
primarily by the rise in imports rather than the more gentle decline in
exports. Oil product exports, primarily of gasoline from the north, have
remained fairly stable as shown in Figure 3.7. However, there has been
a large rise in imports since 1990, moving China into overall oil product
deficit. There have been imports into China from Singapore as a result of

product being moved back from Sinochem's refinery processing deals – indeed in the 1980s this represented virtually all of China's oil product imports.[25] China had originally moved into Singapore processing as a way of keeping its crude prices up in a weak market, by moving some excess volumes into Singapore. In this sense the product exports to China were more a function of crude oil market behaviour than the dynamics of oil product markets. What changed in the early 1990s was that the basis for these exports became purely the inability of the domestic refining system to meet the surge in product demand.

Looking at Figures 3.6 and 3.7 from the vantage point of late 1993 and early 1994, one can see the temptation that there was to project very rapid increases in China's overall oil deficit. The declining path of the net exports of crude oil in particular, was so straight as to create a strong urge to perform interpolations. As we noted in Chapter 1, we believe that straight line interpolations of growth rates can rarely be justified, and here we would add that China is a context in which they can least be justified. However, for the sake of argument, linear interpolations from the viewpoint of early 1994, would indicate a Chinese oil deficit of 1.3 mb/d in 2000, and 2.1 mb/d in 2005. A brief survey of reports in the oil industry trade press in late 1993 and early 1994 would appear to indicate that the linear interpolation figures are very close to the conventional wisdom from industry analysts for 2000, and perhaps tend to the low side for 2005.[26]

If the path of our linear projection of China's oil deficit made from the vantage of early 1994 can then be taken as indicative of conventional wisdom, its key feature as of 1996 is that it is already 300 thousand b/d too high. Interpolations of apparent trends may be useful in circumstances where a system starts from steady state equilibrium, proceeds at sustainable rates without hitting constraints, and where policy and regulatory risk is low. None of these conditions are met in China. As noted in Section 1, the oil sector has not been in a steady state equilibrium but subject to quantity rationing whereby effective demand has fallen short of notional demand. Further, that constraint is being used as a macroeconomic control variable in an economy that has moved in exaggerated inflationary cycles rather than in any steady state trend. The degree to which the constraint binds is a function of the position in the economic cycle, and also to the extent to which the state is prepared to accept the foreign exchange implications of allowing the level of imports to be driven by notional demand.

As the impact of the clampdown in 1994 shows, the potential is large in China for policy changes to overwhelm apparent longer-term trends. In particular, the uncertainty of future import volumes arises more from policy uncertainty than from macroeconomics. Because of the potential for variability in policy, the elasticity of oil demand with respect to GDP growth is extremely poorly defined. For instance, the restrictions of 1994 resulted in oil demand being completely reined in, even with a GDP growth rate of over 10 per cent. Prior to the restrictions, oil demand had increased by 600 thousand b/d, or 30 per cent, in just two years.

The major effect of the restrictions on imports in 1994 was on the gasoline trade. The high inventories in domestic refineries in the north were primarily of gasoline, which was then redirected primarily by rail tank car to the south. As a result, exports to both China and Hong Kong (from where there is substantial re-export into Guangdong province) fell. Domestic reallocation was however less important for gasoil and heavy fuel oil, where the northern surplus was less substantial. The restrictions have then primarily had the effect of demand rationing rather than import substitution for these products.

We noted in Chapter 1 there has often been a failure to ask 'what if' questions in the oil industry. China poses a major 'what if', what are the implications of oil demand continuing to be used as a macroeconomic control variable, i.e. allowing more incremental demand to be satisfied during the down cycles in inflation, and rationing it further in the up cycle? If the macroeconomic boom-bust cycle continues to replay itself, the past cycle implies a net oil deficit of around 0.7 mb/d in 2000 and 1.2 mb/d in 2005. The absolute numbers are to a large extent unimportant, the major implication is that there are plausible circumstances under which the Chinese deficit grows far less fast than many current projections. Indeed, if the oil deficit itself becomes a major policy target, these numbers would err on the high side.

Policy uncertainty is then so great as to account for enormous potential variations in China's oil deficit. The conclusion to be drawn is that the assumption that China will necessarily absorb oil product surpluses from elsewhere is not necessarily warranted. In particular, even on the slow track of refinery construction, the low case for the oil deficit implies that it will be primarily filled by crude oil rather than oil product imports. Policy changes in China are of course the norm rather than the exception, and there is every possibility that circumstances could change in such a way as to allow for a dramatic widening of the oil deficit. However, we contend that it is wrong to consider a rapid widening as a given.

The failure of the potential of Chinese trade to meet expectations in the mid 1990s has had a major impact on the market. A further impact has come from the extremely large changes in the month to month levels of Chinese purchases. Figure 3.8 shows oil product imports and exports by month from 1994 to 1996, and Figure 3.9 shows crude oil imports and exports.

While exports of both oil products and crude oil have shown relative stability, there has been considerable volatility in import levels. Oil product

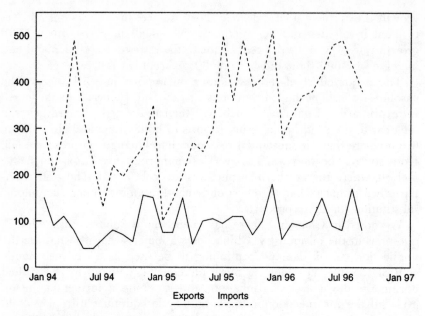

Source: IEA Monthly Oil Report, December 1996.

Figure 3.8:   Chinese Oil Product Exports and Imports by Month. 1994–96. Thousand b/d.

imports peaked at nearly 500 thousand b/d immediately prior to the imposition of restrictions in 1994. Since then they have swung between a low of 100 thousand b/d and a high of again close to 500 thousand b/d. Crude import volumes have shown even greater volatility. Note that the peaks (particularly marked in 1994) in both years have occurred in December, the last chance to use the year's remaining allocations. In 1995 there was a change of 200 thousand b/d or more from the previous level in a majority of the months. China's buying patterns have then tended to be an element adding to oil price volatility. Add the policy uncertainty to the variations in import levels, and a further source of volatility arises, given the potential confusion between temporary increases and those that might herald the start of a less restrictive import regime.

7.   Conclusions

On purchasing power parity terms, China now has the world's third largest economy. In GDP per capita terms, it remains an economy at a low level of development. Its potential for growth in oil demand is

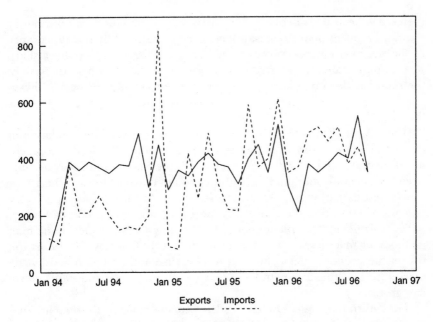

Source: IEA Monthly Oil Report, December 1996.

Figure 3.9: Chinese Crude Oil Exports and Imports by Month. 1994–96. Thousand b/d.

enormous, as it is starting from a very low base of energy consumption per capita. To give some idea of the scale, China with the same oil consumption per capita as the UK would consume 22 mb/d rather than 3 mb/d; with the same as Singapore (domestic consumption only) or the USA, it would consume close to 100 mb/d. Even a doubling of Chinese consumption to over 6 mb/d would still leave it at a very low level of per capita consumption, and the previous doubling of Chinese demand took just ten years. Further, China is currently going through a phase of development in which the scope for substitution for oil by other energy, and in particular coal, is limited. The potential for diesel fuel growth alone as internal trade increases is of an enormous scale. If automobile ownership is allowed to increase unfettered, the scale of gasoline demand growth in such a large country is also of a magnitude now forgotten in the West.

The above represents the case for bullishness about the Chinese oil demand. However, we have stressed in this chapter the extremely strong impact of shorter-term economic policy on oil demand. In particular, strong upwards growth in notional oil demand is not necessarily matched by growth in the policy controlled level of effective demand. A dominant theme has been the discontinuities in Chinese economic and oil industry reform. Frequent policy reversals have been occasioned by the appearance of vicious inflationary pressures, and progress has come in short sharp phases, punctuated by interludes of policy retrenchment where the grip of the state has been tightened. However, the general movement has always been upwards, with each reverse failing to remove all of the previous liberalization.

The relationship between economic development and the regulation of the petroleum industry in China is a non-linear one. Rapid development needs energy, and when the capacity of the state to deliver became stretched, liberalization and the introduction of market forces in the sector were important ways of preserving the boom. But in the face of a boom that was overheating the economy, and in the absence of effective direct policy instruments, petroleum policy has become one of the few remaining methods of macroeconomic control by central authorities. Hence as the rest of the economy has continued to liberalize and be controlled by the market, retrenchment in petroleum policy has taken place and state control has grown.

The return to a *dirigiste* policy has left Chinese policy with some internal contradictions. The market solution to reducing the pressure of demand is to allow prices to rise and vary with market conditions. But using oil as a control variable has resulted in fixing prices and instead rationing quantities, forcing some of the excess demand to be manifested in grey markets that seek to bend the restrictions, and in markets for import allocations. In short, the necessary result of attempting to control prices and quantities simultaneously has been a growth in unofficial markets. In controlling inflation there can be no substitute for responsible fiscal and monetary policy, and measures that ban markets, that are merely reflections rather than causes of inflationary pressures, represent no long-term solution. The correct policy must be to allow the boom to be reined in by higher energy prices, and the development of sustainable fiscal and monetary policy measures, not by rationing quantities. Currently Chinese petroleum policy is torn between a desire for a market solution, particularly by regional authorities, and a desire by the central authorities to use the sector as a policy instrument. While the two forces are both at work, the stop-go cycle in policy is bound to continue. With that cycle in operation, the buying pressure from China will tend to alternate between a trickle and a flood with the consequent impact on market volatility.

The influence of China on the markets while these cycles continue, is likely to be destabilizing. The sharp swings in China's call on petroleum imports, as shown in Figure 3.9, imply the presence of a volatile source of demand impacting on the margin of the market. The potential sourcing of Chinese crude oil and product imports involves the Asian, Middle East,

West African and Mediterranean markets, with knock-on effects onto the North Sea and Rotterdam. China is likely to become a major force in all these markets, not necessarily because of the scale of its purchases, but more because of the inter month volatility in the scale. Other factors being equal, a buyer's influence at the margin is necessarily magnified by the variance of its purchases. Reading these changes has now become an important part of effective trading, and Chinese inflation statistics and political developments have joined the array of information which the market must reflect.

Faced with a doubling of demand in ten years, almost every oil industry in the world would show large infrastructural dislocations and imperfections. In that context the Chinese system has actually stood up reasonably well to the stresses. However, we have identified a series of pressure points in the system. The first is geographical dislocation, with a strong north–south split emerging. The system in the south is incapable of coping without imports, which serve as a more efficient solution than transportation of any surplus from the north. Secondly, the refinery system is in urgent need of upgrading, particularly in reforming and de-sulphurization units. Given these two factors and the capital requirements involved, there may well be a continuing potential niche for foreign investment in upgrading and in joint petrochemical and oil complexes, if not in the construction of grassroots coastal refineries. However, we have noted the slow pace and frequent obstacles to new joint ventures in refining, the cost overruns and the variability of the regulatory climate.

Despite these major obstacles, the prospects and potential returns to foreign involvement in the downstream still appear to be greater than those in the upstream. Chinese production continues to rise, although past experience suggests that it is unwise to accept the more optimistic projections arising from inflation of the potential reserves in the Tarim basin. Large prospects may remain, particularly in the west, but the lack of pipeline infrastructure over the huge distances involved would appear to present an enormous capital and logistical challenge. In total, China is now set on the path of being a significant crude oil import market (even if it proves to be less significant than many current projections), and upstream development and governmental demand suppression can only influence the scale rather than the state of import dependency.

For physical oil trading companies, China does represent a major opportunity. A large market with many coastal entry points, requiring astute use of lightering and blending, as well as a network of storage and personal connections, appears to be the ideal circumstances for the growth of niche traders focusing on a particular trade. As yet the system is not suited to regular large movements in integrated channels, but rather represents a series of many small trading opportunities. Infrastructural dislocations and specific regional characterizations present the opportunity for niche trading, and those are the salient features of the Chinese petroleum sector. Trading in the West has over twenty years been driven less and less by the logistical mentality of the old-fashioned supply men, and more by the frictionless operations of developed markets. In China logistics still dominate the mechanics of trade, together with a volatility and complexity in government policy and policy implementation which has long since ceased to be a major factor in US and European oil markets.

Notes

1. The spiral analogy was coined in Jude Howell (1993), *China Opens its Doors: The Politics of Economic Transition*, Harvester Wheatsheaf, Hemel Hempstead.
2. Susan L. Shirk (1994), *How China Opened its Door: The Political Success of the PRC's Foreign Trade and Investment Reforms*, Brookings Institute, Washington.
3. See Myo Thant, Min Tang and Hiroshi Kakazu (eds) (1994), *Growth Triangles in Asia : A New Approach to Regional Cooperation*, Oxford University Press, Oxford for the Asian Development Bank.
4. The data is drawn from William H. Overholt (1993), *China : The Next Economic Superpower*, Weidenfeld and Nicholson, London.
5. See David S.G. Goodman and Gerald Segal (eds) (1994), *China Deconstructs: Politics, Trade and Regionalism*, Routledge, London.
6. See Todd M. Johnson, (1986), 'The Structure of China's Petroleum Administration', in Fereidun Fesharaki and David Fridley (eds), *China's Petroleum Industry in the International Context*, Westview Press, Boulder Colorado, for the original assignment of refineries by ministry.
7. Alan Troner and Sarah J. Miller (1995), *Energy and the New China: Target of Opportunity*, Petroleum Intelligence Weekly, New York.
8. Yoshiki Ogawa (1995) 'Present Situations and Problems of Overheating Chinese Oil Markets', *Energy in Japan*, September 1995, no. 135.
9. Refinery joint ventures are considered in Section 5 of this chapter.
10. China was granted WTO observer status in July 1995. Full membership was denied, primarily due to the perceived lack of openness in Chinese markets.
11. See Central Intelligence Agency (1977), *China: Oil Production Prospects*, CIA, Washington.
12. Leslie W. Chan (1974), *The Taching Oilfield: A Maoist Model for Economic Development*, Australian National University Press, Canberra, gives a case study of the Daqing field during the 1960s.
13. Vaclav Smil (1988), *Energy in China's Modernization: Advances and Limitations*, M.E. Sharpe Inc., New York.
14. A detailed description of Tarim basin development is contained in Keun-Wook Paik (1995), *Gas and Oil in Northeast Asia, Policies, Projects and Prospects*, The Royal Institute of International Affairs, London.
15. *Energy Economist*, January 1996.
16. Government estimates of about 150 billion barrels of oil and gas reserves in the Tarim(i.e. 50 per cent more than the official estimates for Kuwait), as reported

in Kang Wu and Binsheng Li (1995), 'Energy Development in China. National Policies and Regional Strategies', *Energy Policy*, vol. 23, no. 2, should perhaps be treated with more than a modicum of caution given the past record of Chinese reserve announcements.

17. *China Oil and Gas Report*, July 1995.
18. CNPC runs the Daqing CPF, Liaohe, Dushanzi, Yumen and Gansu plants, together with about 20 smaller units. The Jilin Petrochemical plant is run by the Chemical Industry Ministry, and the Qingdao refinery is owned by local government. About half a dozen smaller units are run by other local authorities.
19. This table was compiled from a large number of sources, including details of Sinopec refineries in *Oil and Gas Journal*, and press reports in *China OGP* (Xinhua News Agency), *Petroleum Argus, Petroleum Economist, Energy Compass* and *Petroleum Intelligence Weekly*. Interview evidence has also been used.
20. A. Troner and S.J. Miller (1995), *Energy and the New China: Target of Opportunity*, op.cit.
21. For a treatment of the mechanics of establishing and running joint ventures in China see John Child (1994), *Management in China During the Age of Reform*, Cambridge University Press, Cambridge.
22. In some cases however a high export share has been accepted, for instance the 90 per cent provisionally agreed in the Yukong joint venture.
23. The most high profile withdrawal to date has been the cancellation in 1995 of Elf's interest in constructing a refinery in Shanghai.
24. The only current plans for a grassroots refinery not involving foreign capital are CNPC's proposal for a 200 thousand b/d in Pengxian. Given the capital constraints on CNPC noted in the last section, we would consider this project most unlikely without a major policy change.
25. Sinochem's processing of oil in Singapore is considered in Chapter 6.
26. To give but one example, one source reports a US research centre prediction of a 1.0 mb/d to 1.3 mb/d deficit in 2000, and a major international oil company implying a deficit of over 2.5 mb/d by 1999. We presume the latter is a misquote. See Petroleum Economist (1994), *China: Its Energy Potential*, Petroleum Economist, London.

CHAPTER 4

JAPAN: THE RE-EMERGENCE OF MARKET FORCES

1. Introduction

Japan is the largest oil market in Asia and the second largest in the world after the USA. However, it has over a long period in effect been insulated from the rest of the international oil industry. In particular, the impact of international market signals on the structure and organization of the domestic oil industry, has been extremely weak. After a long period of tight government control, a continuing programme of deregulation has begun to change the nature and impact of the state's control and persuasive influence. This chapter argues that liberalization of the Japanese oil market involves an element of the reincorporation of Japan into the regional market. The capacity for insulation from regional price signals will be, while still present, reduced. The effective link with international market prices involves a radical restructuring of the Japanese oil refining and marketing industry.

The next section of this chapter considers the current structure of the Japanese oil industry, and its complex pattern of ownership. The following two sections are concerned with the relationship between the government and the industry. Section 3 considers the past history of control, starting in the 1930s and continuing to the mid-1980s. The experience of the liberalization programme after this period is the subject of Section 4, which also contains conclusions on the nature and impact of that process.

2. The Japanese Oil Market

Japan's major source of potential economic weakness, at least in its own perception, has always been its resource dependence. Table 4.1 gives an overview of the use of oil, showing that domestic production is extremely slight, leaving 99.7 per cent of the total crude oil demand to be met by imports. Oil product demand reached a new historic high in 1994, exceeding the previous record levels of over 5.5 mb/d reached in both 1973 and 1979.

The scale of the direct price effect combined with energy efficiency and substitution away from oil occasioned by the 1970s oil price shocks was

Table 4.1: Japanese Production and Use of Crude Oil. 1981–95. Thousand b/d.

Year	Domestic Production	Refinery Throughput	Utilization Rate Per Cent	Non-Refinery Use	Total Product Demand
1981	8	3620	60.9	329	4695
1982	8	3358	56.5	294	4395
1983	8	3241	56.9	323	4390
1984	8	3336	67.1	327	4620
1985	11	3099	62.3	272	4435
1986	13	2965	59.9	280	4495
1987	12	2885	60.9	265	4500
1988	12	2963	64.4	360	4805
1989	11	3148	69.2	387	5005
1990	11	3437	75.5	454	5305
1991	15	3653	79.1	469	5410
1992	17	3882	80.3	444	5540
1993	16	3983	80.2	347	5455
1994	15	4167	82.0	426	5765
1995	15	4169	80.4	357	5780

Sources: Petroleum Association of Japan, and *BP Statistical Review of World Energy*.

so large that it took twenty years of Gross Domestic Product (GDP) growth, a fall in dollar prices after 1985 and a sharp appreciation of the yen to restore the 1973 demand level of 5.46 mb/d.[1] Weak oil prices have combined with the weak dollar to drive domestic acquisition prices for oil in yen sharply downwards. Yen appreciation has produced an exchange rate effect whose downward pressure on domestic prices has been greater than the impact of the 1986 oil price collapse.

Table 4.2 shows the evolution over time of the cost of oil imports to the Japanese economy. Comparing 1995 with 1982 (the peak year for the yen cost of oil) gives an idea of the dimension of the exchange rate effect. By 1995 the dollar price of oil had fallen by 47 per cent, while the dollar had lost 62 per cent of its value against the yen. The compounded effect was an 80 per cent fall in the nominal yen value of oil from 8571 to just 1699 yen per barrel. The equivalent effect in a dollar based economy would have involved the 1995 oil price falling below $7 per barrel. Again in nominal yen terms, crude oil prices had fallen to less than half their level immediately after the 1973 oil price shock.

The proportion of Japanese GDP expended on the crude oil import bill has thus fallen sharply. The total yen cost of crude imports has fallen by 77 per cent between 1980 and 1995 in nominal terms. Over the same period nominal GDP rose by 95 per cent. Put in other terms, in 1980 crude oil imports represented 454 hours of Japanese GDP per year, by

Table 4.2: Japanese Crude Oil Imports 1977–95. Volume, Total Cost and Unit Dollar and Yen Cost.

Year	Crude Imports mb/d	Import Cost $billion	Average Cost $/b	Exchange Rate ¥/$	Average Cost ¥/b
1977	4.79	23.6	13.48	270.11	3641
1978	4.66	23.4	13.77	211.92	2918
1979	4.85	33.5	18.93	219.07	4147
1980	4.37	52.7	32.97	236.14	7786
1981	3.92	53.3	37.29	219.10	8170
1982	3.66	46.3	34.66	247.28	8571
1983	3.57	40.1	30.77	237.28	7301
1984	3.67	39.4	29.36	236.66	6948
1985	3.38	34.6	28.07	239.97	6736
1986	3.25	19.5	16.42	176.12	2892
1987	3.18	20.7	17.78	145.89	2594
1988	3.30	18.9	15.60	127.93	1996
1989	3.53	21.5	16.71	137.47	2297
1990	3.88	31.6	22.29	141.51	3154
1991	4.06	30.2	20.37	134.22	2734
1992	4.25	30.1	19.36	126.52	2449
1993	4.35	28.0	17.65	112.15	1979
1994	4.60	27.6	16.45	102.24	1682
1995	4.55	30.0	18.05	94.13	1699

Source: Petroleum Association of Japan.

1995 this had fallen to just 53 hours.[2] This measure is now at its lowest since 1952, and at around half the level in the economic boom of the 1960s, during what is normally thought of as the age of cheap oil. While it would be wrong to suggest that oil imports are no longer an important concern in political perceptions, these figures do show that the exposure of the Japanese economy to any given magnitude of oil price shock has been very greatly reduced.

The large quantity of crude oil that was shown in Table 4.1 as being for non-refinery use, primarily represents the direct burning of unrefined oil in power generation. Burning crudes such as Indonesian Minas tend to be used as a swing fuel, along with Low Sulphur Waxy Residue (LSWR), at times of peak seasonal electricity demand. For instance, in 1994 a hot summer reduced the availability of hydroelectric power, and increased the demand for electricity for air conditioning above normal levels. As a result non-refinery use of crude oil reached as high as 611 thousand b/d in the August of that year. Like LSWR, direct burning crudes can also be used to blend with fuel oil to reduce the overall sulphur content so as to meet the required specifications.

Refinery runs of crude oil have been increasing since 1987 (as was shown in Table 4.1). The increase was most noticeable during and after the Kuwait crisis, when ministry permission was given to increase runs in an attempt to cut back on oil product imports. One result of this has been that Japan's reliance on the Middle East as a source of crude oil has increased. Table 4.3 shows the composition by origin of Japanese crude oil imports. The proportion coming from the Middle East remained between 67 and 71 per cent throughout the 1980s, but has risen steadily throughout the 1990s to exceed 78 per cent in 1995. In the period after 1987, when crude oil imports began to rise, Japan's dependence on the Middle East for incremental imports has been total. Between 1987 and 1995 total imports rose by 1.404 mb/d. Over the same period, imports from the Middle East rose by 1.46 mb/d. Within Japan's Middle East imports there has been a significant switch from Saudi Arabian to UAE crude oil, which is generally lower in sulphur content.

There are forty refineries in Japan with primary distillation capacity, which are shown together with their configurations in Table 4.4.[3] The refining companies are listed in descending order of total distillation capacity, and the identification numbers used relate to the location of the refineries as shown in Figure 4.1. The system is notable for the relatively small average size of refinery of just 120 thousand b/d. To provide comparisons, the smallest of the four Singapore refineries has a capacity of 230 thousand b/d, which is surpassed by only three Japanese refineries. Combining the capacity of these three plants results in a figure only slightly greater than that of the largest single Korean refinery. The degree of upgrading in Japan is below the average for Asia, with a reliance on fluid catalytic cracking. Only a small number of refineries have any hydrocracking capacity.

Figure 4.1 shows that the refineries are predominantly located close to the major urban areas near the south coast of Honshu, in particular Tokyo, Yokohama, Nagoya and Osaka. Considering the other main islands, Hokkaido and Shikoku have two refineries, and Kyushu one. The Okinawa plants are essentially a separate system, located far closer to Taiwan and China than to the rest of Japan. The Korean refineries shown in Figure 4.1 are detailed in Chapter 5.

In showing twenty-six separate refining companies, Table 4.4 does not give a full account of the pattern of control. The structure of ownership of Japanese refining companies is a highly complex network of holdings and crossholdings. To illustrate, Figure 4.2 shows the primary ownership shares for the industry (there are many more small holdings running across companies, and other companies not shown with significant holdings in more than one company). Of the twenty-six companies in Table 4.4, the figure involves all but two (the total crude distillation capacity is shown for each refiner).

Table 4.3: Japanese Crude Oil Imports by Source 1981–95. Thousand b/d.

	1981	1982	1983	1984	1985	1986	1987	1988	1989	1990	1991	1992	1993	1994	1995
Iran	137	230	395	257	246	221	231	189	293	387	380	373	374	454	394
Iraq	60	66	9	14	72	163	104	171	217	144	0	0	0	0	0
Saudi Arabia	1364	1310	1007	1001	594	444	566	480	472	700	939	961	955	899	896
Kuwait	158	52	58	82	44	105	180	140	167	129	14	149	191	193	222
Neutral Zone	173	196	205	220	221	201	151	141	199	166	90	177	182	200	245
Qatar	146	134	132	220	199	163	114	176	213	229	250	254	290	290	295
UAE	541	495	542	562	727	734	595	650	737	823	1062	1049	1094	1218	1239
Oman	161	130	177	232	302	273	212	294	255	235	296	279	245	314	289
Other Mid East	0	0	0	0	0	2	0	0	15	5	12	13	21	29	34
MIDDLE EAST	2739	2611	2525	2588	2406	2306	2154	2251	2567	2818	3043	3254	3355	3598	3614
Africa	95	47	45	25	43	20	18	29	12	18	11	10	46	22	26
China	181	177	187	221	221	227	249	269	250	270	242	252	246	249	234
Indonesia	627	564	503	483	389	392	435	453	458	482	499	436	391	418	362
Brunei	87	85	83	80	55	46	49	46	36	45	52	48	56	49	53
Malaysia	82	75	63	109	113	137	90	80	82	84	84	81	92	87	94
Vietnam	0	0	0	0	0	0	0	0	27	45	53	70	77	87	86
Mexico	71	98	137	160	142	193	176	175	166	146	148	97	75	83	77
Venezuela	42	26	21	16	10	7	8	8	8	8	9	9	7	8	2
Others	45	16	15	2	3	15	4	6	5	2	4	2	5	5	3
TOTAL	3968	3699	3581	3688	3418	3352	3195	3331	3614	3941	4182	4318	4397	4667	4599
% Middle East	69.0	70.6	70.5	70.2	70.4	68.8	67.4	67.6	71.0	71.5	72.8	75.4	76.3	77.1	78.6

Source: Petroleum Association of Japan.

Table 4.4: Japanese Refineries. 1996. Ownership, Location and Capacities. Thousand b/d.

Refiner	No.	Location	Crude Capacity
Idemitsu Kosan	1	Chiba	250
	2	Aichi	160
	3	Tokuyama	120
	4	Hyogo	140
	5	Tomakomai	130
Cosmo Oil	6	Chiba	240
	7	Yokkaichi	155
	8	Sakaide	140
	9	Sakai	110
Nippon Petroleum Refining	10	Negishi	385
	11	Muroran	170
Tonen Corporation	12	Kawasaki	240
	13	Wakayama	166
Mitsubishi Oil	14	Mizushima	250
	15	Kawasaki	75
Nippon Mining	16	Mizushima	200
	17	Chita	95
	18	Funakawa	6
Showa Yokkaichi	19	Yokkaichi	240
Koa Oil	20	Marifu	127
	21	Osaka	104
Kashima Oil	22	Kashima	180
Fuji Oil	23	Sodegaura	162
General Sekiyu	24	Sakaii	156
Kyokuto Petroleum	25	Chiba	143
Kyushu Oil	26	Oita	136
Showa Shell Sekiyu	27	Kawasaki	120
	28	Niigata	40
Seibu Oil	29	Yamaguchi	120
Taiyo Oil	30	Ehime	89
Tohuku Oil	31	Sendai	100
Nansei Sekiyu	32	Okinawa	100
Kygnus Sekiyu Seisei	33	Kawasaki	70
Nihonkai Oil	34	Toyama	60
Toa Oil	35	Kawasaki	65
Okinawa Sekiyu Seisei	36	Okinawa	110
Wakayama Petroleum	37	Kainan	50
Toho Oil	38	Owase	35
Nippon Oil	39	Niigata	26
Teikoko Oil	40	Teiseki	4
TOTAL			5261

Sources: Petroleum Association of Japan and various.

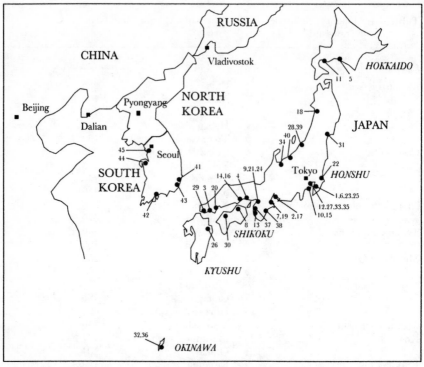

Figure 4.1:   Location of Refineries in Japan and Korea.

The complexity of ownership shown in Figure 4.2 is simply a microcosm of much of the Japanese economy. The major groupings are *keiretsu*, giant industrial groups often consisting of thirty or more corporations engaged in a broad sweep of sectors in the economy.[4] They are distinguished by a relatively loose confederation of companies, in contrast to the tight central control exhibited by the pre-war *zaibatsu*.[5] Among the 'Big Six' *keiretsu*,[6] Mitsubishi, Mitsui and Sumitomo have large holdings of oil assets, with the major ones being shown in the figure.

Many of the holdings of the *keiretsu* are mutually offsetting, as not only do the parent groups own each others' shares, but their own corporations also have offsetting shares running between them. It has been estimated by Huber[7] that about one-half of all *keiretsu* holdings are mutually offsetting. The system consists of a series of almost closed loops, within which a large proportion of dividends simply move around the system without ever leaving it.

There are several reasons for this complex pattern, but as Huber points out, the most forceful arguments see it, beyond representing a series of symbolic relationships, as creating an enormous and all-embracing system

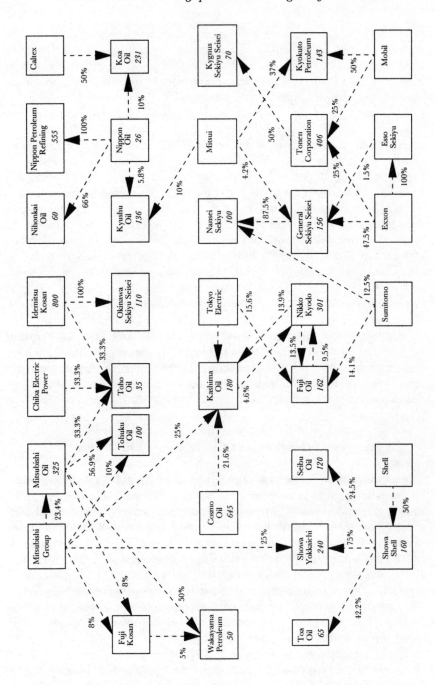

Figure 4.2: Structure of Ownership in the Japanese Refining Industry.

of peer review. The corporations are separate, but mutually accountable, and the crossholdings help to solve some of the principal-agent problems in managerial behaviour that might otherwise arise. Alternatively, others have seen the arrangements in terms of financial integration and providing greater managerial security from takeover.[8]

In Section 4 of this chapter we will talk of a restructuring of the Japanese refining industry. It should be noted at this point that the mechanisms for achieving this are very different from the US or European systems. The severity of the restructuring of refining in the 1980s in Europe and the USA was led primarily by weaker companies leaving the industry, and larger companies closing their least efficient plants, in a series of separate decisions. Those decisions were certainly interrelated, in the sense that the outcome for each firm was affected by the decisions of others, but the industry in both regions was made up of a large number of firms, in the main highly independent of each other. There were no significant crossholdings or long chains of ownership such as those shown in Figure 4.2, nor could such linkages involving all but one company be drawn even for the UK's relatively small industry, yet alone that of the USA.

The Japanese refining industry is more of an organic system (operating as just one aspect of a larger system). As Figure 4.2 makes clear, restructuring would need to be accomplished as part of that system, not as a series of individual exit or remain decisions as would shape the outcome in other regions.

3.   A History of Government Control over Oil

The oil industry in Japan first came under a tight regulatory regime in the 1930s. The original reason for state intervention into oil came not from strategic considerations, but from a consensus between government and industry on the undesirability of *laissez-faire* and unbridled competition. While other imperatives were also certainly at work, the primary impetus came from within the Japanese oil industry itself. The industry's own fear of the impact of competition, and the collapse of all attempts at self-regulation, led it to be a major lobbyist in favour of control by the state. This aspect is a major characteristic of Japanese regulation, policy has been to a great extent accommodatory and a result of consensus and compromise, rather than being overtly *dirigiste*.

The early Japanese industry consisted of a series of small domestic companies, and two large foreign concerns, the Shell subsidiary Rising Sun, and Standard Oil. Gasoline price wars broke out in the 1920s, with the two international firms bringing in products from California and

undercutting the output of domestic Japanese refineries.[9] The uncompetit-
iveness of Japanese refineries in the face of imports is an issue we will see
again below in the changed circumstances of seventy years after the
original price wars.

The competition was temporarily stopped by a market share and price
maintenance agreement in 1928, giving Rising Sun and Standard 60 per
cent of the gasoline market. The agreement was short lived, as was a
similar arrangement in 1929. Finally a government approved cartel was
formed, and was ratified in August 1932, under which Standard and
Rising Sun received 55.5 per cent of the market.[10] A cartel is of course
only completely effective if there are barriers to entry that can be
maintained. There were none, and the cartel was soon undercut by 40
per cent by Russian gasoline and other products imported by Matsukata
Kōjirō. With no accommodation with Matsukata proving possible, the
industry descended into what was essentially chaos. Japan had the cheapest
gasoline in the world, and also the most unprofitable oil companies.

Order was restored through the enactment of the Petroleum Industry
Law of March 1934. The industry would certainly have come under
regulation eventually, as there was a generalized move towards government
control as typified by the Import Industries Control Law of 1931.[11]
However, in contrast to the resistance to control in some other industries,
most notably electricity generation, the regulation had been actively
solicited by the oil companies. The 1934 Petroleum Industry Law laid
down a framework of control, of which many features still survive.
Importation and refining of oil were made subject to licensing, with
minimum inventory holding requirements being set, originally at six
month's of requirements. In addition, price controls and quota management became part of the public domain. Responsibility for the industry
was vested in the Ministry of Commerce and Industry, the predecessor of
the modern Ministry of International Trade and Industry (MITI).

With the 1934 Petroleum Industry Law (PIL I), oil moved into a
regime of tight regulation. While the major motive for control may have
changed over the decades, note that to a large degree this initial
manifestation was due to the perceived deficiencies of the competitive
market. The original law was seen as a sanitization of an industry that
had shown an inability to reach any stable competitive solution. The
alternative view would be that competition had not run its full course, and
that a restructuring of the industry through competitive pressures was
needed. However, given the then prevailing move towards greater
government control for strategic purposes, the regulatory solution was the
only one likely to be chosen.

With the occupation of Japan by the USA, regulation was impacted on
by general competition policy. The major manifestation of this was the

abolition of the giant centralized *zaibatsu*, and the repeal of most regulation, including PIL I and other oil industry legislation. However, the other main facet of occupation policy with regard to oil was the internal debate on whether to reform or to effectively destroy the industry. While refineries using domestic oil on the north coast of Honshu were reactivated, those on the Pacific coast were closed down and a direct import ban on crude oil introduced. In its first phase the attitude of the authorities tended towards the destroy option, most clearly shown in the Pauley report of 1946.[12] Pauley recommended the closure and removal of most of the import refineries, leaving only a rump industry consistent with the aim of reducing Japanese industrial power to the level of two decades previous.

Despite the bluster of the early occupation view of the industry, the focus turned towards its reformation. The reasons for the more accommodatory policy are fairly obvious; with the advent of the Cold War destroying Japanese industry was not only a lesser priority, it was patently counterproductive. The focus had shifted to trying to recreate Japan in a more American image in terms of competitive structure. Further, given the existing economic drain on the USA of the Marshall Plan in Europe, the destruction of the industrial base would simply prolong the time and extent of the US aid contribution to Japan.

From being an implicitly undesirable industry because of its military and economic strategic significance, refining soon became explicitly desirable for the exact same reasons. The need for economic reconstruction combined with growing Cold War tensions (made concrete with the onset of hostilities in Korea in 1950), led to a series of liberalizing measures. Most importantly, in the second half of 1949 crude oil imports were allowed and the Pacific coast refineries began to reopen. Following a relaxation of the rules concerning the involvement of foreign capital, direct encouragement was given for Japanese firms to acquire (predominantly American) outside capital. A series of such agreements were signed, many of which (as shown in Figure 4.2) are still reflected in the current structure of ownership. For example, Tonen tied up with Stanvac (now Mobil and Exxon separately), Showa with Shell, Nippon Oil with Caltex, and Mitsubishi Oil with Tidewater. Foreign companies now had a major stake, and in return the Japanese firms effectively achieved vertical integration given the tie-ups with companies which held the bulk of industry upstream reserves.

The oil regulation regime that emerged at the end of the American occupation in 1952, was essentially very light. Throughout the 1950s the only major direct control instrument that the government had was the indirect control over the volume and distribution of oil imports arising from the allocation of foreign exchange. With no comparable legislation to the Petroleum Industry Law in force, the degree of regulation remained considerably lighter than the immediate pre-war period. However, the tendency that was shown before PIL I, for an unregulated Japanese oil industry to lose profitability through extreme price wars, did not reemerge.

A series of factors account for this new relationship between profitability and the degree of direct industry regulation. First, the nature of foreign involvement had changed. While there were some arrangements between foreign capital and Japanese firms before the war, the basic relationship was highly competitive. By contrast, the links between foreign and domestic capital formed in the late occupation period were primarily direct owner-ship shares, and these involved most of the major domestic companies. A major source of competition had then been removed, especially as the foreign firms were not prone to high levels of competition between themselves in a foreign market. Secondly, the structure of the industry had changed. Before the war Nippon Oil had a dominant position among domestic firms, with a large market share which tended to attract entrants into the industry given the then relatively low barriers to entry. After the war, market shares were closer to a sustainable oligopolistic equilibrium, and in addition the costs of entry were now considerably greater. Finally, even in the absence of direct industry regulation, the indirect effects of foreign exchange allocation were enough to stifle competition. Price wars had tended to arise primarily due to undercutting by oil product imports. Without a free market for foreign exchange, the equivalent tendency for domestic prices to converge on the import price was greatly reduced. In addition, the pre-war import price was often a 'dump' price beneath domestic marginal cost. Post-war, and particularly in the Cold War period, there was no large-scale availability of excess oil from a near source with which to threaten the market, even if the foreign exchange constraint proved to be non-binding. In particular, there were no discounted Russian supplies for any potential new Matsukata to draw on.

There was then in fact little pressure for any successor to PIL I in the 1950s. PIL I had, in part, arisen as a response to a widely held perception that the unregulated oil market had produced suboptimal results. With no such apparent market failures, the equivalent pressure from within the industry was absent. There was also no real consensus within government on the role and importance of oil. While MITI had swung several times between oil and coal as the preferred fuel, no definitive statement on the role of energy in economic growth, and of the government's role in the energy sector had been made. In terms of management of the sector, the indirect effects of foreign exchange allocation also appeared to give the state all the leverage over the sector it wanted.

There is a certain tendency for an increase in the direct regulation of the oil sector often to be associated with liberalization in other sectors. As we shall see in the next chapter, this is true of China in the 1990s, but it was also true of Japan in the early 1960s. Two main factors combined to lead to a return to direct regulation as encapsulated in the Petroleum Industry Law of 1962 (PIL II). The first was a change in the foreign exchange regime. As part of the conditionality for joining the international GATT trading system, in 1960 the IMF ruled that Japan must liberalize its foreign currency control mechanisms. At the level of the macroeconomy the restrictions had effectively run their course, given the strength of Japan's international trade position. But macroeconomic liberalization had an effect at the microeconomic level of the oil industry in necessitating a switch to more overt government control. As the major control instrument for the sector was now to be removed, MITI, perhaps not surprisingly given the performance of the sector before PIL I, sought protection for it against competitive forces and any movements of foreign capital.[13] The optimal response was thought to be more state intervention and regulation. Because the trade regime was to be liberalized, the oil industry then needed to be under heavier direct control, given the fear of the impact of competitive forces in the sector.

The second factor behind the creation of PIL II reinforced the worries about foreign exchange liberalization. After the many swings of the 1950s, a clear energy policy of basing economic growth on oil emerged, and the perception of the economic and strategic importance of the industry was thus magnified. The tone was set by the Arisawa report of 1961, which also provided one of the incentives for the growth of refining in Singapore with the creation of a view that Japan was to be the dominant export market for Singapore. Arisawa recommended two options for state response, the creation of a significant state owned interest in the downstream sector, or a general petroleum industry law. In 1961, the idea of a state company to follow in the footsteps of Mattei's ENI held some considerable nationalistic appeal.[14] Mattei had shown that there was an alternative to the international oil majors, and it was the comparative strength and competitiveness of those same majors that was a major fear in regard to the fragmented and weak Japanese sector. However, while Kyodo Oil was formed as a marketing, but not refining company, there was to be no large-scale state presence in the downstream.

If the state was not to directly participate in the industry, and competition was, through experience, considered highly undesirable, the only remaining option was to impose a new tranche of regulation. This found expression in the Petroleum Industry Law of 1962 (PIL II). The new law gave the government a wide set of direct powers over the industry, very similar to the scope of PIL I. Under PIL II refiners need permission to operate and to expand or change their configuration, and must report production plans. Importers and marketers also need permission to operate, and again must submit plans. In addition, MITI was given powers to fix prices, and to annually formulate an overall plan for oil supplies for the next five years.

The Arab–Israeli conflict of 1967 led to further regulation, as supply security raised its head as a primary reason for heavy regulation. Compulsory stockpile requirements were introduced, and the potential fragility of Japan's supply lines became more of a worry. If, *de facto*, Japanese firms had achieved vertical integration through their relationship with foreign companies, the same political developments that threatened to deintegrate the oil majors was now threatening to deintegrate the Japanese industry as well. With the wave of OPEC nationalizations in the wake of the first oil shock this threat became real, and the Arab oil boycott also reinforced the fear of oil dependence.

After world prices rose due to the 1973 oil shock, the price structure for Japanese oil products was allowed to dislocate. This dislocation lies at the heart of the structural weakness of the industry, and is the reason why deregulation is necessitating structural change. The problem lay in the differential treatment of gasoline prices, which were not seen to be politically sensitive as compared to gasoil, and in particular, kerosene prices. With cold winters and the reliance on kerosene for home heating, particularly in rural areas, kerosene prices serve as the same political barometer as do gasoline prices in the USA. Likewise, just as a US administration would be relatively untroubled by a rise in the relative price of kerosene, gasoline prices have been considered to be relatively free of major political feedback effects in Japan.

Given this dichotomy, gasoline prices were not only allowed to rise, but to rise enough to compensate for the loss in profit implicit in moderating increases of other oil products. Once this principle for relative consumer pricing was in place, the dislocation became greater over time. As a result, the profitability of the Japanese oil industry became heavily skewed towards gasoline. With imports confined to refiners, and refiners unwilling to compete on gasoline prices (particularly as the dislocation in prices was effectively a result of government, industry and public consensus), there was no market mechanism in place to realign prices.

This situation had two further effects. First, it effectively cut off Japan from the world market in terms of prices. True, Japanese companies were important players in the Singapore market. However, the reverse dependence was weak. Changes in Singapore relative prices and price levels did not lead to strong effects in the Japanese market. Throughout the 1980s, Rotterdam prices in Europe helped drive industry behaviour, and the industry restructured in response to the signals being sent by Rotterdam prices. By contrast, signals from the Singapore market had little impact on Japanese operations, while in Singapore itself they forced restructuring and attention to cost minimizing and the full use of economies of scale. One could say that part of the restructuring the Japanese industry must do in the 1990s is left over from the 1980s, when the degree of responsive change in Japan was far less than in other regions. That restructuring did not fully happen, simply because Japan was in effect insulated from the full extent of changes in the world market.

The second effect was the lack of incentive to minimize costs. Profit was primarily being generated by just one oil product, in which there was little effective competition, and where almost complete pass through of costs to the end-user was possible. Inefficiency was then encouraged at all stages of the refining process, and also in marketing. Add a further heavy layer of government regulation, and the quantum difference between modern US or European service stations and their Japanese counterparts becomes explainable. The former tends to be self service and high volume, has a convenience store and other activities attached, and has very low staffing levels. With the development of electronic transfer of funds and prepayment systems, they can operate for 24 hours a day, completely unattended at night. The latter will tend to very low volume, very heavily staffed (with government regulations on numbers and on the necessary qualifications held), and with a minimum of other add-on activities.

The Japanese oil industry entered the 1980s with an industrial structure made static by the disincentives for and direct barriers to competition embodied in PIL II and its implementation. The control of domestic competition was at least codified in law, protection from foreign competition was less transparent. Without codified controls on imports, prices still remained far from the potential border price. The controls on imports were less direct, in questions of foreign trade *all* the iceberg lay under the water.[15]

4.   Liberalization of the Oil Industry

A programme of deregulation of the industry began in 1986, in both its domestic and international aspects. On trade issues, it is symptomatic of the degree of indirect control used by government, that liberalization could be said to begin with the drafting of controls on imports where there had previously been no explicit controls. The extent to which implicit control had been exercised is also shown by the fact that this liberalization still left a highly restrictive regime.

The Provisional Measures Law for Selected Petroleum Product Imports (PML) of January 1986 was a temporary law (intended to end at the start of April 1996) designed to encourage imports of gasoil, kerosene and gasoline. Potential importers needed to have three characteristics. Two of these, namely the requirement to maintain mandated days of supply through oil stockpiles, and the capacity to adjust product qualities, certainly represented an entry barrier, although not a completely insuperable one. The third condition however was totally restrictive, importers had to have capacity to refine crude oil in Japan. This was justified on supply security grounds, i.e. importers needed to be able to replace any quantities lost through war, natural disaster or other *force majeure*. The supply security argument implicit in PML does however bear very little scrutiny. The major potential source for imports at that time would not have been an area prone to political or economic instability, but Singapore. Singapore would also have had the capacity to make up for any truncation of supplies from more volatile areas. Further, the conditions under which supply insecurity could affect oil products and not crude oil are hard to envisage.

Confining importation to refiners did remove the spectre of free import competition, which, due to the experience of the 1920s and early 1930s, we would argue was still a dominant factor behind policy. If the refining industry was fully competitive itself, the restriction would not have stopped import price convergence. The controls on the industry were such that internal competition through access to the import market never became a reality. However, even in domestic regulation, a liberalization process was at least beginning.

In November 1986 the government created a committee to consider future deregulation. The report of the Basic Oil Industry Issues Study Committee was published in June 1987, and recommended a series of liberalizing measures that were brought into effect over the next five years. These included the end of the quota system for gasoline production in 1989, a lessening of controls on new retail stations in 1990, and the removal of direct administrative controls on refinery runs in 1992. While liberalizing measures, none of these affected the relative isolation of the industry with respect to the international market. In particular, there was no move towards any convergence of domestic wholesale prices onto an import price.

As the expiry of PML came closer, the way became clear for a more meaningful liberalization of the oil trade regime. The dislocation of oil prices in the wake of the 1970s oil shocks had been maintained through to the 1990s, and, in particular, profits had remained very heavily skewed towards gasoline. The extent of this, and the degree to which the industry was cushioned from making efficiency savings, is shown in Table 4.5, which provides a breakdown of the retail price of gasoline in USA and in Japan prior to liberalization.

The table shows that while the Japanese industry is considerably less efficient, it is also far more profitable in gasoline production and retailing. In Japan refining costs and wholesale costs are shown to be 50 per cent

Table 4.5:   Gasoline Cost in the USA and Japan. 1992. Yen Per Litre.

	Japan	*USA*
Crude oil cost	15.2	14.7
Refining/Wholesale cost	9.3	6.2
Refining/Wholesale margin	17.3	1.7
Wholesale price	41.8	22.6
Distribution cost and margin	20.2	3.7
Tax	59.7	10.2
Retail price	121.7	36.5

Source: MITI estimates.

greater than the USA, with a profit margin at these stages ten times larger. With a distribution cost and margin more than five times higher in Japan, the total downstream costs and margins (net of taxes and crude oil costs) amount to 46.8 yen per litre compared to just 11.2 yen per litre in the USA. The implication is that higher profit margins, combined with inefficiencies and additional costs due to regulation, amount to some 35 yen per litre, more than half the retail price exclusive of tax. A comparison with other Asian refiners confirms the high cost nature of the operations of the Japanese industry. Using 1993 data, refining costs alone amounted to about $2.20 per barrel in both Singapore and Korea, while the equivalent cost in Japan was $6.80 per barrel.[16]

The modifications to the import regulation regime consequent on the expiry of PML represent the greatest structural change in the Japanese market in fifty years. The new regulations took effect at the start of April 1996, and removed the refining requirement for potential importers.[17] For the first time, independent retailers had the option to conclude direct deals with foreign refiners, with the immediate potential source being Korea. However, the start of the new regime in April 1996 in itself represented no 'big bang' of liberalism. The volume of imports in the five months after the expiry of PML averaged just 32 thousand b/d, about 3.5 per cent of total supply and 10 thousand b/d less than the corresponding period in 1995. All the initial adjustment to the rules was made by direct price rather than import quantity adjustments. Further, this adjustment had been completed by, rather than started in, April 1996.

To illustrate this adjustment, in Figure 4.3 we have shown the domestic cost and profit margin for gasoline. This is calculated as the retail yen price of gasoline in Tokyo, net of tax and also net of the international spot market price for gasoline delivered into Japan. The latter is used as a shadow price, i.e. a proxy for the ex-refinery value evaluated at international prices.[18] The resulting difference as shown is then the aggregation

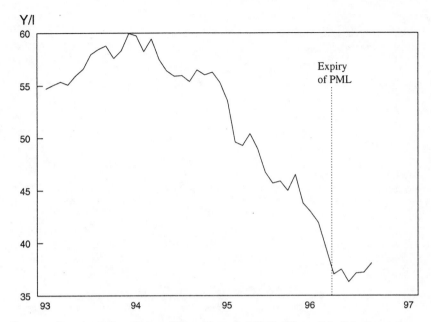

Source: Petroleum Association of Japan data, and *Platt's Oilgram Price Report.*

Figure 4.3 : Japanese Domestic Cost and Profit Margin for Gasoline. January 1993 to August 1996. Yen Per Litre.

of wholesale, distribution and marketing costs and margins, plus the additional refining margin resulting from the difference between ex-refinery values at domestic rather than international prices.

The fall in margins shown in Figure 4.3 began at the start of 1995, before stabilization occurred at the time the new import regime came into effect in April 1996. The total fall over the period shown amounted to about 18 yen per litre, representing about one-third of the original value of the difference between net of tax retail and international cargo prices. About 90 per cent of this adjustment came from falling retail prices, the rest from a period of increasing international dollar prices combined with yen depreciation against the dollar. The tax inclusive retail price is shown in Figure 4.4. Prices fell throughout 1993 and 1994, but as comparison with Figure 4.3 shows, not by enough to pass on the full extent of falls in the international price in yen terms. The fall in price was precipitous through 1995, before stabilization was reached at 107 yen per litre at the time of PML expiry. Two main questions are raised by the adjustment prior to PML, namely the mechanism that induced it, and whether it is a full adjustment to international prices.

The adjustment began before liberalization, because the expiry of PML

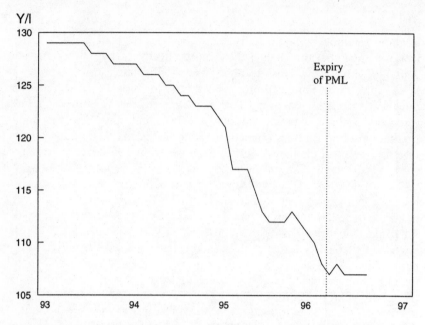

Source: Petroleum Association of Japan.

Figure 4.4:   Retail Price of Gasoline in Tokyo. January 1993 to August 1996. Yen Per
Litre.

made import competition and potential entry into the Japanese market a
credible threat. It thus inititiated competition between incumbent firms,
which, through those firms cutting into some of their cost overhangs and
large margins, brought prices down to levels where the threat of large-
scale import competition was removed. The credibility of new import
flows was made clear by a series of deals signed between (non-refining)
Japanese retailers and Korean refineries in 1995 and 1996 prior to
liberalization. The new importers have primarily been agricultural collect-
ives such as Zennoh on Honshu and the Hokuren Federation on Hokkaido,
which take a significant share of the rural market. While the volumes
concerned were relatively small, such deals were enough to achieve a
strong motive force towards domestic price reductions in order to minimize
the scope for further undercutting by imports. In response, in the months
leading up to PML expiry, refiners began to change their pricing formulae
for sales into the wholesale market. In particular, for the first time
international product prices and product freight rates entered formulae
that had previously been driven by crude oil costs and exchange rates
alone.

A potential entry threat emerged from Japanese supermarkets. In contrast to European markets, the large-scale entry of supermarkets into gasoline retailing in Japan had effectively been blockaded. With liberalization, entry became achievable, most effectively by setting up joint ventures with a *sogo shosha* (trading house).[19] If relative prices were to be favourable, these joint ventures would turn to imports, exerting further downward pressure on prices and a greater degree of domestic market response to changes in international market conditions. The first supermarket gasoline station opened in June 1996 in Nagano prefecture, resulting in an immediate fall in local prices from 105 yen per litre (close to the Tokyo and national averages), down to 95 yen per litre.

The credibility of import flows, and the threat (and realization) of entry, is not enough in itself to assume that the answer to our second question (whether the adjustment to international opportunity costs is a full one) is necessarily in the affirmative. The liberalization was only partial in that entry barriers still remain, leaving some scope for further price falls. The wholesale price of gasoline in the six months after liberalization stabilized at about 26 yen per litre net of tax, about one half of its value at the start of 1995. Over the same six months, the yen delivered cost of imports, based on Singapore prices, averaged about 16 yen. That 10 yen difference (as seen above, and not coincidentally, equal to the local impact of the opening of a supermarket retail site), effectively represents the cost of the remaining barriers to entry.

The barriers to entry after April 1996 come in several forms. We would identify four main types, although the following list is not intended to be exhaustive. First, the stockpiling requirements, while less onerous for a new entrant than before liberalization, still represent an opportunity cost for capital. It is also discriminatory, given that a domestic refiner can count the quantities held in the production process or necessary for smooth operation (i.e. minimum operating inventories) towards their obligations. The amount of inventories held purely because of regulation is thus lower for a refiner than an importer, and hence the associated opportunity cost of capital is also less. Secondly, entry to the retail sector is not allowed without having a certificate showing proof of supply from a wholesaler, i.e. a *moto-uri*. As this effectively means that the *moto-uri* can decide the pace and composition of any entry or potentially blockade it, it represents a barrier to entry.[20] Thirdly, while wholesale prices are transparent, there is also a system of selective and non-transparent rebates from wholesale prices in operation. This is a potential entry deterrent, since it allows the possibility of predatory pricing in areas where entry could occur. Finally, Japanese quality specifications may provide a barrier. We enlarge on this latter point further below.

Despite the presence of some barriers to entry, we have seen that import liberalization brought about large falls in gasoline prices, and greatly reduced the extent to which the Japanese market can move out of line with the international market. The consequences of this were profound, given what had been the skewed nature of profits in the Japanese oil industry. The reduction in prices, and consequent removal of profits, lead to a strong motive for cost reduction and rationalization within the industry. One aspect of rationalization is changes and consolidation in ownership structure. The first such change came in December 1995, prior to liberalization, when Caltex announced a decision to pull out of its major Japanese refining asset, Nippon Petroleum. Caltex sold their 50 per cent share to Nippon Oil, already owner of the other half of the company. As shown in Figure 4.2, while this still leaves Caltex with a stake in Koa Oil, the sale represents a major withdrawal of foreign capital from Japan. Given at that time Caltex were about to bring their new Thai refinery on stream, and expand the Korean refinery in which they have a stake, the move can be seen as being symptomatic of the decline in profitability in Japan compared to other areas in Asia.

While European and US oil companies had gone through severe restructuring during the 1980s and early 1990s, involving large reductions in staffing levels, Japanese companies had maintained, by international comparison, very high staffing levels. Cutting staff by attrition and by a hiring freeze is perhaps the easiest way to cut costs. Such a method reduces the need for making redundancies, particularly difficult given the still dominant Japanese ethos of life-long single company employment, and produces a savings figure that can be quantified long before it is actually realized. The first impact of liberalization prior to its actual arrival came in this form. A series of companies announced large staffing cuts through natural wastage in 1995, for example Cosmo Oil announced plans to lose 700 jobs over three years, while over the same period Japan Energy planned to lose 800. In both cases this represents nearly 20 per cent of the workforce. The succession of restructuring plans (including among the companies with foreign capital Showa Shell) continued through 1996, with the 20 per cent figure dominant in announced plans.

It should be noted that the Japanese approach to what in the West would be termed downsizing, operates in a far less confrontational manner. In the USA and Europe a greater proportion of staff savings are met through outright redundancy, and executive salaries have tended to rise sharply post restructuring. In Japan there is a greater emphasis on natural wastage, and the pressure on executive salaries is normally downwards, for example the Japan Energy programme involved in a 10 per cent cut in the top salaries. It should also be noted that because of the sectoral breadth of many companies involved in the Japanese oil industry, there is also a scope for transfer to other divisions. This enables the pain of oil industry restructuring to be partly absorbed through reduced hiring in other sectors, in a way that was not a possibility in the US and European industries.

The question remains as to whether there is enough scope in natural wastage to absorb all of the adjustment required by liberalization. Given the degree of excess costs in the industry, the natural wastage savings are useful but insufficient to close the efficiency gap alone. Additional savings arise from consolidation of offices, more effective use of information technology and selective withdrawal from the least profitable markets. Further cost reductions are possible from the pooling of distribution facilities, or from an expansion of swap agreements (currently far less prevalent in Japan than in Europe), whereby freight costs are minimized by companies exchanging the rights to physical supplies in different locations.

Despite the scope for cost savings, the problem at source remains that Japan has too many small high cost refineries, and is confronted with competition from more cost effective refiners abroad. Given this, the ultimate long-run solution would be expected to involve consolidation in the sector, with the closure of some small refineries, and the revamping and expansion of the survivors.

Liberalization in Japan has been a process of moving away from explicit controls and explicit protectionism. However, as in other countries, liberalization does not imply a complete withdrawal of government from the market, nor does it preclude the use of either implicit controls or more subtle explicit controls. Indeed, we saw in the previous section that Japanese regulation has at several points in the past swung between explicit and implicit mechanisms. We will use one form of control as an example.

Earlier in this section we mentioned the question of oil product quality specifications in the context of barriers to entry. These specifications are primarily motivated by air quality and other environmental concerns. However, there is a case to be made that they also represent a new indirect form of regulation that can be used to provide insulation for a domestic oil industry. If foreign refiners are not able to meet a specification, then import competition is removed. If they can only achieve the required quality at additional cost (e.g. through blending operations), then that cost provides a buffer for the domestic industry.

Japan's quality specifications are very tight, and continue to become more restrictive. As an important example, from 1999 the allowable benzene content in gasoline sold in Japan falls from 5 per cent to 1 per cent. With the current refinery structure in other countries this would rule out a large raft of potential import sources. Those foreign refiners would then face the choice of incurring the cost of upgrading to meet the new standards, or foregoing the future ability to export to Japan. We have seen that import competition has impacted on the Japanese market through the potential, and not the actuality of large-scale imports. Given this, the likelihood would be that foreign refiners would not consider expenditure on upgrading solely to allow them to meet the Japanese specification, as a viable proposition.

Environmental standards can then have consequences for the scale of potential import competition, and thus have price and industrial structure implications. Oil products have a multiplicity of characteristics, and the possibility of fragmentation of markets due to widely different environmental specifications for several quality dimensions in each product, provides scope for the maintenance of what are, in effect, non-tariff barriers. In a world where price and quantity regulation can complicate external trade relationships, the ability to achieve equivalent effects through environmental quality standards may prove to be a useful addition to a government's regulatory armoury.

Notes

1. A similar pattern was shown in the USA, where oil demand also regained its 1973 level in 1994. However, 1978 remains the peak year for US demand.
2. The equivalent figures for the US crude oil import bill are 211 hours of GDP in 1980, and 51 hours in 1995. The difference between the US and Japanese import bill in GDP units has thus disappeared.
3. Table 4.4 shows facilities with primary distillation capacity. In addition there are units with no distillation or cracking that perform other functions. For example, Nippon Petroleum Refining has reforming and LPG recovery facilities on Okinawa and also in Yokohama.
4. A distinction can be drawn between *kigyo-shudan* (a loose knit group of companies), *kinyu-keiretsu* (a broad group financed by the same financial institution) and *keiretsu* (a dominant firm with often vertically integrated subsidiaries). See Hiroyuki Odagiri (1992), *Growth Through Competition: Competition Through Growth*, Oxford University Press, Oxford. While acknowledging the impreciseness of '*keiretsu*', and that in particular the firms we are referring to are primarily *kigyo-shudan*, we use the broader definition of *keiretsu* in this chapter.
5. For an account of the development and dismantling of the *zaibatsu*, see Hidemasa Morikawa (1992), *Zaibatsu: the Rise and Fall of Family Enterprise Groups in Japan*, University of Tokyo Press, Tokyo.
6. Takatoshi Ito (1992), *The Japanese Economy*, Massachusetts Institute of Technology Press, Cambridge, Massachusetts.
7. Thomas M. Huber (1994), *Strategic Economy in Japan*, Westview Press, Boulder, Colorado.
8. Paul Sheard (1994), 'Interlocking Shareholdings and Corporate Governance in Japan', in Masahiko Aoki and Ronald Dore, *The Japanese Firm: Sources of Competitive Strength*, Clarendon Press, Oxford.
9. See Laura E. Hein (1990), *Fuelling Growth: The Energy Revolution and Economic Policy in Postwar Japan*, Council on East Asian Studies, Harvard University Press, Cambridge, Massachusetts.

10. Richard J. Samuels (1987), *The Business of the Japanese State: Energy Markets in Comparative and Historical Perspective*, Cornell University Press, Ithaca, New York.
11. Chalmers Johnson (1982), *MITI and the Japanese Miracle: The Growth of Industrial Policy, 1925–75*, Stanford University Press, Stanford, California.
12. R.J. Samuels (1987), *The Business of the Japanese State*, op.cit.
13. Richard J. Samuels (1987), *The Business of the Japanese State: Energy Markets in Comparative and Historical Perspective*, Cornell University Press, Ithaca, New York.
14. The ENI example was attractive more in its symbolism than in its strategic implications. An integral part of Mattei's strategy was to displace the majors in the upstream as well as in the downstream. The idea of an upstream challenge outside Japan did not play a major role in the Japanese policy debate in the early 1960s.
15. This is shown most clearly in the case of the independent retailer Lion Oil, who found that delivering Singapore gasoline into Japan brought about indirect pressures on their financing. See Samuels (1987), *The Business of the Japanese State*, op.cit.
16. Hideaki Fuji (1996), 'Why Japan's Oil Refining Costs are so High? A Factor Analysis on Oil Refining Cost Gaps Among Japan, South Korea and Singapore', *Energy in Japan*, March 1996.
17. In addition a change in how the stockpiling requirements are accounted reduces the barrier to new entry inherent in the 70 days of supplies inventory conditions. That barrier is however still significant.
18. We have used the Platt's quote for c+f Japan unleaded gasoline (effectively the Singapore fob price plus freight) converted into yen.
19. The first of these joint ventures has involved Marubeni and Mitsubishi. The tie-up with a *sogo shosha* is primarily motivated by the need to have a supply certificate from a wholesaler, as discussed below.
20. At time of writing MITI is considering removing the proof of supply from *moto-uri* as an entry condition for post 1999 entry. See *Platt's Oilgram News*, 10 September 1996.

CHAPTER 5

STRUCTURAL CHANGE IN FIVE ASIAN OIL SECTORS

1. Introduction

The previous two chapters have discussed Japan and China, the largest consumers of oil in Asia. In this chapter we consider India, Korea, the Philippines, Thailand and Taiwan separately in the next five sections.[1] Their common feature is the presence of a deregulation programme, but with different speeds of liberalization and starting from a range of degrees of government regulation and direct involvement. They provide examples of the full range of regulatory experience. In India we show that regulation was a very gradual process, with a ratcheting up of state control over a long time period. The reversal of the process may be on a faster track, but the structures of the past are still taking some time to unwind. Korean liberalization has involved the lifting of heavy state controls, and unleashed strong but inward looking competitive forces. In the Philippines regulation and state participation in the main grew quickly in the 1970s, and liberalization has been comparatively fast. In Thailand, large-scale state involvement started even later, liberalization has been even quicker, but has been flawed by a tendency towards post liberalization indirect interference in prices. In Taiwan, state control was total from the time of the exile to the island of the Kuomintang, and the process is a dismantling of the previous system, although perhaps only moving from a protected monopoly towards a protected duopoly. A final section offers some conclusions.

2. India

The oil industry in India is the oldest in Asia. Drilling in Assam began as early as 1866, and the first field was found in 1889. This field, Digboi, close to the border with Burma has the distinction of being the oldest continuously operated oilfield in the world.[2] The refining industry also began early, with a small temporary unit being started in 1893 before the commissioning of the Digboi refinery in 1901. However, Assam oil was never prolific, and Digboi itself peaked in 1945 at a little over 7 thousand b/d. The industry that evolved was almost completely dependent on oil product imports. By independence in 1947, the market was supplied primarily by Caltex, Standard Vacuum (with the Indian assets given to Exxon in the division of Standard Vacuum in 1960), and Burmah Shell, the joint marketing company of Shell and Burmah Oil.

Independence saw the start of a process of increasing state control that continued in a relatively linear fashion for more than thirty years. The economic policy of the Nehru government was essentially centralist, but did not endorse any sharp increases in the degree of public control. The policy is best described in the Industrial Policy Resolution of 1948, which placed petroleum into a category of six key industries where the aim was for the state to undertake any new ventures. It was a policy of creeping nationalization, leaving the existing private sector to whither on the vine while all industrial dynamics were to be carried out by the state. Policy grew more centralist in the early 1950s, and a further statement of industrial policy in 1956 produced further controls on foreign capital, and laid the foundations of a system of tight price and profit regulation.[3]

Despite the 1948 resolution that wished to freeze private sector involvement, there was no state apparatus or public sector company in a position immediately to take a leading edge role. The first stage in the development of the industry post-independence placed the private sector in the leading role. The next step in the evolution of the industry, the growth of a significant refining sector, was left to the private sector. Three refineries came on stream in close succession, Standard Vacuum started operations in 1954, Burmah Shell in 1955 and Caltex in 1957. However, after this final flourish by the private sector in a leading role, state involvement began to grow, and the component elements began to form.

The second post-independence stage of the Indian industry was one of the development of the state sector, with that sector playing an increasingly important role. However, foreign capital still had a function in taking minority shares in new developments. The first major public sector operation was the creation in 1955 of what became the Oil and Natural Gas Commission (ONGC), with responsibility for the search for oil in the upstream. In 1958 Indian Refineries was set up to construct and operate public sector refineries, and in 1959 the Indian Oil Company was established to market and distribute products, as well as to manage part of the import trade. Note that the model used for the creation of the public sector is fairly unusual, there was no integrated national oil company but a separate concern in each of three stages of the industry. Greater integration was achieved in 1964 with the merger of the refining and marketing interests to form the Indian Oil Corporation (IOC).

In the refining industry, after Caltex's refinery came on stream in 1957, all new refineries were state led. In the 1960s the Madras and Cochin refineries came on stream, both with a government majority interest, with Amoco and the National Iranian Oil Company having 13 per cent shares in Madras, and Phillips having 26 per cent of Cochin. Six further refineries built between 1961 and 1982 used no foreign capital, although they involved the use of technical collaboration from Romania and Russia. The state was dominating new developments, and the next stage was to remove Western involvement completely.

The highly gradualist nature of the growth in state control is perhaps best exemplified by the movement of Burmah Oil's holdings in Assam (including the Digboi refinery) into the public sector. The state took a one-third share in what was constituted as Oil India in 1959, which increased to a half in 1961. Finally, the state took full control, but only in 1981 bringing to an end a twenty-two year transition process for the company. The incorporation of the rest of the private sector was faster, but still gradualist and piecemeal.

Worldwide the growth of public control and regulation of oil was to a large extent a function of the oil price shocks of the 1970s. In India the process of creeping state control was already at work, and arguably would eventually have been completed in any case. However, the first oil price shock certainly occasioned an acceleration of this process, and moved the Indian industry into the third stage of its development, that of the assimilation of the private sector into the public.

In 1974 the state took control of 74 per cent of Exxon's Indian interests, with an option on buying the remaining assets in 1981, which was actually exercised in 1976. From Exxon's point of view this could be seen as the logical culmination of years of inexorable increases in regulation, which had made their operations increasingly difficult long before the first oil price shock. Indeed, as noted by Wall,[4] India had been becoming a very marginal operation in its attractiveness in the context of Exxon's overall operations because of the remorseless increase in state intervention. Having created Hindustan Petroleum through the purchase of Exxon's assets, in 1976 Bharat Petroleum emerged as the result of the purchase of Burmah Shell assets. The Caltex operations were assimilated into the state sector next, to be followed by Burmah, Amoco and Phillips in the early 1980s.

There had been no sudden nationalization of the Indian industry, and no sharp discontinuity in policy. Economic nationalism had manifested itself in a creeping process, making operations more difficult for foreign capital and then picking the companies off one by one. Given the extremely gradualist nature shown in history of the evolution of state control and regulation, it is perhaps hardly surprising that the reversal of the process should follow the same pattern, with economic nationalism still playing a major role.

By 1991, when the government of P.V. Narasimha Rao came to power, Indian oil policy was showing signs of severe stress. The heavy subsidization of kerosene and LPG in particular was creating an enormous drain on what was an already tight fiscal situation. Oil demand was rising rapidly, and the domestic refining industry was having difficulty in keeping pace. Severe infrastructural and logistical problems existed in the downstream of the industry. In the upstream production was falling, and investment was insufficient to help mitigate the decline. The industry was faced with capital deprivation, increasing imports of both crude oil and oil products, and the system was becoming an increasingly intolerable burden on the public sector. Within the context of general economic liberalization the industry was a prime candidate for reform.

The reform process has proved to be as piecemeal and gradual as had been the expansion of the state sector. However, on the three broad fronts of entry barriers, pricing and privatization, progress has been made. As we have seen above, state companies gradually took over the refining industry, with further private sector involvement being precluded. In 1992 this barrier was removed, a raft of new refinery projects involving both domestic and foreign capital have been approved, and, particularly in the smallest of the existing refineries, some state sales of shares have been possible. Table 5.1 shows the fourteen refineries in operation in 1997, together with their current government ownership share and capacities. The identification numbers shown relate to the location of the refineries as shown in Figure 5.1.

Table 5.1: Indian Oil Refineries. 1997. Ownership, Location and Capacities. Thousand b/d.

Refiner (Government Share)	No.	Location	Crude Capacity	Thermal Cracking	Catalytic Cracking	Hydro-Cracking
Indian Oil Corporation	1	Koyali, Gujarat	195	20	20	20
(91%)	2	Mathura, Uttar Pradesh	160	20	0	0
	3	Begusarai, Bihar	70	22	0	0
	4	Medinpur, West Bengal	60	10	0	0
	5	Guwahati, Assam	20	6	0	0
	6	Digboi, Assam	10	1	0	0
Hindustan Petroleum (51%)	7	Mahul, Bombay	110	0	12	0
	8	Visakhapatnam	90	0	20	0
Madras Refineries (54%)	9	Madras	130	10	12	0
	10	Narimanam	10	0	0	0
Bharat Petroleum (66%)	11	Maharastra, Bombay	135	0	35	0
Cochin Refineries (55%)	12	Ambalamugal, Cochin	95	20	25	0
Bongaigaon Refinery (75%)	13	Bongaigaon, Assam	30	2	0	0
Mangalore Refineries (13%)	14	Mangalore, Karnataka	60	0	0	10
TOTAL			1175	111	124	30

Sources: Various.

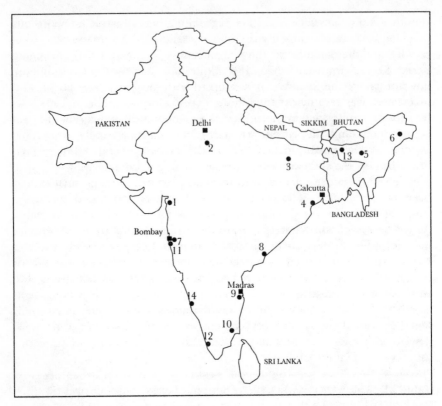

Figure 5.1:   Location of Indian Refineries.

The private sector shares in the first thirteen companies shown in Table 5.1, arose from a partial privatization programme during the Rao government.[5] The key point to be made is that the process is partial, and the state retains a controlling interest. In the case of IOC (and also of the upstream ONGC) this could be put down in part to domestic capital market constraints and political constraints on allowing large-scale foreign ownership at an early stage. The longer-term aim remains for the state to divest 49 per cent of both the giant corporations, with the possibility of a New York listing. However, it should be noted that the sale of 49 per cent is likely to be considerably less attractive to potential investors than the sale of 51 per cent, and that extra 2 per cent still represents a rubicon in terms of political economy. Given the above history of state control, it has proved extremely difficult for India to consider crossing that rubicon. In the case of the smaller companies, the domestic capital constraint argument does not hold. The motives behind privatization are more directly seen as revenue raising to reduce the public sector deficit, perhaps deriving some benefit from benchmarking, but without reducing the primacy of state control, at least in the earlier stages of liberalization.

The control structure for the Indian refining industry works through a mesh of quantity and price controls administered by the Oil Coordination Committee. Pricing controls have resulted in distortions in relative pricing, with gasoline prices maintained far in excess of other products, and used to subsidize kerosene in particular (kerosene subsidies exceed all food subsidies combined). Using 1994 prices, the controlled price of gasoline stood at about six times that of kerosene, a similar structure to that noted in Japan but of a far more severe degree. While of a smaller degree, gasoil (the key fuel in the strong transportation growth associated with the development of the internal Indian market) has also been effectively subsidized through the gasoline price. Beyond this strong cross-subsidization, prices are also set according to rate of return calculations. Refineries are allowed to recover costs, and then achieve a 12 per cent rate of return on equity (calculated post tax). Despite mechanisms built in to try to reduce the complete disincentives this system would provide for cost saving or innovation if unmodified, compared to free market pricing there are still strong disincentive effects. In addition, a complex system of freight adjustments results in there being in effect (with some minor regional variations) an all India administered price. This failure to incorporate any true measure of (often high) transport costs into prices results in a further degree of strong cross-subsidization, primarily away from urban areas in favour of interior rural areas. The system is extremely expensive within the context of India's overall government expenditure, with the state incurring the costs of any increases in international prices. After any such increases, refiners are compensated through the Oil Pool Account, which as of 1996 amounted to a subsidy of close to $2 billion.

The first significant move away from this structure came in 1993 with the adoption of a parallel marketing scheme in kerosene and LPG. Rather like the privatization programme to date, this is a very partial measure, allowing the development of a free market at the margin while leaving the distorted intra-marginal official price setting mechanism unchanged. In economic terms the parallel market price, given the rationing created by the low official prices, is the same concept as a black market price, with the distinction that the parallel market should be sourced through imports rather than through the diversion of supplies from the official market.

The experience of the parallel market scheme has been mixed. To the extent that kerosene imports have loosened the extent of rationing and provided a demonstration effect to consumers that the administered price is absurdly low, there has been some positive effect. On the negative side, the parallel market has been hampered by logistical problems, lack of access to storage for bulk breaking of cargoes, and problems with fair access to pipelines. While state companies have been given instructions to facilitate access to storage and pipelines, this is far away from guaranteed access through regulation of natural monopoly facilities and distribution networks. This has constrained the kerosene trade and made entry to the LPG parallel market all but impossible. The one area that needs regulation to achieve fair competition then remains unregulated, while the core of the system retains tight regulation. Poor infrastructure and lack of access to what there is, have severely impacted on the potential for competition. For example, because of the problems with bulk breaking a standard cargo, traders have often resorted to 'piggy-backing' kerosene, that is to say they arrange for a small cargo to be used to fill space in a cargo of gasoil entering India to fill an IOC tender. The volumes of kerosene that can enter the market are accordingly constrained.

It should also be noted that the parallel market price is *not* a free market price. Given that it is the price at the margin into a rationed market, and the flows involved are not enough to proxy what would be the unrationed price at the margin, it is in excess of the undistorted market price. Given that, primarily due to logistical factors, the bulk of the parallel market is in urban areas and its existence allows more from the regulated market to flow into rural areas, indirectly the result is further cross-subsidization of rural areas by urban areas. The political problems with lifting kerosene subsidies, particularly in rural areas, then remain a key constraint.

While we would argue that the parallel marketing scheme has so far not borne major fruit, there were other significant liberalizing measures in 1993. As well as kerosene and LPG, the IOC monopoly on imports of paraffinic naphtha and low sulphur waxy residue (LSWR) was also removed. It has been the LSWR market that has showed the first real competition, as state companies have vied to supply the fuel to utilities. However, it is again significant that foreign traders have mainly kept out of the competition for business with the utilities, primarily due to the same infrastructure access problems referred to above.

We have seen above that economic nationalism was at the heart of Indian policy from independence. As such it still has some political popularity. The process can not be described as purely a smooth but slow evolution towards reform and a growing role for foreign capital. There are still political risks capable of causing some discontinuities and reversals. The clearest example of this came in August 1994, with the cancellation by Maharashtra state of a $2.8 billion power station that was to be built by Enron. The Hindu nationalist parties Bharatiya Janata (BJP) and Shiv Sena that had taken power in Maharashtra used strong anti-foreign capital rhetoric in cancelling a deal towards which Enron had already incurred large sunk costs.

At one level the Enron cancellation could be seen as a major deterrent to foreign investors, especially as the BJP gained more influence in national politics in the 1996 election (and indeed achieved a very brief spell in national government). Further, the use of the economic nationalist line by politicians had achieved some popularity. At another level, the Maharashtra decision was not so dramatic a symbol. First, the deal itself may not have been very typical, there had been no competitive tender and suggestions of some special circumstances had abounded.[6] Secondly, as we have already seen, there is often a considerable gulf in India between political rhetoric and political practice. In particular, the picture of liberal forces facing an illiberal opposition does not hold. In terms of implementation of outward looking economic liberalism, BJP-controlled states in India have often been far more effective than central government. After renegotiation, the Enron deal went ahead, but its progress has demonstrated that a degree of political risk remains.[7]

We would not wish to overemphasize the importance of political risk in India. To a large extent there is political consensus on reform, and the differences in emphasis can be seen in practical terms as being about the pace of reform. After the defeat of the Rao administration in 1996, the Janata Dal centred United Front minority government led by Deve Gowda was left with a series of difficult unfinished reform tasks. The outgoing government had planned the process on a six-year timetable. By 1998 price subsidies on oil products were due to be removed, by 2000 domestic distribution of products were to be liberalized, and controls on crude oil prices removed. Finally by 2002 restrictions on exports and imports were due to be fully removed. The Deve Gowda administration soon found that even this slow timetable was optimistic, with the first element in particular being problematic. Some parts of a series of increases announced in July 1996 had to be rescinded after widespread public protest, resulting in the government acknowledgment that the lifting of subsidies would need to be phased over an extended period. Deregulation may remain the ultimate policy goal, but the timetable is no longer fixed and likely to become very extended.

There is a noticeable difference between countries with fast and linear oil sector reform, and those using a slower pace. In the former, deregulation has been accompanied with an infusion of both foreign and domestic capital into refining, often, as we will see below in the case of the Philippines and Thailand, resulting in at least a period of short-term overcapacity. In the slower countries one is more likely to see a long queue of proposed projects. Such has been the case in both China and India. By 1996, the list of refinery projects approved by the Indian government reached fifteen. Some involved independent Indian capital, others Middle East capital through Saudi Aramco or the Kuwait

Petroleum Corporation. Some, for instance a proposal involving Shell and its former Indian self now embodied in Bharat, involved the return of international oil capital displaced during the 1970s. Were they all to be built, India would have more excess refinery capacity than the rest of Asia combined. In reality they will not all be built, as the intention to build an Indian refinery is rather different than an intention to build a, say, Thai refinery where intentions have tended to become realities. In most cases, the refinery plans should be seen as more of a statement of intent to commit to building a refinery should the liberalization plans hold their course, rather than as any actual commitment. In the medium term enough problems remain on issues such as retail network access, pipeline access, port logistics and crude procurement to keep most of the projects firmly in the category of possible developments for some time to come.

The degree to which India will become a 'sink' absorbing the products from excess capacity in other countries is a major variable in assessing the implications of the fast growth of refining capacity in Asia. As we have noted above, the pace of new refinery construction is bound, *inter alia*, to the progress of the reform process. However, to give some idea of the scale of the possible 'sink', some rough orders of magnitude can be derived.

Consider first the supply side. Expansion plans for existing refineries seem to imply an addition of some 250 thousand b/d of capacity by 2000, with the major expansions (around 60 thousand b/d) due at IOC's Koyali and Digboi plants, the Hindustan Petroleum Visakhapatnam plant, and the Mangalore plant. This produces a figure for additions of some 500 thousand b/d by 2000, which can be taken as a lower bound for additions. On the upper bound, allowing for a possible further 300 thousand b/d in new capacity produces a range for total capacity in 2000 of about 1.6 to 1.9 mb/d, with a large tranche of projects feasible after that point should reform progressing to plan and regulation of the natural monopolies in the infrastructure be in place.

There are also major uncertainties about the rate of demand growth in India, related not only to the pace of economic growth but also to the extent to which prices are allowed to adjust outside regulation, and the extent to which infrastructure and rationing constraints are reduced. We would place reasonable demand estimates for 2000 in the range of 2 mb/d (medium growth, moderate liberalization success) to 2.3 mb/d (faster growth, success in developing infrastructure and in liberalizing).[8] Taking supply and demand together produces a range for excess demand of between 0.1 mb/d and 0.7 mb/d, with a midpoint of 0.4 mb/d probably representing a fair estimate. Given that the equivalent figure in 1996 was 0.5 mb/d, even the upper estimate implies that the degree to which India can act as a sponge for excess capacity elsewhere should not be overstated. Indeed, the conditions that generate the upper estimate are precisely those which facilitate a large number of refinery projects early in the next decade, thus suggesting that any large deficit could be short lived given the endogeneity of refinery capacity. It should also be stressed that the appearance of any deficit is extremely sensitive to whether any more than two new refineries are constructed before 2000.

There has been a split in opinions within the international industry between the relative wisdom of and prospects for investment in China and India. From this section and Chapter 3 we can note the different character of reform in the two countries. Chinese reform can move extremely fast, but is prone to reversals and subject to considerable political and economic uncertainties. Reform in India is extremely slow, but has at least been moving forward, and the uncertainties relate to timing rather than objectives. Invoking an analogy drawn from Aesop might prove too much of a temptation for many.

3. Korea

The oil sector in Korea represents an extreme in the world context. The growth of oil demand has been faster in South Korea than anywhere else, making it the world's sixth largest consumer of oil in 1996. Fast demand growth has been accompanied by even faster growth in refinery capacity, with absolute increments so large and so fast that they have had strong repercussions on the profitability of refining across Asia as a whole. By 1995, Korean oil demand had nearly doubled over the previous five years, and quadrupled over the previous ten years. The dynamics of this extremely fast growth have been highly non-linear, and have been greatly magnified by dislocations and discrete once and for all structural changes. To illustrate, consider the evolution of oil and total primary energy demand as shown in Table 5.2.

There are two very distinct periods shown in the data. From 1980 to 1987 real GDP increased by 7.8 per cent per annum, with oil demand rising by only 2.7 per cent per annum. From 1987 to 1995 real GDP increased by 7.9 per cent per annum, about the same annual rate as in the earlier period. By contrast, oil demand accelerated to 16.9 per cent, and exceeded 20 per cent in each of the first three years of the 1990s. In the first period the share of oil in total primary energy fell consistently, with its incremental share being low, while in the second oil made up 85 per cent of the increase in primary energy demand.

These two distinct time periods relate closely to the period of the effects of Korean structural adjustment, followed by a period of structural changes and dislocations in energy use brought on by rapid economic growth. In the wake of the second oil shock prices increases were passed

Table 5.2:   Korean Energy Demand and Real GDP Growth. 1980–95.

Year	Oil Demand th b/d	Primary Energy th boe/d	Growth Rates, Per Cent Real GDP	Oil Demand	Primary Energy	Oil as % of Primary Energy	Oil as % of Incremental Energy
1980	485	795	-2.2	-1.0	3.2	61.0	-
1981	480	835	6.7	-1.0	5.1	57.5	-
1982	480	835	7.3	0.0	0.0	57.5	-
1983	500	910	11.8	4.2	8.9	54.9	26.7
1984	505	990	9.4	1.0	8.8	51.0	6.3
1985	535	1050	6.9	3.0	6.1	49.5	25.0
1986	590	1185	11.6	9.6	12.8	48.1	37.0
1987	620	1325	11.5	6.1	11.8	45.7	25.0
1988	740	1485	11.3	19.0	12.1	48.5	71.9
1989	855	1615	6.4	16.0	8.8	51.7	88.5
1990	1040	1830	9.5	22.8	13.3	56.0	88.4
1991	1255	2060	9.1	20.5	12.6	60.0	91.3
1992	1520	2315	5.1	21.1	12.4	64.6	102.0
1993	1675	2525	5.8	9.7	9.1	65.0	69.0
1994	1840	2735	8.4	10.1	8.3	66.0	78.6
1995	2010	2980	8.9	9.0	8.9	67.4	69.4

Sources: BP *Statistical Review of World Energy*, IMF *International Financial Statistics* and own calculations.

on into retail prices, producing a strong price effect on demand.[9] In addition energy efficiency and conservation were encouraged by both direct regulation and fiscal incentives for switching to more energy efficient technology, and in particular towards substitution away from oil. As shown by Table 5.2, the result was that energy demand grew considerably slower than GDP, and oil demand grew slower than total primary energy.

By 1988 several factors had combined to reverse this process. First, the oil price collapse of 1986 helped demand to recover and reduced the incentive for further substitution away from oil. However, more significant was a change in the sectoral use of energy. In the domestic sector, and particularly in rural areas, incomes had risen as far as to induce a switch from traditional fuels and domestic coal towards cleaner and more efficient oil products. Car ownership began to expand significantly leading to high growth rates for gasoline demand. The growth in domestic demand increased the internal transportation of goods and thus diesel demand grew sharply. Electricity demand grew faster than the rate of commissioning of new capacity, which was becoming heavily biased towards coal, nuclear and gas. As a result, the only way of meeting electricity demand was to increase throughputs at existing oil fired plants, and bring back extra oil fired capacity that had been taken off stream in the early 1980s. Taking all these factors together, energy demand began to rise faster than GDP, and as shown in Table 5.2, the incremental demand was very heavily biased towards oil.

The strong demand growth since 1987 has then been due in large part to temporary factors. The change in domestic utilization is a once and for all structural shift in demand away from more traditional but dirtier and less convenient fuels. As such the impact of this shift has begun to decline, and in the longer term some oil would be displaced from the domestic sector by gas and electricity. Likewise, the surge in oil demand for electricity generation was temporary, arising from a mismatch between the actual and desired power sector configuration given the long project times involved in bringing new plant on stream. The demand was due more to over pessimistic projections on electricity use rather than any positive and intentional substitution towards oil. Oil has therefore acted as a swing fuel, and over time new base load capacity will be biased away from oil. In total, post 1987 growth rates are unsustainable given that some of the major factors behind them are now either disappearing or beginning to work to erode oil's share. Future percentage demand increases will be heavily damped compared to the explosive growth of recent years, given that a significant part of this growth has been due to temporary factors. Table 5.3 shows the composition of Korean demand since 1990. Note the exceptionally strong rates of growth for naphtha due to petrochemical expansion, and also for gasoline as car ownership expands.

Deregulation in Korea has gradually brought prices to a direct linkage

Table 5.3: Composition of Korean Oil Demand. 1990–5. Thousand b/d.

	1990	1991	1992	1993	1994	1995	Growth 90–95 % p.a.
LPG	98	118	148	163	173	182	13.2
Naphtha	130	180	266	296	338	360	22.6
Gasoline	65	79	97	116	140	163	20.2
Jet/Kerosene	108	105	125	155	175	216	14.9
Diesel/Gasoil	266	314	349	378	403	447	10.9
Fuel Oil	279	336	384	399	438	446	9.8
Other	27	32	40	39	40	41	8.7
Bunkers	43	57	71	84	91	102	18.9
TOTAL	1016	1221	1480	1630	1798	1957	14.0

Note: The source used for this table does not include fuel used in the refineries. This accounts for the difference from the oil demand figure in Table 5.2.

Source: IEA, *Monthly Oil Report*, various issues.

with the Singapore market. During the 1980s some products were freed from the system of direct government price fixing – jet fuel in 1983, asphalt in 1988 and naphtha in 1989. Fuel oil in power generation was also freed in 1990. All of these were products where the final consumer was an industrial firm rather than an individual consumer. The basic consumer products, gasoline, kerosene and diesel, were left to government price fixing. As an intermediate stage of liberalization, prices were tied on a monthly basis to crude oil prices (in Korean won) from January 1994. In September 1994 the linkage was changed to the previous month's Singapore price.

The tax component in Korean retail prices is fairly large, particularly for gasoline where tax constituted about 60 per cent of final price in 1996, and was then increased by 20 per cent at the end of 1996. However, changes in those prices are now fully linked (albeit with a month's lag) to the Singapore market. The final stage of Korean liberalization will allow full liberalization of prices for all petroleum products. This was originally due at the start of 1997; instead, with less than a month to go, the government stepped back and imposed an interim system. Under this, all retail price changes would have to be given to the government for approval in advance, for a six-month period before full liberalization in July 1997. The change shows that Korean policy is often not robust, and is subject to change at short notice. There remains a lingering suspicion of a competitive system, and liberalization does not envisage leaving circumstances where Korean firms are uncompetitive in relation to outside capital (indeed the retail business is not due to open to foreign entry until 1999 at the earliest).

Just as Japanese industry is dominated by the *keiretsu*, the bulk of Korean industry is made up of a series of large multi-sector conglomerates known as *chaebol*. All Korean refineries are either wholly owned or have at least 50 per cent shares held by *chaebol*. Similar to the *keiretsu*, each *chaebol* is in effect a small economy, such is the degree of sectoral coverage in their activities. Where the similarity breaks down is in the central control of a *chaebol*, as opposed to the more decentralized and looser form of *keiretsu*. A *chaebol* tends to be fairly tightly controlled by a single entrepreneur, and act more as a unified decision-making unit.[10] In this regard *chaebol* are closer to the pre-war *zaibatsu* than to any organizational form found in the modern Japanese economy.

Vertical integration in the oil industry is normally defined as an alignment of production, refining and marketing assets. Under this definition the *chaebol* are vertically integrated, but their degree of integration is at a greater level, what could be termed as 'super-integration'. As well as production (in five countries), exploration (in fifteen countries), refining and marketing, within the oil industry they own petrochemical plants, are engaged in construction of plant and pipelines, and the production of oil tankers and service industry equipment. The integration extends beyond the retailing of energy to the consumer as they produce the capital goods that use energy products; automobiles and other consumer capital goods, power stations and industrial generation systems.

The degree of integration within the *chaebol* represents formidable competition within the oil industry. As an example, when countries with oil and gas potential have begun to liberalize, access is often suddenly available not just to the oil sector but to the whole economy. A conventional oil company focuses on the oil sector alone, while the *chaebol* can use oil as just one facet of a programme for entry into the economy as a whole. A *chaebol* can not only put together a package for the operation of, say, a new refinery. It can construct it, together with associated roads, ports and other infrastructure, and place the whole enterprise within multi-sectoral development. In upstream oil development the *chaebol* are often prepared to bid aggressively, as part of the value of an upstream contract can be in helping to facilitate entry into other sectors of the host economy. In short, the *chaebol* with their 'super-integration' have become a major force within the industry. While it is too simplistic to portray *chaebol* and *keiretsu* as being locked into ferocious competition, it is certainly true that the increased competition following the growth of the *chaebol* has become a major feature throughout the Asia-Pacific region and often beyond.

However, the major impact of Korea on the oil market stems not from the increasing internationalism and foreign success of the *chaebol*, but from the expansion of the domestic Korean industry. Korea is embarking on a massive programme of refining capacity expansion. Table 5.4 shows the crude distillation capacity of the five South Korean refiners at the start of 1995, and that which will pertain in 1997. The identification numbers for each refinery relate to their location as was shown in Figure 4.1. As we detail further below, the increments shown in Table 5.4 are by no means the end of the expansion process in Korean refining. The current involvement of foreign capital is limited. Shell had been involved with the Hyundai plants but withdrew in 1977. Gulf withdrew from the Sunkyong plant in 1980, Unocal from the Hanwha (formerly Kyung In Energy) plant in 1983, and the National Iranian Oil Company from Ssangyong in 1979. Currently the only foreign capital represented is Caltex (50 per cent share of the LG Caltex refinery) and Saudi Aramco (35 per cent share of the Ssangyong refinery).

Table 5.4 demonstrates an expansion in crude distillation capacity over two years of some 820 thousand b/d, a total which is roughly equivalent to the combined size of the three largest Singapore refineries. The Sunkyong refinery is already the largest of the more than 700 refineries

Table 5.4:   Korean Refineries. Crude Distillation Capacities. 1995 and 1997. Thousand b/d.

Refiner	No.	Location	Crude Oil Capacity	
			1995	*1997*
Sunkyong	41	Ulsan	610	810
LG Caltex	42	Yocheon	380	600
Ssangyong	43	Onsan	300	500
Hyundai	44	Sosan	110	310
Hanwha	45	Inchon	275	275
TOTAL			1675	2495

Source: *Petroleum Argus* and various.

in the world, even before expansion brings it to the extraordinary capacity of 810 thousand b/d. To put this figure in context, taking the largest refinery in each of France, Germany and the UK and combining the distillation capacity of the three still falls some way short of the expanded Sunkyong facility. Current post 1997 plans involve a further 300 thousand b/d in 1998 for Hyundai, followed by another 200 thousand b/d by 2000.[11] Ssangyong has also planned an additional 80 thousand b/d, and Hanwha 50 thousand b/d. Taking these into consideration, plus making an allowance for capacity creep,[12] Korean refinery capacity could reach 3.2 mb/d by 2000. A probable scenario is then as follows. The refiners are not capital constrained, particularly as they are part of *chaebol*.

The removal of state controls on refinery capacity and market share in the retail market, has then led to an extraordinary surge in refinery capacity. Indeed, capacity has expanded far faster than the pure profits and loss economics of refining could justify. We would suggest that the apparent overshooting of Korean capacity can be explained by a series of other factors. First, the expansions are part of a battle of market share among the *chaebol*, and are being driven more by that than immediate profit margins, particularly as the ability to effectively cross-subsidize across activities is a natural feature of the scale and scope of the *chaebol*. Secondly, the expansions are also an attempt to block further entry by those *chaebol*, and most notably Samsung, who are not currently involved in oil refining. Thirdly, there may be a fear that government regulation over market share may return in the future, especially, as noted above, given that policy has tended not to be completely robust. If shares were then to be ossified at that point, there is an incentive to build up share, and therefore refinery capacity, as soon as possible and regardless of short-term profit implications.

The loosening of government control has led to a refinery building programme so large that it has overhung the entire Asian market. A key question is then to what degree demand increases will erode Korea's potential refining capacity surplus. To provide a plausible range for the evolution of oil demand, we make the following assumptions for the major oil products. In electricity generation fuel oil use is kept in check by continued substitution by nuclear, coal and in particular gas in new plants. Naphtha increases are modest in terms of the recent past, given that petrochemical capacity has already expanded. Distillate demand continues to increase faster than GDP, with the bulk of the increase arising from internal commercial transportation. Gasoline demand increases fastest given a continued increase in car ownership and leisure travel. We use slow, medium and fast rates of GDP growth (4, 6 and 8 per cent respectively), and a set of income elasticities by product.[13]

The use of this exercise for our purposes is to demonstrate that, given plausible assumptions, there will be at least a short-term surplus of refining capacity in Korea in 1997, considering just the expansions reported in Table 5.4. With a (by recent Korean standards) modest rate of real GDP growth of 6 per cent, the 1997 surplus is about 300 thousand b/d; with 8 per cent growth (similar to that achieved between 1980 and 1995), it is still 200 thousand b/d. For the moment, assuming that no further capacity comes on stream, by 1999 the surplus has been removed by the medium growth rate. The fast growth rate removes the surplus in 1998, and produces a demand figure for 2000 of just below 3 mb/d, and thus a significant refining deficit. Under the slow growth forecast (the least likely *a priori* among the three), no refining deficit appears until 2001.[14]

Surplus capacity in Korea might then appear to be a relatively short-term phenomenon. However, there are other processes at work that could maintain the surplus for longer. In particular, government tax policy should never be taken to be exogenous. Increases in taxation can choke off the increases in demand, and weaken the impact of GDP changes on oil demand. The possibility that higher demand and a rising crude oil deficit could lead to significant taxation changes has already been demonstrated in Korea. In early 1996 import tariffs on both crude oil and oil products were raised significantly (from 3 to 5 per cent). The major point to be made is that refinery surplus calculations should never be divorced from considerations of both government regulatory and fiscal policy, nor should the feedback effects between them be ignored.

There is also one structural feature of Korean demand that suggests that there would be significant excess capacity, even in equilibrium. It should be noted that Korean oil demand is extremely seasonal, with a very large swing between the high and low demand within a year. The peaks are in December and January, the trough between May and August. In 1995 the peak month demand was some 0.9 mb/d above the May to

August average, in 1993 and 1994 it was 0.8 mb/d. In 1995 the swing was then about 45 per cent of the average annual demand, and resulted in a maximum monthly demand of nearly 2.6 mb/d. The seasonality could be met by a combination of inventory smoothing and changes in utilization rates. However, given the implicit costs of these, plus an open trade system, in addition to some production smoothing by inventories, the optimum refinery solution involves planning for a level of demand greater than the annual average, and having a potential exportable surplus during the months of low demand.

If Korea is to have a significant refining surplus until 2000, particularly during the seasonal periods of relatively slack demand, the question is raised as to where surplus production could be sold. There are two relatively minor outlets. First, when Korean reunification is effected there will be a new market where oil demand has so far been suppressed. Secondly, economic growth in the Russian Far East marks the Vladivostok region as a potential market. China is possible, but given that the demand is mainly into the south of the country, Korea has no large logistical advantage. The most obvious market, as implied by Figure 4.1, is Japan. However, we noted in the previous chapter that the adjustment to Japanese deregulation has been through prices and not quantities. Japan is unlikely to absorb any large proportion of Korea's excess, and as a result that excess will have to go further south into China, losing freight advantages, and in market terms appear somewhat distressed.

Regardless of the size of oil product surplus, one feature that does emerge from the above is the growing importance of Korea as a crude oil importer. Table 5.5 shows the composition of Korean crude oil imports between 1990 and 1994. Over this period, with a near doubling of imports, Korea has moved from being a smaller importer than Singapore, to importing about 50 per cent more. Supplies from the Middle East have made up 78.6 per cent of the increment, with 10.8 per cent of this increment coming from within Asia, mainly from Indonesia, China and Australia.

Environmental regulation of oil product quality is tight in Korea. Gasoil for use as automotive diesel was mandated to have a maximum 0.2 per cent sulphur content from 1993, with the same specification for gasoil as an urban heating fuel. Fuel oil sulphur content has been limited to a maximum of 1 per cent in most major cities, with more areas being added to the requirement every year. The number of desulphurization units has run behind the demand for low sulphur products, with only some 30 thousand b/d of desulphurization in place at the end of 1994. By 1997 this capacity should have reached around 115 thousand b/d, but Korea still remains short of low sulphur products and long on high sulphur products.

As a result of the shortfall in desulphurization capability, there has

Table 5.5:   Korean Crude Oil Imports by Source. 1990–5. Thousand b/d.

	1990	1991	1992	1993	1994	1995
Saudi Arabia	108.5	366.0	423.8	472.4	486.6	620.9
Iran	93.9	144.6	167.4	208.4	226.9	187.2
UAE	131.3	122.8	144.3	157.7	133.1	198.1
Oman	170.8	140.2	201.4	190.6	130.7	145.8
Yemen	14.8	8.2	2.8	15.0	97.0	71.7
Kuwait	48.7	0.0	39.3	80.4	74.6	69.1
Qatar	18.1	22.3	39.6	39.5	39.3	29.8
Neutral Zone	20.1	5.5	20.9	16.5	13.4	0.0
MIDDLE EAST	606.2	809.6	1039.5	1180.4	1201.8	1322.6
Indonesia	51.2	91.0	103.9	95.0	102.6	82.9
Malaysia	56.0	71.5	69.6	75.0	54.8	34.1
Brunei	31.5	33.5	39.5	35.5	36.5	29.0
China	19.8	20.7	34.0	32.8	35.8	37.7
Australia	1.9	4.6	0.0	5.0	10.6	13.6
Papua New Guinea	0.0	0.0	4.5	13.8	3.3	0.0
Philippines	0.0	0.0	1.4	1.1	0.0	0.0
Thailand	1.8	3.1	2.3	0.0	0.0	6.9
ASIA	162.3	224.4	255.1	258.2	243.7	204.2
Nigeria	0.0	0.0	10.1	7.5	16.6	51.4
Angola	0.0	0.0	0.2	4.7	11.8	17.8
Cameroon	3.7	3.2	2.1	5.3	11.3	7.9
Gabon	3.4	0.0	0.0	8.0	5.1	17.5
Congo	0.0	0.0	0.0	0.0	0.0	2.5
WEST AFRICA	7.1	3.2	12.4	25.4	44.8	97.1
Ecuador	4.7	35.0	52.4	48.5	55.0	50.5
Mexico	8.5	3.3	2.5	0.0	0.0	0.0
LATIN AMERICA	13.2	38.3	54.9	48.5	55.0	50.5
Algeria	1.8	2.4	8.8	2.3	2.1	0.0
Egypt	15.5	7.0	13.8	15.5	14.3	6.6
Canada	6.4	7.5	2.7	0.0	2.5	0.0
Russia	0.6	1.6	4.5	5.5	5.8	6.8
TOTAL	813.0	1094.0	1391.7	1535.8	1569.9	1687.8

Source: Korea Petroleum Association.

been a simultaneous export of high sulphur products and import of products with a quality sufficient to meet requirements. For example, in 1994 Korea exported 108 thousand b/d of fuel oil (excluding bunkers), but imported 97 thousand b/d of lower sulphur fuel oil. Likewise, exports of 94 thousand b/d of gasoil coexisted with imports of 48 thousand b/d of lower sulphur gasoil. The apparent exported surplus of 57 thousand b/d from the two products combined masks a total trade flow of 347 thousand b/d, the gearing arising from the problems that the domestic industry has in meeting sulphur requirements. Given this, even an apparent balance between Korean refinery capacity and product demand can leave the potential for considerable flows of simultaneous exports and imports. It also leaves a role for traders in arbitraging between high sulphur excess and its lower sulphur deficit.

4.   The Philippines

The oil sector in the Philippines is now emerging from a period of heavy distortionary regulation. Historically, the economic climate had been comparatively liberal until the first oil shock in 1973. After that, under the authoritarian regime of Ferdinand Marcos, the amount of regulation and government intervention grew sharply. The result was that most foreign companies simply left. In the early 1970s six had been active in the retail market, Shell, Caltex, Getty, Mobil, Gulf and Exxon. When democratic government returned in 1986, only the first two were still present. The government response to the company exits in the 1970s had been simply to tighten regulation, and to take accommodatory action to fill the gap they left. A national oil company, the Philippines National Oil Corporation (PNOC), was established in 1973 and took over the running of what had been Exxon's refinery. Gulf also left in 1973, and PNOC added the Gulf retail assets to their portfolio.

Conditions that would have been conducive to fast effective liberalization, at first sight appear to be met in the Philippines. The transition from autocracy to a democratic regime was associated with a general approval of economic reform. The attitude towards foreign capital was accommodatory, and such capital had always been present in the oil sector.[15] The national oil company served relatively few vested interests, and there had been no history of a long-term consensus in favour of state ownership, nor any history of creeping nationalization as in India. In the general energy industry, chronic power shortages and frequent brownouts increased the public receptiveness to reform in the sector. As noted above, the presence of the state in the downstream of the oil industry was primarily by default rather than by imposition. The infrastructure of the industry was relatively good, and the general economic policy was highly liberal. However, one key condition was not met. Relative price adjustments were not easily accomplished; while the country had an inflationary history, it did not have the hyperinflationary experience and relative price insensitivity that could facilitate rapid price changes. Indeed, regulation had created a system where, not only did retail prices rarely change, when they did the process was a very highly political one. A climate had been created where, not only had the acceptance of fluctuations in oil prices as a normal feature been lost, any changes had been made a matter for immediate recourse to political appeals. As we see below, this aspect has proved to be the major barrier to even faster reform than that which has been achieved. The problem of depoliticizing oil prices has become central to the reform process.

Domestic oil prices have been regulated by the Energy Regulatory Board (ERB). The main mechanism of control has been the Oil Price Stabilization Fund (OPSF), which was set up in 1984 as a method for insulating the domestic market from swings in world prices. This operated as follows. Every two months the three refiners (Petron, Caltex and Shell) submitted figures to the ERB for their actual (i.e. realized) costs of crude imports in the prior two-month period. An overall average across the three refiners was then calculated, and the oil company margin on oil products adjusted according to the movement in the average crude cost, and transfers made through the OPSF. Thus when oil prices were low the refiners contributed to the OPSF, when they were high they received transfers from the OPSF. The OPSF thus obviated the need for frequent changes in the regulated price in response to changes in world crude oil prices. As we have suggested above, in terms of consumer psychology this was not necessarily a helpful aspect.

With regulated prices set too low, the OPSF would eventually be exhausted, or were they too high it would increase over time, together with the opportunity cost of the capital tied up in it. The required second part to the price system was thus the method by which the regulated prices changed. This became a highly politicized mechanism, whereby the case for changes had to be made in (often lengthy) public hearings. Theoretically, any exhaustion of the OPSF would trigger a regulated price increase through the hearings. In practice, the fund often went into deep deficit (i.e. funds were owed to refiners by government) while price increases proved difficult to enforce. When a perceived political bar to full adjustment was in operation, the only solution was for the government to put monies into the OPSF, in effect a direct subsidy of retail prices. By the start of 1996 the OPSF had fallen into deficit by the equivalent of about $350 million. The ERB agreed to an increase in prices of just over 0.5 pesos per litre, less than half the rise suggested by the refiners, with an accompanying transfer of some 10 billion pesos (nearly $400 million) by the government to the OPSF. The difficulty in adjusting prices was then causing an appreciable drain on already stretched public finances, and what was intended as price stabilization around a proxy for the market average, was in fact subsidization below that average.

Originally the ERB used refiner cost recovery in setting the regulating prices, allowing full recovery of capital, crude, marketing and refining costs. Incentives for cost reduction and efficiency were thus removed.

From 1993, the ERB relied on a rate of return method for setting the regulating price, which leaves some incentives for cost saving. This allowed a rate of 8 to 12 per cent (the actual rate at any point in time being discretionary) on the refiners' asset base plus an allowance for the operating capital used. The rate of return analysis thus produced a figure for allowable income, and the regulated prices set accordingly. There was no requirement that prices should be set in such a way as to avoid cross-subsidization, and indeed fuel oil and LPG tended to receive such implicit subsidies.

The government's general liberalization plans had two main strands, the deregulation of the industrial structure and the privatization of the bulk of the state's assets in the sector. Given the problems with untying the Gordian knot of the regulated pricing structure, the programme began with state divestment before deregulation. The privatization of Petron, the downstream component of PNOC, took place in 1994. The state was left with 40 per cent of the company, 20 per cent was in the hands of private shareholders (around 500 thousand in total, including a significant stake for employees and dealers), and the remaining 40 per cent (after some competition, primarily from Malaysia's Petronas) was sold to Saudi Aramco for about $0.5 billion. The privatization was a success, bringing an infusion of capital (and the operational freedom to use it), access to international capital markets, effective vertical integration in regard to crude sourcing,[16] and an injection of management with an international perspective.

Having achieved privatization, the focus switched to removing control structures. Full deregulation is due in 1997, with both parts of the Philippine Congress having passed the enabling legislation in 1995.[17] With its implementation the role of the state shrinks to what can be considered the minimum achievable. In other words, the impact of state becomes primarily confined to the application of general, rather than specific, competition law, the control of quality standards, and also fiscal intervention. The deregulation involves the lifting of controls of imports, the dismantling of the price control mechanism, elimination of cross-subsidies, and less control on construction and entry within the sector. As an interim measure before full price liberalization in March 1997, the first phase of price reform took effect in August 1996, with the end of the previous implementation of the ERB system for setting prices. Instead, a price ceiling was defined by the ERB, using a formula based on Singapore market prices over the previous two months.

For the existing refiners, long buttressed by regulation from the incentive to be efficient or act competitively, liberalization has meant the choice between responding to those imperatives or withdrawing. The liberalization has also involved the competitive pressures of potential new entry into the downstream. As a result, it has been accompanied by a raft of expansion and upgrading by existing refiners, together with a raft of building proposals by new entrants. In early 1995 the Shell STAR (Shell Tabangao Asset Renewal) refinery, within the site of the existing Shell plant, became the first full new facility to be constructed in the Philippines since 1962. Table 5.6 shows the current four refineries, with a set of possible developments. All the provisos we used in describing the avalanche of new refinery projects in India hold, although it is more than possible that the list will grow.

Table 5.6: Refinery Capacity in the Philippines. Thousand b/d.

Refiner	*Location*	*Capacity 1996*	*Potential Capacity*	*Putative Completion*
Petron	Bataan	155	180	(1998)
Shell	Tabangao	40	40	
Shell (STAR)	Tabangao	110	110	
Caltex	Batangas	72	72	
Petron/Ssangyong	Bataan	-	200	(2000+)
Asian Dragon	Surigao	-	65	(1999)
Kaibigan	Mindanao	-	140	(1999)

Sources: Various.

As already noted in the case of India, the list of potential refinery projects at any one point of time tends to be a highly movable feast. However, even with the acknowledgement that such a list tends to be outdated as soon as it is recorded, Table 5.6 does illustrate two features. The first is how, just as in India, Asian capital is being attracted by oil liberalization. Asian Dragon is a joint venture between Thai Petro-chemicals Industry and the Philippine conglomerate Chem Holdings, while Kaibigan is an Indonesian company. The second feature is the strong possibility that the Philippines could have surplus capacity, and hence become an export refiner. With such a geographical advantage in relation to proximity to the South China market, export refining in a liberalized industry could have considerable economic potential.

To give some idea of the possible scale of the exportable surplus, oil demand in 1996 was about 345 thousand b/d. The new Shell facility lifted total capacity above this level and removed the country's net product deficit. There is a continuing process of substitution away from fuel oil in power generation, leaving demand increases very heavily skewed towards gasoline and distillates. Even the most optimistic demand growth projections only bring the total to some 450 thousand b/d by 2000. By contrast, on the basis of Table 5.6 we would estimate a bare minimum refinery capacity in 2000 or soon after of 450 thousand b/d with a maximum of over 700 thousand b/d. The Philippines may then have an exportable surplus into the next century, and during the first decade that surplus has the potential to be substantial.

With full market liberalization, the above is enough to make clear that the profitability of the Philippine industry will depend on the South China market, the most obvious vent for surplus. However, there is one further major factor at work. Liberalization has also attracted investment in making the Philippines a major centre for strategic trading storage facilities. The withdrawal of the US navy from the Subic Bay base, some 100 kilometres from Manila, brought the abandonment of a substantial military oil product storage facility. The facility is some 1100 kilometres from Hong Kong and Guangdong province, i.e. closer than the distance between Singapore and Bangkok. As such Subic Bay represented a major opportunity for any trader with the resources to use it.

The opportunity was taken by the US company Coastal, who are significant traders in the Singapore market. For an investment amounting to $100 million, Coastal took a 25-year lease effective from April 1994 on 2.4 million barrels of storage. Beyond the straightforward use of storage for trading optimization, and as a trading tool, Subic Bay represents the possibility of blending and bulk breaking, and at a far more economic cost than acquiring or using facilities in Singapore. Independent storage in the Philippines was increased later in 1994 when Chemoil Asia took a 15-year lease on 1 million barrels of storage on the island of Mindanao in the south of the country.

There are some potential competition issues that arise from the presence of large-scale independent storage, primarily in depressed markets. The question at core is one of which part of the market ultimately contains discounted, and sometimes distressed, cargoes. Storage has a strategic role and source of value added in terms of bulk breaking and blending operations. It also of course has the economic role of being a buffer between supply and demand. The issue is then the location of 'swing' storage. This is likely to be spread between oil at sea, refinery storage and independent storage throughout Asia. However, if, perhaps due to constraints elsewhere, the swing is predominantly taken by storage in one location, there is a potential danger. If freight charges to other locations are relatively discouraging, there is a threat that supplies will move into the domestic market during periods of excess supply in the region as a whole. This does not happen out of refiners' storage, since changes in their storage are already derived from the balance of their own supply decisions and the market demand conditions. However, it is possible out of storage held by agents that are not involved in domestic refining. In such circumstances the domestic market would be playing a swing role for Asia as a whole, with a saturation of supplies precisely when market conditions are already weak. Were there to be such a concentration of regional adjustment in one country, a clear case arises for correctional protectionism of the domestic refining industry.

The above is posed as a question rather than a statement of inevitability. Given that we do not have access to data to test the proposition, we have no evidence that this could be the case in the Philippines or any other area of Asia. The point to be made is that policy makers in the presence of large-scale independent storage do need to have a position on whether there are any asymmetric adjustment effects in operation.

The one element of government regulation that will be tightened in virtually all countries is the regulation of oil product quality. In the Philippines the specification for sulphur content in fuel oil is due to be reduced to 3 per cent from 3.5 per cent in 1996, with sulphur in gasoil to be reduced from 0.7 per cent to 0.5 per cent. The process is extremely unlikely to stop there, and further tightening to perhaps 2 or 2.5 per cent and 0.2 per cent for fuel oil and gasoil respectively seems inevitable. As we will see below, particularly in the case of Taiwan, comparatively the new specifications represent less of a problem for the industry in the Philippines than many current proposals in other countries.

5. Thailand

The Philippines moved from liberalism through a period of tight regulation, which is now being unwound, resulting in a highly competitive industry, all within a period of some 25 years. Thailand's regulatory regime was never quite so overtly restrictive, but has described the same circle of liberalism to control in the wake of the first oil shock, and then back to less regulation in the 1990s. Thailand's involvement with the international industry began early. In the same journey that led to the purchase of Shell's first facility in Singapore (documented in Chapter 6), Marcus Abraham came to Bangkok to build a facility. Part of the cargo from the first tanker journey through the Suez Canal was unloaded there. Standard Oil started operations in Bangkok in 1894, so both Shell and Exxon have maintained a presence in Thailand for over a century. However, as in Singapore, despite the long history, oil refining is a relatively recent industry, with the first refinery having come on stream in 1964.

In contrast to this long involvement of foreign companies in Thailand, the national oil company, the Petroleum Authority of Thailand (PTT) is a relative newcomer. Just as the first oil shock helped spawn PNOC in the

Philippines, PTT was also partly a product of the international dislocations of the 1970s. PTT was founded in 1978, and the system it joined was oligopolistic and displayed relatively few competitive tendencies. PTT had a share of virtually every asset within a downstream industry subject to price regulation, and entry barriers to new marketers removed the chance of any further sources of competition.

Stabilization (and, as in the case in the Philippines, effective subsidization) of retail prices was achieved through the use of the Oil Fund, instituted in 1974 as a direct response to the first oil shock of the previous year. Price regulation had several tiers and considerable complexity, but the basic structure was the setting of refinery gate prices, plus tax and a regulated fixed marketing margin, plus or (and normally) minus the contribution of the Oil Fund. More flexibility tended to be shown in the setting of refinery gate prices, leaving the Oil Fund to be the major cushion between the consumer and changes in international prices.

Oil price deregulation was first explicitly put on the political and economic agenda as part of the Sixth National Economic Five-Year Plan, which began in 1987, and within that the Sixth Energy Plan. This provided a blueprint for gradually sweeping away most of the distortionary controls over the industry.[18] A succession of liberalizing measures ensued, primarily the removal of constraints on refining capacity, on imports, and on the entry of new companies into both refining and retailing. A structure was thus created prior to any price liberalization that aimed to create elements of the competitive forces that had previously been totally absent, be that competition between incumbent firms, through entry or through imports.

Price liberalization provided the last major piece of the reforms, and was implemented in August 1991. The prices of all oil products were freed, with the exception of LPG (current plans involve the full deregulation of LPG prices by the end of 1997). The deregulation caused a surge of competition in the retail market, with the number of retailing companies rising to over twenty, and within two years the number of retail stations increased by about a quarter. Among foreign capital besides the refiners, BP, Conoco, Mobil, Kuwait Petroleum and Cosmo are all now represented in the retail market.

On the eve of liberalization in 1991, Thailand had three refineries. The first was operated by Exxon with a 12.5 per cent Thai government holding. The second, Thai Oil, had a 49 per cent PTT stake, with Shell and Caltex holding minority shares. The third, Bangchak Petroleum, had PTT and the Thai government as main shareholders. The industry was then split between four main companies (Exxon, Shell, Caltex and PTT) in most downstream operations, with PTT having a protected monopoly on sales of fuel oil for power generation.[19] The present refining capacity in Thailand is shown in Table 5.7. The industry expanded significantly in

1996 with the coming on stream of two large new plants in the Map Ta Phut industrial area of Rayong, Shell's Rayong refinery, and Caltex's Star refinery. PTT has a share of 36 per cent in both developments.

Table 5.7: Refinery Capacity in Thailand. 1997. Thousand b/d.

Refiner	Location	Capacity
Exxon	Sri Racha	157
Bangchak	Bangkok	120
Thai Oil	Sri Racha	207
Shell	Rayong	145
Caltex	Rayong	130

Sources: *Petroleum Argus* and industry sources.

The addition of the new refineries has taken total capacity up from about 485 thousand to 760 thousand b/d. With domestic demand running just below 700 thousand b/d, Thailand moved, albeit temporarily given the pace of demand growth, into an oil products surplus, primarily of gasoline.

Even before the sharp increase in refining capacity consequent on the addition of the two new refineries in 1996, Thailand had been becoming a significant crude oil import market. Table 5.8 shows the source of imports between 1990 and 1995. The bulk of Thai demand is for, within the context of Middle East crude oils, the relatively lighter and lower sulphur grades from that region, most particularly those from Oman and the UAE. Within the sharp increase in the total volume of imports, the percentage dependence on the Middle East has been fairly constant, with about a third of incremental demand being met by Asian, and Malaysian in particular, crude oil.

Additional grassroots refining projects are currently either under discussion or in the early stages of implementation. A joint venture in Rayong between Thai Petrochemical Industries (TPI) and the Kuwait Petroleum Corporation has been considered with between 200 and 300 thousand b/d of capacity. With progress on this refinery being slow, TPI has also commissioned the building of another 150 thousand b/d plant in Rayong, due for completion by the end of 1998. A Thai independent company, Sukhothai Petroleum, has also been given government permission to build a 125 thousand b/d plant. The removal of entry controls, plus the imperatives of booming demand, have then already led to a supply side response, and the progress on additional projects seems to indicate that the response will continue. In particular, Thailand is becoming a large crude oil, rather than oil product, importer. Indeed, given the lumpy

Table 5.8:   Thai Crude Oil Imports by Source. 1990–5. Thousand b/d.

	1990	*1991*	*1992*	*1993*	*1994*	*1995*
Iran	11	18	5	4	4	15
Iraq	8	0	0	0	0	0
Kuwait	15	7	7	2	1	0
Neutral Zone	2	7	35	55	44	30
Oman	14	18	35	50	73	112
Qatar	6	4	0	0	0	12
Saudi Arabia	36	23	25	5	5	6
UAE	29	41	43	68	100	98
Yemen	0	0	0	0	0	18
MIDDLE EAST	121	118	151	185	227	291
Brunei	22	21	28	33	38	39
Malaysia	56	57	67	90	94	108
Indonesia	0	4	3	6	6	7
Australia	2	2	14	6	3	5
Papua New Guinea	0	0	2	4	2	3
Angola	0	0	0	0	2	3
Nigeria	0	0	0	0	0	1
TOTAL	201	201	265	323	373	458
% Middle East	60.3	58.8	57.1	57.2	61.0	63.6

Source: Thailand Department of Energy.

nature of refinery additions, there remains scope for Thailand for limited periods to become a net oil product exporter, just as it has in the wake of the Shell and Caltex additions.

The general Thai policy in favour of private ownership has led to a structural reorganization of PTT in advance of privatization to facilitate eventual listing on the domestic equity market. The central policy management of PTT now oversees four groups, natural gas, oil refining, petrochemicals and gasoline. Partial privatization, involving the flotation of one or more operating units could be possible by 1998. PTT also faces the ending before 1999 of a series of sales monopolies to state firms, although the guarantee of the 80 per cent of fuel oil sales not covered by tender has not as yet been lifted.

In the main, the remaining distortions not removed by the 1991 liberalization are being reduced and removed, and prices have tended to follow the Singapore market closely. However, the tax structure is still causing a distortion manifested in the problem of smuggling, primarily of gasoil, which remains a potent political issue. Official estimates of smuggling indicate large volumes, 2 billion litres in 1995,[20] equivalent to some 30 thousand b/d, more than 10 per cent of domestic diesel demand. With past efforts to suppress smuggling producing results that were patchy at best, one is forced to the conclusion that reweighting the balance between product taxes and retail price taxes would be more effective than direct action by navy, customs and police.

There has been considerable pressure on improvements in oil product specifications, especially given the extreme pollution in Bangkok, primarily caused by expanding numbers of diesel fuelled vehicles combined with a transport infrastructure stretched beyond breaking point. The sulphur in gasoil was reduced from 0.5 to 0.25 per cent in January 1996, and the National Energy Policy Committee has approved a plan that involves substantial further tightening. Under this plan the sulphur content of high speed diesel will be reduced to 0.05 per cent nationally by January 1999, with implementation in Bangkok in early 1997.[21] The specifications for fuel oil for power generation have also been tightened, with the previous 3.5 per cent sulphur limit replaced with 2 per cent in Bangkok and 3 per cent elsewhere.

The removal of direct price regulation does not always imply that regulation can not reoccur in a different form. We saw that the setting of regulated prices was a political issue in the Philippines before deregulation. Thailand entered into price deregulation five years earlier than the Philippines, and manifests the next stage of the politicization of prices. The threat of further regulation now comes from political pressure for consumer protection, in the face of any perceived failure in the competitive mechanisms. The opening up of competition prior to price deregulation was aimed at removing the need for explicit controls on oil company profit margins. In addition, a series of other measures designed to improve and monitor the operation of the market had been put in place. Transparency had been increased by the obligation for retail stations to display price boards visible from the street, retail price maintenance by wholesalers was implemented to prevent individual retailers from exploiting any local monopoly status,[22] and two government departments were given the task of price monitoring.

The potential problems arise when the margins earned are deemed to be too high for political expediency. When that level is actually defined in the political arena, the result is indirect price regulation, since wholesalers will realize that if they breach those margins direct regulation is likely to follow. If the political articulation is for a level greater than that which would obtain in a competitive market, all is well and good. Should it be below, then problems ensue. Thailand appears to be in this situation, with the acceptable level of margins being a political issue. To provide but one example, in 1996 the Thai Commerce Ministry, backed by the Parliamentary Committee for Energy, articulated plans for the direct regulation of margins, with their suggested 13.5 per cent return on investment attracting a counter proposal of 15 per cent from PTT.[23]

Price deregulation had been based on the premise of a structural solution to facilitate competition. Its underlying logic had been the following. If margins were to be temporarily high, then competition should bring them down. If it did not, then there would be two alternatives. First, the definition of 'too high' may be wrong, the margins could have been at the competitive level. Secondly, there might be some structural factor that was preventing the margin from being eroded. The corollary to this is that unless a structural block can be identified, then margins must be at the competitive level. The policy implication was then that the structure of competition needed to be monitored, not the level of profit margins directly. The return to a political discussion of what represents the acceptable level of margins (even if unaccompanied by explicit regulation) is then a highly retrograde development, and a return to the old ethos. From the point of view of the oil industry, the moral is perhaps that the end of a liberalization process tends not to bring the end of potential government involvement in prices.

6.   Taiwan

The national oil company of Taiwan, the Chinese Petroleum Corporation (CPC), had a domestic monopoly from the time when the Kuomintang government established itself on the island in 1949, until 1996. Until then, the position of CPC within the industry had been one of complete dominance. Currently there are two refineries in Taiwan, the giant Kaohsiung plant with 570 thousand b/d of crude capacity, and the recently expanded Taoyuan plant with a capacity of 200 thousand b/d. With domestic oil reserves negligible, the refineries are almost entirely dependent on imports, with more than two-thirds coming from the Middle East, primarily from Saudi Arabia. Taiwan's oil industry policy in the 1990s has had three main facets, the removal of CPC's monopoly, privatization, and strict specifications on oil products. We consider each facet below.

Potential entry creating (potential) competition with CPC became reality with the announcement of a new refinery project by the Formosa Plastics Corporation. After years of delay before approval was granted and delays due to site problems and subsidence, completion is due in 1999. Of all the single potential refinery projects in Asia, this is the most significant in terms of its implications for trade flows. The design capacity is about 430 thousand b/d for a complex of a refinery attached to an ethylene plant, situated in Mialiao in the south of the island. The project is impressive in terms of engineering and capital equipment, comes with the equally impressive cost of about $10 billion, and would immediately make Formosa

Plastics a significant force in both Asian oil and particularly petrochemical markets. It also changes the oil products balance of Taiwan dramatically.

One might be tempted to think that the addition of refinery capacity equal to 60 per cent of the current level of demand might dampen the enthusiasm for other projects. When a single giant refinery such as Kaohsiung is such an imposing presence on the island, building another large one might be brave, but considering further projects could be seen as a little excessive. However, while there are the usual provisos about their coming to fruition, there are other projects. In particular, a Taiwanese petrochemical company, Tuntex has planned a 140 thousand b/d refinery, and CPC has considered a joint venture with a group of petrochemical companies with around 200 thousand b/d of crude distillation.

It should be noted that scepticism in regard to the additional Taiwanese projects is less justifiable than it might be in respect of some proposed refineries in other countries. These plants are primarily an expansion of Taiwan's petrochemical industry.[24] The economics of joint oil refining and petrochemical complexes are different from those of a stand alone refinery, and the potential impact on regional refining economics can be thought of as being to some extent a negative externality from the comparative strength of petrochemical markets. The combined crude capacity of the three projects we have mentioned (one firm, two more speculative) is close to 800 thousand b/d. However, a significant proportion of the output is to fulfill the derived demand from the associated petrochemical plants, and, while the impact on the exportable surplus is potentially severe, it is not the full extent of capacity additions. A correlated warning is not to be too surprised by the extremely high rate of apparent Taiwanese oil product demand growth when the complexes come on stream.

Having unsuccessfully attempted to blockade entry, the normal rational strategy for an incumbent monopolist is to agree an accommodation with any new entrant, hoping to avoid all-out competition and to blockade further entry. The strategy of CPC with the advent of the competition of Formosa Plastics follows the textbook to the letter. Following a meeting between the chairmen of the two companies, a CPC spokesman has been quoted as saying the following: 'We are hoping the government will implement some regulation to prevent an all-out price war.'[25] In particular, they wished to confine the importation of oil products to the two refiners. It is possible that the two companies had noticed the impact of import competition in Japan, and had decided that using the old Japanese system in Taiwan would be their best form of protection.

Rather than merely confirming some validity in modern industrial economic models of entry deterrence, the above quote is very telling. The reader will note the implicit equivalence drawn in the statement between import liberalization leading to border price convergence and 'all-out price war'. From the Japanese experience of confining imports to refiners, it becomes clear that such a policy in Taiwan would render some of the motives behind the general liberalization redundant. In particular, the limitation would prevent any benefit accruing to the customer, and provide disincentives for cost saving and efficiency in the domestic market. It could lead to a situation where a potentially important exporter of oil products had such a distorted domestic price structure that (were they allowed to be created by non-refiners) import flows would still be profitable. However, the position of the two companies is also a signal that they realize that the formulation of the import regime is the critical issue in the whole liberalization process. As we note in the final section of this chapter, a degree of protective regulation for domestic refiners may be the only way to buttress profitability from what could be a relatively poor outlook for refining margins. There is a clear policy trade-off between encouraging competition and wishing to maximize the value of a process of gradual stock market release of CPC shares. The resolution of the question of the import regime is then likely to be affected by the view taken of the path of international market conditions.

Privatization of CPC is planned as a gradual process, with eventual complete divestment the ultimate aim. The timetable towards this, as of mid-1996, was the following. Shares equating to 10 per cent of the company were to be listed on the domestic stock exchange in July 1997, and further tranches released each succeeding year. On this timetable, by 2000 the state share will have fallen below 50 per cent, at which point the intention is to sell the entire remaining ownership share en bloc. The obvious candidates for the purchase would be Middle East national oil companies. In terms of privatization plans for state oil companies in Asia, this is the most radical and complete. The original timetable began to slip in June 1996, when the Legislature voted down the original proposal. At time of writing no clear new timetable has emerged, although the first shares are now unlikely to be listed before 1998.

The tightening of oil product specifications in Taiwan has been especially severe compared to elsewhere in Asia. From July 1996 power generators have only been able to use fuel oil with a sulphur content of 0.5 per cent or less in Taipei, Taichung and Kaohsiung. New trade patterns are likely to be opened up by regulation. In particular, the domestic inability to produce sufficient quantities of such a low sulphur fuel oil might, for instance, force the generators to import LSWR from Indonesia to blend into higher sulphur fuel oil to meet the new requirement. The new specifications are equally tight for other products. Leaded gasoline is due to be phased out completely by 2000, a far more ambitious strategy than has been attempted in the USA or Europe. The sulphur content of gasoil will fall to 0.05 per cent in 1998 compared to the previous 0.25 per cent, again an extremely ambitious timetable compared to those laid down elsewhere.

Taiwan's requirement is then for greater flows of low sulphur crude oils to help meet the tightening of the fuel oil and gasoil specifications. While maintaining a predominantly Middle East slate of imports, CPC is having to cast a wide net for low sulphur supplies. Throughout 1995 a succession of trial cargoes of low sulphur crude were taken into Taiwan from exporters as far afield as Argentina and the North Sea. Until very recently the suggestion that oil trade flows from Norway to Taiwan could be viable would have been thought rather ludicrous. However, it is a measure of the power of the supply side factor of a surge in low sulphur crude output, and the demand side impact of regulatory changes in sulphur specifications, that such flows are now possible.[26]

7. Conclusions

While the general observation that the oil industry is becoming more liberalized throughout Asia can be made, closer examination reveals that there is little homogeneity in this experience. Some form of liberalization process is generally under way, regardless of the degree of economic development or political orientation. The process occurs in countries where there has been a sharp, even revolutionary, political change and in those where there has been none (witness the Philippines and India). Sometimes oil industry liberalization has been clearly linked to general economic liberalization, sometimes (and most notably in China) the two processes have been negatively correlated. The speed of implementation ranges from the fast (Thailand and Philippines) to the extremely slow (China and India). In some cases the regulatory structure being stripped away is comparatively recent, in others (especially Japan and India) it has a very long history.

In the two decades after the Second World War, a consensus on the position and responsibilities of the state in the economy emerged outside of the centrally planned economies. Stemming largely from the theoretical base of Keynesian economics combined with the wartime experience of increased central control, a clear policy framework was formulated. The government had a role in active demand management of the macro-economy, and intervention in key industries was taken to be not only acceptable but often desirable. While this tendency was less marked in the USA, it still played a major part in the formation of economic policy. In Europe the consensus was dominant. In the UK the high water point was perhaps reached in 1967 with the publication of a government white paper on nationalized industries. This advocated the use of marginal cost pricing, i.e. necessarily involving subsidies in declining cost industries such as those which made up the bulk of the public sector, and reinforcing a view of the sub-optimality of any pure market solution.

However, the tide was changing, at the academic level at first, but then increasingly at the political level. The rise of what can be termed new classical economics saw an emphasis on the supply side of the economy and on the correct functioning of markets, rather than on demand management. Arising from the work of Lucas in particular, consideration was given to the impact of private expectations, and from these models it emerged that demand management was not just ineffective, it was actually destabilizing. A view was forming that the correct role for government was in prudent fiscal and particularly monetary policy, using these variables in a targeted and observable fashion in such a way as not to generate inflationary expectations. Within this view was also the idea that the public sector should be reduced in importance, and impediments to market clearing solutions should be removed. The concept of the economic role of government moved from active management of the demand side, to a primary emphasis on correcting market failures on the supply side.

Reinforcing the new macroeconomic thought was a change in microeconomics. Issues such as agency came to the fore, implying that state enterprises had implicit efficiency losses due both to a lack of incentives and a failure of control mechanisms. Industrial economics began to focus on barriers to entry, revealing that the welfare of losses of monopolies could be reduced by the presence of potential entry. All these developments also led to a renewed emphasis on the benefits of competition, and on the efficacy of the price system. This implied that the state's role should not be to subsume markets, but to allow them to operate fully and indeed create new quasi-markets in areas where there were missing markets. Throughout the public sector, including education and health policy, free market economics was pushing towards the idea of full internal markets and internal transfer pricing.

The conventional Keynesian view was then coming under theoretical attack, but the impact of the oil price shocks proved to have greater immediate impact. Demand management had apparently failed, and, caught up in stagflationary cycles of rising unemployment and rising inflation, governments increasingly turned towards new thinking. In some cases the change was apparently involuntary, for instance the UK government in economic crisis in 1976 turned to the IMF for loans, and as part of the conditionality attached put into place a policy which was essentially an implementation of a crude form of monetarist rather than new classical thought. However, in all cases the focus was put on inflation, and with that change new classical elements began to enter policy.

While the theoretical agenda was shifting away from strong state controls and wide state ownership elsewhere in the economy, developments in the oil industry were moving matters in completely the reverse direction. This was due to three main factors, all in some way associated with the 1970s oil price shocks. First, the sudden increases in prices created an emphasis on supply security, and heightened government appreciation of the effects of volatile energy prices. Both considerations led to a sharp increase in state intervention and the degree of regulation, even in countries where the oil policy regime had previously been liberal and accommodatory. Secondly, the growth of non-OPEC output increased the number of countries with significant domestic production, and brought the policy problems not only of how to manage oil revenues in the macroeconomy, but also how the new oil sector should be organized. In virtually all cases this was brought about by the creation of national oil companies, or by significant expansions in existing state institutions. Thirdly, the emphasis in many countries had turned to resource national-ization, and at a political level foreign capital became a target. While typified by the nationalization of assets in OPEC countries, this oil nationalism was by no means purely an OPEC phenomenon, nor even was it confined to countries with oil production rather than just refining assets. In many countries (including as we have seen within Asia, in India, the Philippines and to some extent Thailand), major oil companies found their assets nationalized or under threat of nationalization, or they found a worsening operating climate where the degree of new restrictions forced them to reconsider their operations. The oil price shocks may then have helped move the agenda away from state control at the macroeconomic level, but they had led to the reverse effect in the oil industry. In general, by 1980 the industry had become significantly more regulated, and direct state participation had been greatly increased.

Liberalization of state controls over the 1980s and beyond has in most industries, and particularly in the utilities, been focused on the reduction of state ownership. While privatization has certainly played a role in liberalization of the oil industry, the major thrust has been towards the removal of existing layers of regulation. This regulation has taken a wide variety of forms. Thinking primarily of the downstream oil business as described in this chapter, we can distinguish eleven main varieties of government regulation of oil. This is not an inclusive list, but its elements when in operation tend to be the dominant ones in terms of the structure of the domestic industry.

(i) Government Controls on Entry or the Maintenance of Statutory Monopolies
The state may often license or otherwise control entry into refining, storage, retailing and trading. Often entry into these activities is prohibited

due to the presence of a statutory monopoly, normally held by state companies.

(ii) Constraints on the Movement of Foreign Capital

The involvement of foreign companies in the downstream may be limited. This limitation *in extremis* could be the complete exclusion of foreign capital, otherwise it can take the form of a stipulation on the minimum domestic capital element of any venture. Further restrictions arise if there is a minimum domestic element for any resource used in the production process.

(iii) Direct Price or Quantity Controls

A regulation of prices means that the setting of retail prices is not a commercial decision. This can be in the form of mandated prices or price caps. There can also be indirect controls implicit in general competition policy designed to avoid either predatory or exploitative pricing, but these are designed to correct departures from a competitive market outcome, while direct controls tend to preclude that competitive outcome. Altern-atively, competition can be precluded by quantity regulation, for instance regulating refinery runs of crude oil or fixing either absolute quantities or market shares in the retail market.

(iv) Minimum Inventory Constraints

Legal requirements exist to hold inventories higher than those that would be held under strictly commercial considerations. These constraints tend to be based on supply security criteria, and, if they are binding, would imply that governments feel that the degree of risk aversion held by private industry is socially sub-optimal. Inventory requirements represent a barrier to entry, and, since they constitute holding capital in a (from a commercial perspective) sub-optimal form, reduce the overall returns to industry capital and thus discourage investment.

(v) Constraints on Construction

There are of course constraints on the construction of refineries and retail gasoline stations caused by local planning permission procedures. Here we refer to constraints placed at a central level on the total number of facilities allowed. For example placing quotas on the number of service stations by company will tend to reduce competition.

(vi) Constraints on Integration

Integration may be constrained by having different criteria for entry into, say, retailing than into refining. Existing refiners who are unable to retail will tend to find their performance impaired, just as difficulties in entering retailing may stop further entry into refining. Other barriers might involve constraints on the asset market so that vertical integration through merger or takeover is impaired, or directly precluding entry into certain stages of the production process.

(vii) Constraints on National Oil Company Investment

If company investment can only be funded through general state funds, rather than through retention of company revenues or through borrowing on a commercial basis, under-investment tends to take place. Likewise the maintenance of internal transfer pricing which is not market based will tend to result in misallocation within the company together with associated efficiency losses. Commercialization of the company can be achieved by full or, with some provisos, partial privatization, or by the development of one-step removed management structure with separate company accounting, internal financing and access to capital markets, and the implementation of full internal transfer pricing.

(viii) Import and Export Restrictions

Governments may ban imports or exports of certain products, or place restrictions and conditions on importers. Import bans are motivated by balance of payment and foreign exchange considerations, as well as being protectionist towards the domestic refining industry. Export constraints are more likely to be motivated by supply security concerns, or occasionally, as in the case of Alaskan oil, have an element of protectionism towards some part of the labour force such as seamen. In some cases (China is the most obvious example), where logistics imply that a country will have both imports and exports of the same product, an export ban can be little more than an indirect import ban. Such controls may also be tariff rather than quota based, e.g. discriminatory tariffs on imported crude oil or oil products. A further form of insulation for domestic producers can arise in the form of differential tariffs for inputs and outputs. For instance, a regime where oil product imports carry a higher tariff than crude oil, provides a form of protectionism for domestic oil refiners. Indeed, if a government feels that, due to the operation of other considerations of national costs and benefits, a degree of protection is justified, then the differential import tariff often provides a highly efficient method for achieving this.

(ix) Foreign Exchange Constraints

Particularly in countries which operate systems of dual or multi-tier exchange rates, companies may be forced to conduct different operations at different rates, usually to their disadvantage. Alternatively, a national company may be rationed not just in its access to capital, but specifically in the amount of foreign exchange it can utilize. One final form of this constraint is the imposition of limitations on the amount of profit earned by foreign capital that can be repatriated.

(x) Rate of Return Regulation

While more common in the regulation of natural monopolies, constraints can be placed more generally on a range of activities, stipulating maximum rate of return to assets. Alternatively (and more commonly in the Asian oil industry) rate of return regulation can be imposed through price regulation which fixes a rate of return on capital in computation of prices.

A contrast should be drawn between government regulation that is designed to correct market failures, and regulation that merely causes market failures. For instance, in the absence of a competitive market, price caps are often used as an alternative to monopoly pricing, or as a sanction arising from competition policy legislation against companies that have engaged in either overt or tacit cartelization. Likewise, if the government really does believe that companies are not sufficiently risk averse, then they would justify minimum inventory requirements as a correctional rather than a distortionary policy. Correctional policies relating to environmental concerns motivate one further type of regulation.

(xi) Regulation of Oil Product Specifications

As other regulations have been relaxed, national product specifications and timetables for further tightening of specifications have in many countries become the most important form of government regulation. While the other forms of regulation are likely to become less common, specification regulation will continue to grow. Social costs are perceived in, for example, leaded gasoline and the sulphur content of oil products, which are not reflected in private costs. To correct for this market failure which would result in product of a suboptimal quality being used, governments can either use the tax system, price regulation or direct regulation. In most cases direct regulation has been preferred, sometimes combined with tax incentives for consumers to switch to the cleaner products earlier rather than later during any phase-out period for the old specification.

The attitude of the new economic orthodoxy to the above is clear; the first ten forms of regulation should be abolished, unless their purpose is correctional in the sense defined above. This implies a downstream with freedom of entry into all activities, refining, storage, transportation, trading and retailing. The only stage where any regulation beyond the provisions of general competition legislation might be needed is transportation, where legislating for third party access and setting rate of return conditions may be necessary if there is a situation of natural monopoly. There should be freedom of movement for capital, and transfer prices should reflect international market prices. State oil companies should be commercialized, and preferably through a programme involving the transfer of assets to the private sector. Under this essentially neo-classical viewpoint, even the government mandated minimum inventory requirements should be scrapped, as the time structure of prices should provide the correct signals about when to build stocks and when to release them on the market.

The now orthodox view of what to do with government regulations is straightforward enough. The problems start in the transition from theory into practical implementation, and the questions of how fast to liberalize, in which order the component steps in liberalization should go forward, and how the political effects of the associated economic restructuring should be handled. There has been a handful of cases outside Asia that governments could draw on as examples. The largest restructuring and effective privatization in oil has been in Russia, but that particular experience is probably not one on which other countries would wish to model their own strategy. However, there is one notable case that serves as an example, that of Argentina.

The Menem government in Argentina followed an approach of fast oil sector liberalization, the 'big bang' approach. When Menem took office many of the elements of regulation listed above were in place. Both crude oil and product prices were heavily regulated, and at levels well below world prices. There were controls on the repatriation of profits, and a system of import and export taxes. Product supplies were subject to quotas, and there were barriers to entry, particularly in retailing. The state oil company Yacimentos Petroliferos Fiscales (YPF), was highly inefficient in all stages, a situation compounded by the lack of any effective price signals. The policy undertaken in response was essentially shock therapy. Price controls were lifted rapidly, less than eighteen months into the new government, and the barriers to entry removed. The import and export taxes were abolished in 1989 and freedom of capital movement was facilitated. YPF was reduced in size and in relative importance, including the sale of major interests in producing oil fields and the removal of its monopoly over crude oil and oil product sales. Finally, YPF was privatized, with major tranches of stock being sold not only in Argentina but also in the USA and Europe.

The underlying conditions that created the ability to implement fast track liberalization are not necessarily immediately reproducible every-where. The government had a strong mandate for change from a popu-lation willing to countenance radical economic restructuring throughout both the state sector and also in regulated markets after years of economic chaos. The element of resource nationalism over oil was far weaker than in many countries. Readjustments of prices and the lifting of price controls were easier in a country with a long experience of annual inflation rates of several thousand per cent, where large adjustments in relative prices were commonplace. Given that the economy as a whole moved from four digit inflation rates to below 20 per cent in just four years, the lifting of oil price controls was not a major political issue in the overall picture. It could also be said that political discontinuities, and in particular the movement from authoritarian to democratic systems, helped the climate for rapid economic change. Finally, the overall policy stance was extremely open to foreign capital, and saw the reduction in state control as a desirable object in itself, not just as a remedy forced by expediency.

These conditions have not held in Asia. While the Philippines provides a close parallel, adjustment was hampered by the politicization of oil prices and the mechanism for changing prices. The inflationary and distributive effects of removing subsidies have been a major problem in several countries, in others it has been ambivalence to, or open hostility towards, foreign capital. The number of Asian countries that on an ideological level could be said to have subscribed to the new economic orthodoxy is extremely small. A degree of liberalism has often been motivated more by capital and sometimes technological starvation of the domestic industry. In some cases the attempt has been made to achieve a greater flow of capital and technology without compromising the role of existing institutions.

There is no general Asian model, and to the greatest extent the liberalization of the oil industry is country specific. On the whole this springs from the oil industry being somewhat separate to most other sectors. In countries with upstream production, the industry often represents the greatest source of rents for the state and for the more powerful interests within the state. The social and economic effects of pricing are greater than for other commodities. Further, in many countries the oil industry has represented in value terms the largest single part of the entire state-owned industrial sector. Where unemployment has been disguised by over absorption of labour by this sector, this means that the social consequences of industry rationalization play even more of a constraining role.

Notes

1. A listing of the refinery structure in the major Asia-Pacific countries not covered in this, Chapter 6, or in previous chapters is given in Appendix 3.
2. For the early history of Indian oil, see S.N. Visvanath (1990), *A Hundred Years of Oil: A Narrative Account of the Search for Oil in India*, Vikas Publishing House, New Delhi.
3. Ibid.
4. Bennett H. Wall (1988), *Growth in a Changing Environment: A History of Standard Oil*

Company (New Jersey) 1950–1972 and Exxon Corporation 1972–1975, McGraw-Hill, New York.

5. The final refinery in Table 5.1, Mangalore Refineries and Petrochemicals Limited's plant in Karnataka state, began operations in March 1996. This is a joint venture between Hindustan Petroleum and a private group (26 per cent of equity each), with the residual share held by other private agents.

6. See *The Economist*, 12 August 1995.

7. The project was given approval by the (transient) BJP national government in June 1996.

8. To make the same provisos repeated throughout this study, these are not intended as point estimates but, in this context, solely to provide a plausible range of magnitudes for the size of any Indian excess demand 'sink'.

9. For an examination of the impact of structural adjustment policies in Korea, see Vittorio Corbo and Sang-Mok Suh (eds) (1992), *Structural Adjustment in a Newly Industrialised Country: The Korean Experience*, The World Bank, Washington, DC, and Il Sakong (1993), *Korea in the World Economy*, Institute for International Economics, Washington, DC.

10. See Il Sakong (1993), *Korea in the World Economy*, Institute for International Economics, Washington.

11. See *Petroleum Intelligence Weekly*, 1 January 1996.

12. The tendency for the capacity of refinery to expand due to debottlenecking and efficiency changes, which can add more than 1 per cent per annum to capacity.

13. Among the main products, we use income elasticities of 0.25, 0.5, 1.5 and 2 for fuel oil, naphtha, distillates and gasoline respectively.

14. The range for these projections of Korean demand in 2000 is a low of 2.4 mb/d to a high of 3.0 mb/d, with a proposition that the low end is less likely than the high. However, even the low end of this range is higher than the estimates reported in Table 2.13, both of which (even with the factors that are likely to slow Korean demand growth) we would consider as being too cautious.

15. Namely Caltex and Pilipinas Shell (which has a 30 per cent domestic capital share with Royal Dutch Shell owning the remaining 70 per cent).

16. In 1994 Petron signed a 20-year agreement with Aramco for most of their crude oil supplies.

17. The two versions were however different, with the major point of divergence being whether crude and import tariffs should be the same (i.e. a tendency to full import price convergence), or whether crude should attract a lower rate (i.e. some element for protection for domestic refining).

18. For a detailed discussion of the components of the Sixth Energy Plan, see Tienchai Chongpeerapien (1991), 'Development of Energy Policy in Thailand' in Shankar Sharma and Fereidun Fesharaki (eds) *Energy Market and Policies in ASEAN*, Institute of Southeast Asian Studies, Singapore.

19. In 1993 the Electricity Generating Authority of Thailand (EGAT) opened 20 per cent of input business to competitive tendering. The first three-year tender was won by PTT.

20. This estimate comes from the Industry Ministry, as reported in *Platt's Oilgram News*, 4 January 1996.

21. *The Nation*, 9 February 1996.
22. Retail price maintenance is in itself a highly anti-competitive process. Its use in Thailand seems designed to keep the competition at the wholesaler rather than the individual retailer level. While this reduces the scope for price competition in areas with many stations (i.e. urban areas), it does allow for the spillover of the impact of urban competition into the prices charged in rural areas.
23. *The Bangkok Post*, 30 March 1996.
24. The Kaohsiung refinery is also very closely tied in to existing petrochemical facilities.
25. Quoted in *Weekly Petroleum Argus*, 12 February 1996.
26. We return to this issue in Chapter 12.

PART III

The Singapore Market

the industry went into a boom-bust cycle, culminating in an extremely difficult operating climate through most of the 1980s. Yet the cycle was broken and the industry strengthened despite many a death elsewhere. Section 4 contains a brief description of the industry that has emerged from this process.

CHAPTER 6

SINGAPORE AS AN OIL CENTRE

1. Introduction

In Part II we noted that, albeit at different speeds, Asian oil markets were gradually being opened up to the impact of market signals. This leads directly to the question of the nature of these signals, and the mechanisms that generate them. In turn, it places the focus of this part firmly upon Singapore, a nation of just three million people, as the primary centre of price signal generation. The basic requirement for the development of any market is the existence of trade, and of the necessary physical industry infrastructure. This chapter considers the growth and current structure of the oil industry in Singapore, as a basis for the development of Singapore as a trading and price setting centre as discussed in the following chapters.

Singapore is today the third largest refining centre in the world, after the complexes in the US Gulf coast and around Rotterdam. However, nowhere else can match Singapore for the sheer geographical concentration of its industry. At points along the southern coast of Singapore the observer can see in one vista four refineries with over 1 million barrels per day of capacity, a power station, large independent storage facilities, a petrochemical plant and the largest oil bunkering port for ships in the world. The industry is remarkably compact given its enormous size.

Virtually all the current Singapore energy industry, including all the refineries, has been built since 1960. Yet the origins of Singapore as an oil centre go back far further. Throughout the world, the development of markets has often been related, if not to historical accidents, at least to the course of political and economic developments and geographical imperatives. The growth of markets often shows hysteresis, i.e. the current structure is a function of its own past. We feel that the history of the oil industry in Singapore does provide clear pointers to an understanding of its current role. The next section looks at the history of the oil centre started in the 1870s, up to the beginning of the domestic refining industry. The strategic position of Singapore led to it establishing an entrepôt role from its situation as a crossroads in trade, and indeed it played an early part in the evolution of the world oil industry as a whole.

Section 3 considers the growth of the refining sector after independence. From demand optimism and the rapid expansions of the 1960s and 1970s, the industry went into a boom-bust cycle, culminating in an extremely difficult operating climate through most of the 1980s. Yet the cycle was broken and the industry strengthened despite many challenges from elsewhere. Section 4 contains a brief description of the industry that has emerged from this process.

For many years a mainstay of the industry was the refinery processing deal, in the simplest terms the renting out of excess refinery capacity. Section 5 considers the functioning of these deals, now rapidly becoming an historical oddity. We see these arrangements as having arisen as a market substitute, that, together with other factors, have begun to die away as a necessary consequence of the development of the markets described in later chapters. Section 6 considers the industry's export pattern, which acts as the physical base for much of the oil trading activity described in the following chapters.

2.   Early History: Singapore as a Strategic Entrepôt

The origins of Singapore's strategic role lie in the early history of the oil industry, and the movement from conditions that proxied an international monopoly to a more oligopolistic structure. With the growth of Standard Oil in the US industry, Rockefeller sought to internationalize the operations of Standard Oil and move more ambitiously into the potentially lucrative markets of Europe and the Far East. International trade in oil had begun almost as soon as the industry itself. Within a year of Drake's discovery in Oil Creek, Pennslyvanian kerosene packed in tins and then in crates was being exported to Europe, Australia and the Far East. The first attempts at bulk transport used wooden barrels, and the *Elizabeth Watts* became the first bulk carrier across the Atlantic in November 1861.[1] However, carriage in barrels proved difficult and expensive, and for thirty more years the case oil trade using tins dominated.

Standard Oil kerosene in its distinctive blue tins soon spread around the world. Cases of American oil would have reached Singapore in small quantities in the 1860s. In the 1870s traders realized that Singapore's strategic location was ideal for a distribution centre. Case oil from the USA was kept in warehouses near the harbour, for distribution around the peninsula. A re-export trade sprang up, and official figures for the export of oil from Singapore exist from 1877. By the early 1880s the re-export trade in Standard's tins was booming. Storage of case oil then began early, but the next step in Singapore's development was the establishment of bulk storage. The origins of this lie in the history of Shell, and of the attempts of its founders, Samuel and Marcus Samuel, to enter the international marketplace. It should be noted however that

Shell's involvement represents the start of a new phase in the Singapore industry, not, as often stated, the birth of that industry.

The competitive position of US oil was undermined by the start of the oil industry in the Russian empire. The first wells were drilled in Baku (in modern Azerbaijan) in 1871, and by the mid-1880s, with that industry primarily controlled by the Rothschilds and the Nobels, Russian kerosene was starting to become internationally competitive. The problem for the industry was finding markets, and in particular in entering the Asian market. Some limited penetration had been achieved with the case oil trade. The first Russian oil arrived in small amounts into India by 1885, and in February 1888 the first full cargo from the Russian port of Batum in cases was discharged in Singapore.[2] Russian kerosene had considerable cost advantages over US grades sent into the Far East from New York via Cape Town. New York oil took four and half months to reach India, while Batum oil could complete the journey in just one month. This not only gave Batum material a freight advantage, the reduction in the time of the forward commitment to be made by intermediary traders should have made it far easier to market.[3] However, without a broad geographical foothold and a network of marketing channels, the impact of Russian oil remained localized and its market penetration was limited.

Carriage of kerosene in cans was not a long-term solution. The structure of the market was altered, in a way that has an echo in changes over a hundred years later that we detail in the next chapter, by the development of large-scale transportation. Only this could gain the necessary economies of scale to make worldwide kerosene distribution possible on a large scale. The solution came from the Nobel brothers through their involvement in the early Russian oil industry. They had soon realized that transportation was the major technological problem they faced, as Ludwig Nobel had told Alfred in 1875, it was 'the great problem'.[4] By 1880 the first Nobel tanker, the *Zoroaster*, was delivered, and by 1884 the Nobels had eleven more. This represented a major technical advance, but the market for Russian oil remained constrained, primarily because of the syndicate's lack of marketing channels and expertise. They needed a marketer, and their solution was to bring in the Samuel brothers.

The insight of the Samuels was that tanker movement alone was not sufficient to gain penetration into the Far East market. What was needed was the ability to move tankers through the Suez Canal to cut the costs and time of transport, and the infrastructure to take the oil into storage from which it could then be distributed. To achieve this, the Samuels worked on three fronts. They lobbied actively for the Suez Canal to remove its restrictions on oil traffic, they began to explore the options for building their own tankers, and they set about building up an Asian network of facilities. To the latter end, two nephews of the Samuels, Joel and Marcus Abraham, were sent east to build a network of storage facilities, Joel to Calcutta to take responsibility for India, and Marcus to Singapore and Bangkok, to look for facilities there and to the east.

On arrival in Singapore in July 1891, Marcus Abraham swiftly identified possible areas to build a storage facility. Two sites were considered, at Bukit Chermin and Pasir Panjang, both close to the city centre. However, there was considerable hostility from local commerce to the idea of any oil operations taking place. Not surprisingly this was fiercest from those engaged in the existing case oil trade, who were quick to stoke fears that bulk storage would lead to explosions, and fires that would threaten both life and commerce. As a result Abraham found himself facing a rising tide of opposition which was making his enterprise virtually impossible. His solution was a small island, the modern day Pulau Bukom then known as Freshwater Island. It had the positive points of deep water and close proximity to the mainland, but for Abraham a mainland site would have still been preferable. However, the great advantage of Pulau Bukom was that it lay outside the control of the port of Singapore, and indeed of any Singapore based authority. To buy the island Abraham dealt with the Colonial Office in Kuala Lumpur, who proved more amenable to the scheme and not subject to the pressure of the Singapore traders. Not for the last time in Singapore's history, an oil installation site was chosen as a result of a favourable regulatory environment, albeit on this occasion an escape from local jurisdiction.

While the nephews continued to build up a network of installations in Asia, back in London the Samuels were making progress with tankers. Despite ferocious opposition from other parties,[5] the Suez Canal in January 1892 announced regulations that would allow oil tankers with certain defined specifications to pass through the canal. It might appear as if the Samuels had been taking a huge gamble on this, since by the start of 1892 their first tanker, the *Murex* was well into construction, and sister ships had been started. In fact they had already agreed specifications with the Suez Canal Company in August 1891, and were thus able to place themselves in a position to use the canal as soon as it was available to oil traffic.

The new regulations came into force on 1 July 1892. On 26 July the *Murex* was launched in Hartlepool, and then proceeded to Batum to load with 4000 tons of oil before passing through the Suez Canal. At Pulau Bukom 2500 tons were discharged, representing the first bulk carriage of oil into Asia, and then the rest was taken to Bangkok. The Samuels built on the advantage of their Suez route. By the end of 1893 their Suez worthy fleet had reached eleven, and the fleet was able to dominate the oil trade through the Canal for many years.[6] They had effectively proved that large cost savings were possible on the basis of improvements in transportation and storage, but they had forgotten the third part of the equation, i.e. distribution. Case oil could be immediately distributed, and indeed the tins were of intrinsic value in themselves.[7] Bulk oil needed some form of container, and with the customers loath to provide their own, the swift addition of a tin factory was made to Shell's installations.

With Shell's facility established, Singapore's importance as a strategic site grew. It nearly grew even more in the mid 1890s. In the first instance of a company seeing logic in centralizing its Far East operations, Kessler, its then head, decided to move the administration for the operations of the Royal Dutch company to Singapore. As Royal Dutch had no operations anywhere outside the Far East, this would mean moving the entire sales and transport administration from the Hague to Singapore. The plan was stopped by fierce opposition from Deterding who argued that information gathering and intelligence were more efficient in the Hague. Deterding, while in a position far below the board, managed to change Kessler's mind.[8] The central sales office remained in the Netherlands, but the history of the oil industry might have been very different otherwise. Deterding would have been relatively isolated as a sub-manager in Singapore rather than being Kessler's right-hand man in the Hague and in a position to take control later. And perhaps Singapore's importance in the oil industry as something beyond an entrepôt centre might have started 65 years earlier than it did. Beyond a point for historical speculation, the incident does show that the logic of basing operations in Singapore followed by many oil and trading companies in the 1980s and 1990s was already present nearly a century previously.

Royal Dutch did however build the second storage facility in Singapore, in 1897, to store some of their Sumatran production. In 1898 Standard Oil opened a Singapore office, and began to build storage on Pulau Sebarok. The competition for Asian trade was now an intense three-cornered fight. There was a strong incentive for some accommodation to be reached, and collaboration in the east between Royal Dutch and Shell began in 1903, four years before the worldwide amalgamation of the two companies. Together with the Rothschilds they formed the Asiatic Petroleum Company (APC), which then took a leading role in the Singapore industry.

The development of petroleum exports from Singapore from the 1870s until 1960 is shown in Figure 6.1 on a logarithmic scale.[9] Note that the start of Shell's operation led to no major sustained boom in exports during the 1890s, its major effect was the substitution of the case oil trade by the bulk oil trade. However, by the end of the 1890s when strong competition between Royal Dutch, Shell and Standard Oil had arisen in Singapore, exports began to grow. The accommodation between Royal Dutch and Shell in 1903 reduced competition, and the export trade was

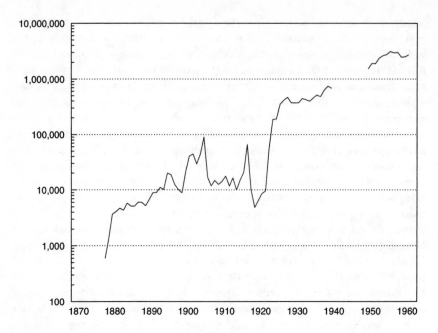

Figure 6.1:   Petroleum Exports from Singapore 1870–1960. Long Tons. Logarithmic
Scale.

adversely affected. Indeed, apart from a temporary expansion during the
First World War, exports did not grow significantly until the early 1920s.
In 1920 petroleum trade out of Singapore was in fact lower than it had
been during the 1890s.

Starting in the 1920s the domestic oil infrastructure of Singapore was
set up piece by piece. Gasoline retailing began in 1922, and fuel oil
distribution to industry from storage at Tanjong Pagar in 1926.[10] With the
start of commercial air traffic in 1930, the aviation fuel market started. As
Figure 6.1 shows, in the early 1920s exports began to pick up again,
finally surpassing their peak of twenty before, and by the outbreak
of war had reached some 700 thousand tons per year. By this time the
trade was split between two main companies, Royal Dutch Shell's APC,
and the Standard-Vacuum Company, a cooperation between Standard
Oil of New York (now Mobil), and Standard Oil of New Jersey (now
Exxon).

Much of the oil storage infrastructure was destroyed just before the
surrender of British forces in Singapore in 1942. The destruction was less
than totally efficient, particularly as it had been left to the last moment for
fear of alarming the population.[11] One would have thought that there
were enough causes for alarm already without worrying too much about

the impact of storage tank destruction, and delay meant that some usable facilities were left for the occupying forces. However, two-thirds of APC's tanks on Pulau Bukom were destroyed. Post-war reconstruction was swift, and, as Figure 6.1 shows, by the end of the 1940s exports far exceeded their pre-war levels. Indeed, a new phase of the industry, and a new source of value added began with the introduction of large-scale blending operations.

In the 1950s the Singapore oil trade started to flourish after the traumas of occupation and the Malayan Communist uprising. Three large tank farms served as a blending, bunkering and distribution centre throughout the region. Singapore also served as a port for regional refineries, particularly in Borneo and Southern Sumatra. Typically these were located on rivers which did not have sufficient draught to enable full cargoes to be loaded. As a result partial cargoes were taken to Singapore, where full cargoes could then be loaded. Even before the start of a domestic refining industry, Singapore's entrepôt role had become highly significant to the domestic economy. In 1950 about 20 per cent of all vessels going through the port were involved with the oil trade.[12] By the late 1950s trade had grown to between 2.5 and 3 million tons per year.

At independence in 1959 there was a further potential source of value added left, and one that was a natural progression from the past. Over eighty years the industry had grown through storage, distribution and then blending functions. The next step, consistent with the new government's policy of industrialization, was to set up a refining industry.

3.   The Start and Growth of the Singapore Refining Industry

With independence, the economic climate for industrial operations had changed. The colonial administration had essentially been highly *laissez faire* with regard to the oil industry. It certainly did not attempt to block the trade, but neither did it provide explicit incentives. Now the industry was faced with an administration that was prepared to give the market mechanism some encouragement. The Singapore government, keen to promote industrial development, offered 'pioneer' status to new developments, which entailed tax free operations for the first five years. Shell became the first pioneer company, making the decision to build a refinery on Pulau Bukom in late 1959. Site clearance began in mid-1960 and the refinery began operations on 26 July 1961.

It is highly probable that there was already an inevitable logic to Shell building a refinery on Bukom. There was a perception that demand would increase throughout the Far East, particularly in Japan, and Shell needed more capacity in the region. Bukom, with its existing infrastructure of tanks and jetties was the logical choice. But the government initiative must have helped cement the decision, and the tradition of governmental incentives to maintain Singapore's position as the key industry location in the region was born.

In the twelve years after Shell began to refine, four more refineries were constructed. The second, a small plant on the mainland at Pasir Panjang, was opened in March 1962 and was owned by the Japanese companies Maruzen and Toyo Menka, primarily to supply fuel oil to a nearby power station. The refinery was bought in 1964 by BP and continued to operate as a simple facility until 1995. Mobil built a plant onshore in the Jurong industrial estate that commenced in June 1966, and in 1970 Exxon started operations on Pulau Ayer Chawan. Finally, in 1973 the Singapore Petroleum Company (SPC), started on Pulau Merlimau. SPC is a joint venture, originally owned by Amoco, Oceanic Petroleum and the Singapore Development Bank (31.33 per cent each), and the Japanese *sogo shosha* C Itoh.[13] From no refining industry at all, Singapore had then grown to five plants within twelve years. This growth was due to the combination of a series of factors.

Faced with the threats caused by the uncertain politics elsewhere in the region, particularly that of nationalization of assets, companies saw Singapore as a more stable environment. While the riots of 1964 were still a recent memory, Singapore post separation from Malaysia had an anti-communist ruling party with entrenched power and support, and with an overtly free market philosophy. Nationalization was a remote threat, as was the insurgency which had beset most other countries in the region. Singapore was indeed 'an oasis in Southeast Asia'.[14] Governmental institutions and politics were also a major factor in encouraging companies to centre their activities in Singapore. Elsewhere in the region they faced restrictive price regulations, excessive bureaucratic red tape and the threat of partial or total nationalization. Indeed, Esso Eastern used all these as justifications for selling their Philippine refining and marketing assets in 1973, while continuing to expand their Singapore capacity.[15]

Throughout the 1960s and early 1970s the perception of a booming market persisted, particularly in Japan. Acceptance by the Japanese government of the Arisawa Report of December 1959,[16] and the policy switches it then led to in the early 1960s, had made the direction of Japanese energy policy extremely clear. The economy was to be fuelled by reliance on imported oil. For the Singapore oil industry the combination of the Arisawa report and the Japanese economic boom meant that they could plan on moving ever greater amounts of product into Japan. While the same companies that built in Singapore also had a presence in Japanese refining (with the exception of BP and SPC), Japan offered a less benign regulatory regime, and higher costs. Helped by the relative commercial environments, not only were new refineries built in Singapore, but Shell continued to add distillation towers to their existing facilities, in the belief that the Japanese economy alone would be able to absorb almost the entirety of Singapore's export potential.

The Vietnam war also had an effect on perceptions in the 1960s, particularly of course for American companies. Refining operations in Singapore would cut the length of the supply chain to the war, without exposure to direct risk, and post-war would leave an increased presence in an important emerging market. In 1970 over 20 per cent of oil exports from Singapore went to South Vietnam, and this was still close to 15 per cent by the fall of Saigon in 1974.

Added to the above factors were the same natural advantages of Singapore which had, as discussed in the last section, made it an entrepôt centre since the 1890s. Singapore offered deep water and extremely good harbour facilities. But most of all its geographically strategic position within South East Asia, close to the major trading routes, was still as attractive as it had been to Raffles when founding the original colony. Singapore had good communications, good labour relations and political stability, and offered an accommodatory industrial policy, and none of the new refinery builders could have had many qualms about their decision.

Shell expanded again in 1973, completing their fifth distillation unit, and in the same year the SPC refinery began operation. But the year also represented the start of a change of economic climate for the industry. The first oil shock and its aftermath brought the consequences of the deintegration of the industry, and the demand implications of the quadrupling of prices. The basic tenet of the leading role of Japanese oil demand had to be abandoned. Apart from the economic implications of higher prices leading to substitution away from oil, supply security moved to the top of the Japanese energy agenda. The essentially pro-oil stance the Arisawa report had created was abandoned, and policy turned against reliance on imported oil.

The SPC complex was the last grass-roots refinery to be built in Singapore. But there was one last major influx into the industry. In 1978 SPC, BP and Caltex set up the Singapore Refining Company, with SPC holding 40 per cent and the other two companies 30 per cent each.[17] On the completion of a major expansion of the SPC facility from 70 thousand to 170 thousand b/d, SRC became the joint holding company for the Pulau Merlimau refinery in 1980. To some extent the expansion of Singaporean refining may have been motivated by dislocations elsewhere in the world, and particularly the nationalization of assets in the Middle East. The importance of political risk in refining had become clear, and the SRC joint venture offered the chance to regain a little of the lost capacity without any significant increase in that risk.

The events of the 1970s and their aftermath were enough to stem any further growth in terms of new plants, as the decade had seen a huge increase in Singapore's capacity. By the end of the 1960s total capacity had reached 180 thousand b/d, and, with the expansion of the SPC refinery into the SRC refinery, in 1980 capacity passed through 1 million b/d. In retrospect it had overshot what was a sustainable level given the conditions of the late 1970s and early 1980s. In fact a sixth refinery had been planned, involving Elf, Total and the Singapore government. Given the appearance of growing excess capacity and weaker profit margins, this project was abandoned. From the point of view of the rest of the industry the abandonment was welcome, and certainly Singapore's problems in the 1980s would have been exacerbated by the presence of even more excess capacity.

In the early 1980s the viability of Singapore refining was affected from both the supply and demand sides of the market. On the supply side the export refineries of the Middle East, particularly in Saudi Arabia and Kuwait, began to bite into Singapore's traditional markets. The Middle East, especially Saudi Arabia, had been a source of significant demand in the late 1970s, but now the area was in direct competition in supply and had a geographical advantage for markets on the west side of India. On the demand side, the second oil price shock of 1979–80 led to a further dampening of consumption. The refiners cut their own runs, and, as is detailed in Section 5, were forced to begin to rely on processing deals to fill some of the excess capacity that arose. Refineries built on the assumption of an unremitting diet of Middle East crude oil, suddenly had to cope with a variety of, often difficult, regional crudes, resulting in a significant fall of effective relative to design capacity.[18]

The capacity to process was not just reduced by the change in crude oil inputs, there were also capacity shutdowns as the industry went into recession. In February 1983 Shell announced plans to close down nearly half of its capacity, bringing it down from 460 to 250 thousand b/d, in the words of the management at the time because 'Singapore's refining capacity is too big for the future'.[19] A view began to prevail that the future was distinctly unpromising, and that the industry was in serious decline. Not only were the Middle East refiners expanding, but new refineries were planned for South East Asia. In the context of the then prevailing demand pessimism, the new capacity was expected to lead to a further erosion of Singapore's position.

In the first half of the 1980s the pressure was quantity based, too much capacity and not enough demand. The oil price collapse of 1986 was to a great extent the start of the turning of the corner. A positive price effect on demand was added to the underlying impact of economic growth. The mechanics of the price fall were driven by the adoption of netback deals by major crude oil producers, effectively giving refiners a guaranteed margin.[20] As a result, while the value of refinery output fell sharply, the value added in refining was less severely impacted.[21] Table 6.1 shows the value of output and the value added of the petroleum industry in relation to the Singaporean industrial sector, and the level of employment generated.

Table 6.1: Industrial and Oil Sector Output, Value Added and Employment in Singapore. 1983–93. Million Singapore Dollars and Number Employed.

| | Output | | | Value Added | | | Employment | | |
	Total	Oil	Share	Total	Oil	Share	Total	Oil	Share
1983	37,222	13,164	35.4	9822	1383	14.1	273,228	3755	1.4
1984	41,078	12,449	30.3	11,106	956	8.6	276,225	3605	1.3
1985	38,495	11,031	28.7	10,702	874	8.2	254,802	3494	1.4
1986	37,259	6990	18.8	11,900	780	6.6	247,732	3367	1.4
1987	46,084	7491	16.3	14,471	728	5.0	277,031	3245	1.2
1988	56,470	7663	13.6	17,918	994	5.5	325,235	3125	1.0
1989	63,626	8765	13.8	19,676	1248	6.3	338,043	3113	0.9
1990	71,333	11,365	15.9	21,607	1662	7.7	352,067	3291	0.9
1991	74,575	11,288	15.1	23,450	2023	8.6	358,723	3725	1.0
1992	77,276	10,272	13.3	24,911	1750	7.0	358,788	3808	1.1
1993	87,212	11,351	13.0	28,312	1986	7.0	354,515	3998	1.1

Source: Singapore Yearbook of Statistics.

The fall in the value of output between 1985 and 1986 was some 36 per cent. The fall in value added was little changed from the fall experienced in 1985, and considerably less than that of 1984. The industry had in fact been going through a deepening recession between 1983 and 1987, with a decrease in the value added contained in output of 47 per cent, as shown in Table 6.1. The sector also shed 14 per cent of its labour force, and its importance within total industrial value added had fallen from 14.1 per cent to just 5 per cent. With the boom in the rest of Singaporean industry since the 1980s, petroleum has never regained its former relative importance. But since the mid 1980s the industry has rebounded to a thriving position in the late 1990s. We can identify three main reasons for the renaissance.

The most obvious factor has been the growth in volumes caused by the underlying economic growth in Asia, helped by a strong price effect after the 1986 price collapse. The excess capacity of the mid 1980s was steadily eaten away. In 1984 the growth of Middle East exportable oil product surpluses was seen as a major threat, combined with the growth of capacity elsewhere in South East Asia. But the demand effect was enough to swamp these factors. Table 6.2 shows the significant increase in the Middle East oil trade surplus between 1984 and 1994 and the increases in the crude oil trade deficit in Asia.

Table 6.2:   Oil Trade Balances by Region. Crude Oil and Oil Products. 1984 and 1994. Thousand b/d.

	1984		1994	
	Crude Oil	*Oil Products*	*Crude Oil*	*Oil Products*
Middle East	+8825	+835	+14,319	+2034
Australasia	-80	-90	-235	+49
South Asia	-270	-150	-705	-462
Japan	-3690	-600	-4667	-807
China	+444	+115	+95	-245
Other Asia	-595	+160	-2272	-537

Note: Singapore is included in 'Other Asia'.

Source: *BP Statistical Review of World Energy.*

The exportable oil product surplus from the Middle East did indeed rise sharply over the period, up by 1.2 mb/d. Refinery runs also increased in Asia, as evidenced by the 3.6 mb/d increase in Asia's total crude deficit. To that extent the pessimists had been proved right. But regional capacity expansion was swamped by demand increases, and the region's net oil product deficit with the rest of the world increased by more than the additional Middle East surplus. Japan's product deficit increased by 207 thousand b/d, South Asia's by 312 thousand b/d, and China slipped from net exporter to net importer, its oil product balance worsening by 360 thousand b/d. But the deficit grew most among the economic tigers in the rest of Asia, worsening by 697 thousand b/d. While Australasia managed to move into surplus, overall Asia and the Pacific needed to buy an extra 1.4 mb/d of oil products from the rest of the world, in addition to the extra 3.6 mb/d of crude oil. Far from declining, Singapore was at the heart of an area seemingly moving into an ever deeper oil product deficit.

However, the return to strength of Singapore was due to more than just the effect of rising regional demand. There was also a strong effect arising from the response of refiners to the problems of the 1980s. Note that in Table 6.1 the value of petroleum output had not returned to its 1983 level by 1993. In contrast the industry's value added had surpassed the 1983 level by 1990. Value added as a share of output was 10.7 per cent in 1983, and had declined to 9.7 per cent by 1987. In 1993 it was 17.5 per cent. The refiners had reacted to their decline by both becoming more competitive in cutting costs, and also through investment increasing value added by upgrading their refineries. New complex units were added, with a consequent improvement in product yields. At the same time that Shell was closing distillation towers on Pulau Bukom, it was building its first hydrocracker outside of the UK. The guiding principle had become 'less, but better'. In total the industry became far more competitive, and far more flexible.

The third reason for the improvement is the decline in third party processing detailed in Section 5. As refiners' own volumes grew and markets developed, the need for such arrangements declined. Refining became a more integrated operation, and it became easier to optimize operations, further increasing the proportion of value added in output.

4. The Modern Singapore Refining Complex

After 1986 the runs through the refineries steadily increased. Figure 6.2 shows the monthly total crude runs from 1987 to the start of 1996, and demonstrates a clear upward trend, with sporadic dips due to refinery maintenance programmes. While runs had been increasing, capacity fell

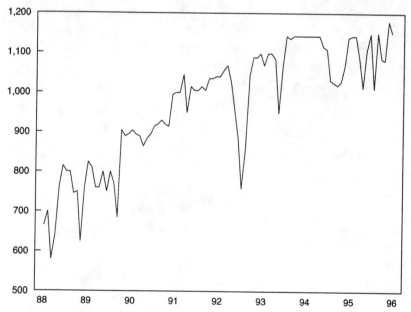

Source: Petroleum Argus Data.

Figure 6.2:   Singapore Refinery Throughputs. 1988–96. Thousand b/d.

from 1984 until 1990, before growing again. In 1993 Shell's latest expansion brought the capacity of Pulau Bukom back above 400 thousand b/d, and for the first time total capacity surpassed 1.1 million b/d. Compared to the rapid increases of the 1980s, capacity had grown by just 100 thousand b/d from 1981 to the start of 1995. Further expansions by Mobil and SRC brought capacity to over 1.2 mb/d by the end of 1995. The difference was that now the refineries were at full capacity with new complex units, and the excess fat of the 1980s had disappeared. The industry thus went through a whole cycle of boom and bust, before reaching its modern position of some strength.

The locations of Singapore's major oil facilities are shown in Figure 6.3, which illustrates a section of the southern coast of the main island. The area shown is very compact, no more than ten miles across, but it has a remarkable agglomeration of installations. Loading of cargoes is effected through terminals (at both refineries and independent storage facilities) and through the three single buoy moorings shown. The western refineries are linked to other installations. Direct pipeline links run from the SRC refinery to a Public Utilities Board (PUB) power station, and

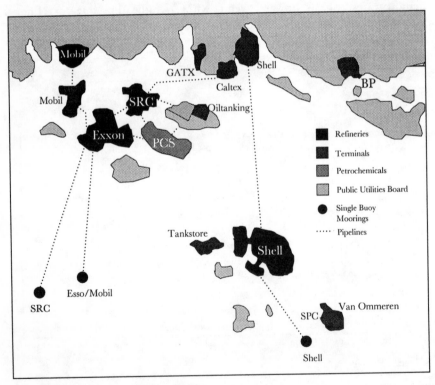

Figure 6.3: Singapore Oil Installations.

from both SRC and the Exxon refinery to the Petrochemical Corporation of Singapore (PCS) installation. Figure 6.3 also shows the locations of independent terminal and storage facilities for crude oil, oil products and chemicals, which are considered in Chapter 7.

The technical characteristics of the four Singapore refineries in terms of their major units are summarized in Table 6.3. A fifth refinery ceased operations at the end of June 1995. BP's small Pasir Panjang facility (as noted in Section 3, the second refinery to be built in Singapore) had been a simple refinery with no facilities for further breaking down the residual fuel oil produced by crude distillation. Its site is shown in Figure 6.3 as a BP storage facility and terminal.[22] The remaining four refineries are all complex, and through a variety of processes can reduce fuel oil yields.

Table 6.3: Capacities of Singapore Refineries. 1996. Thousand b/d.

Refinery		*Crude Distillation*	*Vacuum Distillation*	*Visbreaker*	*Reforming*	*RFCC**	*Hydro-Cracking*
Shell	Pulau Bukom	440	78	60	20	28	28
Mobil	Jurong	275	90	55	58		23
Exxon	Pulau Ayer Chawan	230	40	43	12	-	-
SRC	Pulau Merlimau	280	75	32	14	33	31

* Residue Fluid Catalytic Cracking

Sources: Various.

The form of catalytic cracking found at the Shell and SRC refineries is residue fluid catalytic cracking rather than the less advanced vacuum gasoil fed catalytic cracking utilized elsewhere in Asia.[23] Appendix 2 gives a brief description of refinery units, and from this the comparative advantage of the Singapore complex can be summarized as being in middle distillate production, and particularly in the production of low sulphur products.

In the last section we detailed the major advantages that drew refiners to Singapore. However, the island does have one major disadvantage, namely the lack of spare land for future expansion. Many of the southern islands have been expanded by reclamation, for instance the modern day Pulau Bukom is many times greater in size than the original site. The initial development of the refinery created 25 acres of land, and the area has been consistently increased since. The most ambitious reclamation plan yet in Singapore is the current government plan, that would create one island out of the seven shown to the west in Figure 6.3. The Exxon and SRC refineries would then, together with the Petrochemical

Corporation of Singapore (PCS) petrochemical plant, be on the same island. A second phase would further extend the island towards the south west, with the new land being earmarked for petrochemical expansion.

Petrochemicals are the main focus of the future growth of the complex. The PCS plant began operations in 1984, primarily using capital from the Singapore government and the Development Bank of Singapore, together with Japanese funds. Having launched the project, the government sold its share into the private sector. The company is now equally split between Shell and the Japan-Singapore Petrochemicals Company (the latter being a consortium with the primary involvement of Sumitomo). PCS primarily supplies ethylene, propylene and acetylene to a series of downstream petrochemical companies.

The expansion of the complex, which is due for completion in mid-1997, involves a new naphtha cracker that will double the capacity of the company. Some of the downstream companies will also expand, and new ones will be started. Most notably, a Shell/Mitsubishi joint venture is building a large plant for the production of styrene monomer and propylene oxide, to be used as input to a series of other downstream companies.

We have seen the history of oil in Singapore since the 1870s has been one of gradually obtaining more sources of value added. The expansion of the petrochemicals sector is seen by the government as the next important enlargement of the hydrocarbon industrial complex. They have given backing in the form of the giant land reclamation programme. The resultant Jurong island, linked to the mainland by a new road bridge, will, by a considerable distance, represent the most massive concentration of oil refining and petrochemical complexes in the world. While a long way from the case oil storage facilities of the 1870s, there has been a natural progression in this development.

5.   The Rise and Fall of the Processing Deal

As we noted in Section 3, in the late 1970s and the 1980s, Singapore refiners had substantial excess capacity, and they also had internal company distribution systems within Asia that were not large enough to absorb Singapore's surplus. In Rotterdam or Houston a refiner could simply use that excess capacity up to the point where the margin was no longer profitable, and sell out the resultant excess production in the open market. Singapore refiners did not have that option. In the absence of any developed markets that could be used to sell products and manage the inherent risk, selling out surplus production tended not to be seen as a viable option. This combination of excess capacity and the lack of markets led directly to the use of refinery processing deals. While the evolution of markets led to the rapid demise of large-scale processing in the Mediterranean and the Caribbean,[24] in Singapore it represented more than 20 per cent of crude oil runs as late as 1992.

For the refiner a processing deal can act as a substitute for risk management by essentially locking in a refinery profit margin. The generalized mechanics of a processing deal are as follows. The deal involves specifying four main parameters; the type of crude to be run, the volume, the yield structure and the processing fee. For example, consider the following arrangement, which, while imaginary and not representative of any one particular deal, is still broadly representative of a typical arrangement. Assume a deal for, say, Dubai, for a set volume. The deal might involve specifying five main yield parameters, the percentage yields for naphtha, kerosene, gasoil and fuel oil, which in the case of Dubai will be high sulphur fuel oil (HSFO). The fifth yield parameter is an allowance for refinery losses such as refinery fuel input. Typical yields might look like those shown in Table 6.4, which summarizes the refinery economics of the deal.

Table 6.4: Structure of a Typical Refinery Processing Deal. Per Cent Yields and $/Barrel.

	(1) Yields	(2) Price	(3) Value	(4) Premium	(5) Margin
Naphtha	0.150	19.75	2.96	0.00	0.45
Kerosene	0.165	23.75	3.92	0.00	1.16
Gasoil	0.200	23.25	4.65	0.00	1.30
HSFO	0.450	13.75	6.19	0.15	-1.20
Losses	0.035	-	-	-	-0.59
TOTAL	1.000	-	17.72	0.15	1.12

In Table 6.4, multiplying the market price of each component for a given day during the deal in column (2) by its yield produces its contribution to the total value in column (3). Adjustment may be made for premia from the market quotations. In particular, the heavy fuel oil produced is likely to have a different sulphur content than the specification used for the market quote, and an adjustment will be made. The deal will specify the sulphur content to be used in the calculation, and the method for determining the premia. The gross product worth (GPW) is $17.87/barrel, being the summation of columns (3) and (4).

The yields offered in a processing deal will normally be those from simple distillation. The deal therefore assumes that the processed oil is marginal in the sense that the upgraded units are filled up. In terms of economic theory it is correct to offer simple yields if at the margin the upgraded capacity is fully utilized. This does not of course imply that every barrel of processing deal crude oil merely goes through simple distillation – some will fill up the crackers. In this case the additional value of output simply accrues as rent to the refiner. Pricing efficiency should only lead to the bargaining process resulting in the offering of complex yields in three sets of circumstances. First, if the refinery is so upgraded that even at full distillation capacity there is space in upgraded units. Secondly, if refinery runs were so low that these units also had spare capacity. The final circumstance is where the customer's crude can be considered intramarginal, e.g. if they were the holder of a long-term contract, at a time when there was a large volume of short-term processing. The first two conditions have tended not to hold in practice. Long-term contract holders have normally received better terms, but these have tended to operate through lower processing fees rather than better yields.

Having derived the gross product worth, the net profit or loss under the prevailing structure of prices is simply the GPW minus the price of Dubai minus the fixed processing fee. To illustrate, assume the Dubai price at $16.75 per barrel, and a fee of $1.50/barrel, we get a loss of 38 cents per barrel to which would have to be added the freight fee for getting the crude oil from Dubai to the Singapore refinery gate. An alternative method of expressing the same result is to calculate the margin by product, i.e product price minus crude price multiplied by the product yield, as shown in column (5). Valuing the loss factor at the prevailing crude price and summing produces the same margin of GPW over the crude price.

The lack of markets then provided a reason why refiners would want to offer processing deals. On the demand side both state and private companies had reasons to sign such a deal. There were five main sets of motives. First, regional producers, and in particular Malaysia, Indonesia and China, had crude oil surpluses but a lack of domestic refinery capacity that left them with oil product deficits. These deficits tended to take the form of a refinery structure that was unable to produce enough middle distillate oil products. With oil product demand booming, a processing deal could not only close an overall deficit, but also help balance the composition between domestic demand and domestic product supply. We provide an example of the latter when considering Indonesian policy below.

Secondly, demand also came from Middle East producers. Before 1986, when OPEC operated under an official price system, refining crude and selling products offered an alternative to directly discounting crude oil prices. With the development of refining capacity in the Middle East, and thus of the development of the producers' marketing channels in the Far East, Singapore could also act as an additional supply source.

A motive was also provided by the asymmetric development of markets. In particular, oil product markets have always developed faster than crude oil markets in the region. This has produced trading opportunities, and a demand for spot processing. In acquiring a spot crude oil cargo it would often be easier for a trader to refine it themselves in Singapore and sell on products, rather than trading on the cargo in the limited and imperfect spot crude oil market. With underdeveloped markets that were slow to react, profit opportunities were slow to be eroded. Hence, according to market conditions, there has sometimes been a healthy demand from traders for spot processing contracts of one, two or three months duration.

The fourth source of demand came from companies with distribution channels within the region. If there existed a full set of markets, one of their major functions would be as a source of supply. Without those markets a refinery processing deal could act as substitute. For example, until recently Japanese companies would use processing deals primarily as a way of sourcing for the domestic market. The final motive arose as the Singapore market complex began to develop. Processing could be used as a way of sourcing products to trade with, rather than as a source of supply for distribution channels. A deal could even be used as a way of learning the market by companies entering into Singapore trading. Later, products might be sourced elsewhere, but a processing deal could be used as an easy entry point in developing trading skills in the market.

In the late 1970s and throughout the 1980s, the volume of crude oil refined under processing deals in Singapore represented a third or more of total refinery runs, reaching a peak of some 350 thousand b/d in 1986. The main sources of demand were from the state oil companies of Indonesia, Malaysia and China.

Indonesia was for many years by far the largest processor. In the late 1970s and early 1980s Pertamina, its affiliates and other companies acting for it, were running some 200 thousand b/d through Singapore. Pertamina faced a chronic deficit of gasoil, and its processing activity was primarily motivated by the need to obtain it. A typical Pertamina deal would involve contracts specifying yields of 70 to 80 per cent of gasoil, and zero yield for other products. Refiners would then have to sell out or use in their own systems the naphtha and fuel oil produced, and divert the gasoil from their own production. Pertamina would also, rather than make the transaction part of an explicit processing deal, simply swap crude oil for a smaller volume of gasoil with Singapore refiners.

Faced with large product deficits, Indonesia brought refinery capacity on stream, including an impressive battery of hydrocrackers. However, despite the new domestic capacity Indonesia kept processing in Singapore, albeit at the reduced levels of 80 to 100 thousand b/d in 1985 and 1986. The motive for processing before domestic capacity expansion had simply been to acquire gasoil. Now, the main motive came from the need to secure outlets for crude oil production. Wholesale recourse to the spot market was not an option. Given the thinness of Far East spot markets, the potential downwards pressure that would be exerted on an already weak market was considered too great. The combination of an unwillingness to sell below inflated official prices, and the difficulty of selling spot, combined to make processing an attractive option. However, whereas in the early 1980s virtually all products had been taken to Indonesia, now most were sold back to the refiners or sold on to traders, with product only being brought back to cover for operational problems at domestic refineries.

Indonesia's demand for processing was then first a function of a lack of effective markets to act as a source of supply, and later a function of distortion between markets. Faced with constraints in crude oil markets, they preferred simply to transfer the pressure to discount into the products markets. Secure outlets for crude were maintained, at the expense of making product market trades with poor underlying economics. Processing deals were not only a substitute for the market, but also a means of siphoning off downwards pressure on crude oil markets and realizing the reduced revenues in a less transparent fashion. However, with the move away from official prices and greater reliability in domestic refining, the reasons for Pertamina's processing disappeared. While processing of Indonesian oil continued, by 1987 Pertamina had withdrawn from direct large-scale Singapore refining. It evolved a system of swaps with four affiliate companies. The affiliates were allocated crude oil, which they were free to process in the region or to sell on, in return for supplying the parent company with gasoil, jet fuel and kerosene.

The loss of large-scale Indonesian processing and the growth of Indonesia's domestic refining had no long-term effect on Singapore's viability, given the size of the regional imbalances shown in Table 6.2. The short-run impact was mitigated by new sources of demand for processing, and in particular the growth of Chinese volumes.

Processing of Chinese crude oil in Singapore started in 1980, and resumed in the first quarter of 1982 when Coastal Petroleum ran 1.5 million barrels of Daqing crude oil through the Exxon refinery. In late 1982 China began negotiations for a direct processing deal, originally with SPC.[25] The first deal finally emerged in early 1984 with Shell, and then China started to sign deals with the other refiners. The next year it became the largest processor in Singapore, with a volume of about 115 thousand b/d. A similar volume was maintained the next year, but from that point Chinese processing declined. In 1987 and 1988 it was about 80 thousand b/d, and from 1989 to 1992 it remained in the range of 40 to 50 thousand b/d. By 1993 the volume was down to just 15 thousand b/d with Shell, maintained through to 1996. Volumes fell as China's crude export surplus began to disappear, and they began to rely more on sourcing from the market than by processing. The original reason for processing had also disappeared, i.e. China had moved into Singapore processing primarily to avoid putting direct pressure on its crude prices.

Malaysia, through state oil company Petronas, processed for Singapore for some seventeen years. They started in the late 1970s, and only finally withdrew in 1995. Volumes peaked at around 75 thousand b/d, held through most of the 1980s, and then declined as domestic refinery capacity came on stream. Other countries with excess crude oil have also used Singapore in particular circumstances. The Iran–Iraq war and the loss of Iranian refining capacity, led to Iran processing as much as 100 to 150 thousand b/d in some years. Bangladesh received crude oil from Saudi Arabia on a state-to-state basis, but was not allowed to resell. With no domestic capacity, Bangladesh became a processor at Singapore until 1983, when it became cheaper to buy products on the open market rather than refine the crude from the state-to-state deal.

During the 1990s several factors have combined to lead to a decline of third party processing in Singapore. The refiners, with the exception of SPC which only operates in the Singapore domestic market, have found the demand for products through their own Asian distribution channels increasing and biting into the excess capacity. Furthermore, just as the lack of markets led them into processing deals, the continued development of markets has reduced their need for such deals. Refinery utilization increases in Japan have led to less need to obtain products from Singapore, and again the developing markets now provide an alternative to processing deals. With refinery capacity expansions and surging demand, regional producers have less excess crude oil, and have become more willing to place what remains directly on the market. Finally, the rapid expansion of the Singapore trading community has reduced the potential pool of newcomers who wish to use processing as a learning instrument.

With less motivation for refiners to enter into such deals, and less demand for the deals, third party processing volumes in Singapore have declined sharply. Table 6.5 shows volumes by refiner at selected points, the total and the total net of SPC.[26] By the start of 1995 processing had fallen to just over 10 per cent of total runs, with only 45 thousand b/d outside of SPC. Exxon withdrew from processing deals in 1994, and Mobil and Shell have both cut back to a small fraction of past volumes. The closure of BP's small plant removed the company from processing, as it refines in its SRC share on its own account. This has left third party processing mainly in the hands of SPC, the only company without a sufficiently large distribution system to move their own product. Indeed SPC now tends to obtain its own requirements through buying back from its third party processors.

Table 6.5:   Processing Deals by Refiner. 1990–5. Thousand b/d.

	Shell	*Exxon*	*Mobil*	*BP*	*SPC*	*Caltex*	*Total*	*Non-SPC*
Mid 1990	35	40·	95	45	50	0	265	215
End 1991	45	30	65	15	65	0	220	155
Start 1992	40	25	50	15	65	0	195	130
Mid 1992	40	15	55	15	30	0	155	125
End 1992	40	25	50	15	30	0	160	130
Mid 1993	40	25	35	15	60	0	175	115
End 1993	15	10	25	10	60	0	120	60
Mid 1994	15	0	25	15	85	0	140	55
Start 1995	15	0	10	20	80	0	125	45

Sources: Various.

The decline is shown from the demand side in Table 6.6, which lists the companies undertaking processing at discrete points in early 1989, late 1991 and in early 1995. The major declines are from the exit of Petronas and the scaling back of Sinochem volumes, the exit of traders and Wall Street companies such as Vitol and J Aron, and the overall decline in Japanese processing. The declines in these categories are larger than the overall decline shown in Table 6.5, as there has been some expansion of other volumes. Kuwait's active development of its Far East and South Asian marketing has led to an increase in KPC processing. Chevron has always had a large presence in the region through its share in Caltex, but now, like Texaco, it has sought to create a presence on its own account,[27] and is the only new entrant to processing shown in Table 6.6.

Outside of Singapore, Japan and China are also important third party processing centres. In Japan the use of processing has two main causes, the major being the regulation of the Japanese industry. Export of products from Japan has only been possible for the output of processing deals, indeed they must export more than half of that output. Faced with domestic surpluses, particularly of heavier products, a processing deal is thus the only means available for a Japanese company to gain access to the export market. The second reason is purely logistical, with Japan being a more proximate source of supplies than Singapore for East Asian countries facing product deficits. Korean refiners as well as Taiwan's CPC (Chinese Petroleum Corporation) have been regular processors at the Okinawa refining centre.[28] Again processing acts as a substitute for spot and forward markets, in this case due to the suppression of such export markets by regulation. Third party processing in Japan stood at some 100 thousand b/d in 1995, marginally less than Singapore.

China has also become an important centre, with some 80 thousand b/d of processing deals in 1995. Again the rationale is the absence of a

Table 6.6: Holders of Processing Contracts and Volumes. 1989, 1991 and 1995. Thousand b/d.

Company	1989	1991	1995
KPC	10	7	30
Sinochem	52	45	18
Chevron	-	-	11
Marubeni	15	19	11
Agip	10	10	10
Cosmo	-	25	10
Total	-	8	7
Idemitsu	-	5	5
Mabanaft	-	5	5
Mitsubishi	15	-	5
Petronas	60	30	-
Vitol	20	20	-
Sanseki	-	12	-
Kuo	-	10	-
Yukong	-	8	-
Attock	-	5	-
J Aron	-	5	-
Kyoseki	-	5	-
Toyotsu	-	5	-
Phibro	15	-	-
Iran (NIOC)	7	-	-

Sources: Various.

market because of regulation, with China constraining access to its domestic market except for third party processors. Trading companies, in particular Vitol and Nicor, are the major processors in China. Under such circumstances the processing fee should contain a premium, representing an entry price into a rationed market, just as a Japanese deal represents an exit price. Compared to Singapore, in 1995 Chinese fees were indeed some 50 to 100 cents higher, and Japanese about 20 to 40 cents higher. Other countries in the region have in the past had third party processing in their refineries, in particular Taiwan and the Philippines. However, the only other recent processor has been the Sunkyong refinery in South Korea.

6. Singapore and Asia

Singapore's role is a balancing one. Other countries have a mismatch between domestic demand and their refinery configurations, and this can be balanced, particularly in middle distillates, by an industry with sufficient flexibility. This role has changed over time. Once Singapore was the only major export refining centre in the Asian market, and even its then limited upgrading gave it the advantage. Now it has gained a niche through a concentration on middle distillates and by efficiency gains, also having the most reliable refinery system in the region.

The key feature of the Singapore industry is its concentration on the supply of middle distillates rather than gasoline. In particular, the industry has an array of hydrocrackers with two residue fluid catalytic crackers. Only Indonesia has a greater ratio of hydrocracking to distillation, and it remains an importing country. While the development of Indonesian hydrocrackers certainly removed an important source of demand from Singapore, demand growth in other countries has been able to compensate. Elsewhere the reliance is primarily on vacuum gasoil fed fluid catalytic cracking, leaving many countries relatively gasoline rich but distillate poor.

The Singapore industry is well placed in relation to Asian countries with larger distillation capacity. Japan has huge capacity, but with a large number of refineries the average size is small and so it does not have the same advantages of economies of scale. As we saw in Chapter 4, the Japanese industry has been highly regulated and relatively inefficient, whereas Singapore has become extremely efficient. Finally, Japan's upgrading is not only less extensive in proportion to Singapore, it relies primarily on fluid catalytic cracking. China has by far the most upgraded refining industry in Asia. Catalytic cracking dominates, a mixture of sophisticated modern units together with some more obsolete Soviet designs and home grown variants. The average size of refinery is even smaller than in Japan. The industry is extremely inefficient, and, as was explained in Chapter 3, suffers from some major geographical dislocations. Overall, the structure of its refining industry hints at a country with a product deficit biased towards deficits in distillates. This is indeed the case, and the Chinese distillate deficit has represented a major potential new opportunity for Singapore in the 1990s. South Korea has a large distillation capacity, which is however as yet extremely simple capacity with little upgrading. It also has relatively tight specifications for sulphur limits in products, and such an unsophisticated refinery configuration implies an excess of high sulphur products but shortages of low sulphur, providing a further niche for Singapore.

Singapore's structural advantages have resulted in the export pattern shown in Table 6.7. Diesel (gasoil) is by far the most important cargo export, and has made up nearly half of the increase in exports since 1986. Exports of cargoes of fuel oil have increased little over the same period, with incremental production primarily being absorbed by the ship bunker fuel market, and also by the increase in the cracking capability of the refineries. While gasoline exports have risen faster than any other major

Table 6.7: Singapore Oil Exports by Product. 1986–95. Thousand b/d.

	1986	1987	1988	1989	1990	1991	1992	1993	1994	1995
Gasolines	53.9	53.1	64.9	86.7	101.0	108.6	110.5	124.7	138.0	145.9
Naphtha	61.5	54.5	55.6	61.8	68.1	84.6	95.8	113.1	110.7	105.5
Kerosene	38.1	25.0	30.5	25.4	24.0	29.5	16.0	16.1	11.2	12.0
Jet Fuel	74.7	100.2	105.4	101.1	125.6	122.5	109.1	121.0	133.0	132.6
Diesel	163.8	159.8	187.0	207.5	223.0	229.2	245.4	303.2	358.0	330.8
Fuel Oil	166.0	148.7	156.2	167.0	170.1	158.4	169.3	168.9	181.2	183.8
LPG	7.0	7.4	10.6	9.8	13.0	10.2	9.6	12.3	17.6	17.6
Lubricants	9.9	10.8	12.5	13.5	14.9	12.9	13.6	15.2	16.8	17.4
Bitumen	6.6	5.6	10.4	6.3	7.5	5.3	8.2	11.6	14.1	17.9
Bunkers	n.a	n.a	n.a	174.3	182.0	203.3	205.0	230.7	250.6	230.6
Air Fuelling	n.a	n.a	n.a	13.2	14.5	14.4	17.0	19.7	20.0	21.2
TOTAL (excluding bunkers)	581.5	565.1	633.1	679.1	747.2	761.2	777.5	886.1	980.6	963.5

Source: Own calculations from Singapore Trade Statistics, various years.

product, they remain a relatively minor export compared to the trade in the middle distillates.

The destination of oil exports from Singapore is shown in Table 6.8 which includes the major products, but excludes bunker fuels for ships and aircraft, lubricants and bitumen.

Data on trade with Indonesia are excluded.[29] As a result of the refinery processing deals detailed in the last section, exports to Indonesia were large in the early 1980s, often above 200 thousand b/d. However, for the period shown in Table 6.8 and in following tables, the Indonesian market has been relatively unimportant.[30] Table 6.8 shows some strong shifts in the relative importance of Singapore's oil trade partners. The Japanese market has become considerably less important, declining from 23 per cent of the total in 1986 to 11 per cent in 1995. A key feature has been the emergence of the Chinese market (Hong Kong and China combined).[31] The total exported to the two countries in 1986 was just 114 thousand b/d, which had increased to 306 thousand b/d in 1994 before falling back in 1995 in the face of continuing Chinese import restrictions. Note also the increasing importance of Vietnam which regained a little of its core importance in the 1960s (noted in Section 3), and the South Asian market, in particular India and Bangladesh.

As capacity expansions in other countries come on line, Singapore exporters will need to find new markets for the displaced volumes. The importance of the Chinese, Vietnamese and South Asian markets is then

Table 6.8:   Destination of Oil Product Exports from Singapore. 1986–95.
Thousand b/d.

	1986	1987	1988	1989	1990	1991	1992	1993	1994	1995
Hong Kong	87.0	81.3	99.8	115.0	117.0	115.0	130.0	184.0	210.0	215.7
Malaysia	89.7	87.6	93.5	95.3	106.0	126.0	146.0	134.0	152.0	135.5
Japan	127.3	145.4	156.7	147.0	150.0	123.0	104.0	86.5	97.1	103.3
Thailand	38.5	62.0	96.7	104.0	147.0	127.0	113.0	117.0	95.2	91.2
Vietnam	27.0	22.3	26.6	45.1	17.6	17.9	23.7	91.1	95.9	75.4
China	6.0	9.6	9.1	7.0	11.3	21.6	30.8	33.4	52.8	59.0
Taiwan	8.9	13.7	7.0	8.7	15.3	27.9	33.0	52.8	40.7	41.7
Australia	29.9	21.8	31.9	32.2	24.7	28.6	29.7	22.8	31.5	33.6
Korea	10.7	7.6	11.2	17.5	30.8	25.4	20.3	13.7	28.9	32.3
Bangladesh	6.9	11.9	8.9	13.3	8.6	10.2	11.0	21.1	27.6	28.9
India	5.2	3.8	2.3	6.9	8.5	25.6	28.7	21.8	25.8	25.3
Guam	5.4	5.0	3.7	3.0	13.2	13.7	14.2	16.2	23.6	21.5
USA	37.3	36.5	25.7	25.7	22.2	15.7	20.7	22.5	22.9	16.5
Philippines	9.6	12.2	13.8	16.2	22.5	17.3	14.9	15.3	19.8	13.2
New Caledonia	5.1	2.9	1.6	3.5	4.0	1.6	3.3	4.1	5.5	5.9
New Zealand	2.4	4.3	1.4	2.9	1.4	2.4	3.2	3.0	3.8	5.7
Cambodia	0.0	0.0	0.0	0.0	0.0	0.0	0.0	0.7	3.4	5.3
Papua New Guinea	0.0	0.0	0.1	0.1	0.9	0.6	1.0	3.0	2.5	4.8
Burma	0.7	0.9	1.3	1.0	1.7	1.8	2.1	1.9	2.2	1.9
Sri Lanka	10.4	1.9	0.3	0.4	2.3	1.5	1.2	0.4	1.6	1.2
Seychelles	0.4	0.7	2.2	0.7	0.5	0.8	0.8	1.2	1.3	1.0
Maldives	1.3	2.7	1.5	3.6	0.3	1.7	7.0	4.0	1.1	0.7
Iran	39.8	2.8	5.8	0.0	1.1	3.3	2.1	1.9	0.0	0.0
Others	15.1	11.6	8.9	10.8	18.2	35.0	15.0	6.7	4.9	8.6
TOTAL	565.0	548.6	610.1	659.0	725.0	743.0	756.0	859.0	950.0	928.2

Source: Own calculations from Singapore Trade Statistics, various years.

not just in their current volumes, but in the increasing focus on them as Singapore's other major export destinations cut their immediate oil product import dependence.

Destinations of exports by oil product (gasoil, fuel oil, jet fuel, gasoline and naphtha) in 1986–95 are given in Appendix 5.

The origin of Singaporean crude oil imports is shown in Table 6.9, with Indonesia included.[32] The major structural change is the renewed reliance on Middle East supplies, increasing to 84 per cent of imports in 1995. This degree of Middle East dependence is not out of proportion with other major refining countries in the region. For instance, in Table 5.5 we showed Japanese dependence on the Middle East to be 77 per cent, and the data of Table 5.5 imply the same 77 per cent figure for Korea. However, the scale of the increase has been far greater in

Singapore. Japan has had a 98 per cent reliance on the Middle East for incremental crude oil imports. The equivalent figure for Singapore implied by Table 6.9 for the 1986 to 1995 period is 145 per cent. With the decline in processing deals in Singapore with Indonesia, China and Malaysia, as noted in the last section, the amount of Asian crude oil run has been sharply reduced. China has slipped from being the third most important supplier to Singapore in 1986, to thirteenth in 1995.

The demand for Middle East oil has then been increased by both the overall increase in runs and the decline of third party processing. The major beneficiary has been Saudi Arabia, regaining the market it lost in being a swing producer in the first half of the 1980s,[33] which has increased

Table 6.9: Source of Singapore Crude Oil Imports. 1986–95. Thousand b/d.

	1986	1987	1988	1989	1990	1991	1992	1993	1994	1995
Saudi Arabia	49.8	105.5	194.5	280.5	320.1	402.8	493.9	483.9	510.0	433.6
UAE	43.1	68.4	100.2	119.5	169.5	166.0	79.9	127.5	131.3	170.2
Kuwait	141.3	80.9	11.2	25.4	29.8	0.0	23.5	88.3	80.9	122.4
Malaysia	87.5	99.6	126.4	104.1	102.7	93.8	103.5	100.5	67.4	56.9
Qatar	17.2	40.7	37.9	14.5	6.0	8.6	28.1	17.5	15.4	47.3
Iran	135.4	71.5	68.3	57.0	39.9	49.3	59.0	90.2	80.9	43.5
Yemen	0.0	0.0	4.4	8.2	0.0	2.6	4.2	0.8	25.4	28.7
Vietnam	0.0	0.0	4.2	2.5	10.7	22.6	19.9	23.2	19.6	26.0
Indonesia	75.5	46.0	8.0	26.8	27.0	12.0	12.5	16.3	80.0	25.1
Brunei	21.9	17.3	16.2	9.0	12.8	19.6	30.7	27.2	29.2	24.7
Oman	4.3	24.9	5.2	23.6	42.9	56.2	30.2	36.4	10.6	20.2
Egypt	6.3	4.5	3.6	11.1	11.2	0.0	9.9	17.5	17.9	19.6
China	129.7	95.3	101.2	71.1	72.5	62.0	40.1	20.0	12.5	16.5
Australia	5.3	17.5	16.6	12.1	33.3	21.2	9.7	20.9	19.4	12.6
Nigeria	0.0	0.0	0.0	0.0	0.0	0.0	0.0	0.0	1.9	5.2
Gabon	0.0	0.0	0.0	0.0	0.0	0.0	0.0	0.7	2.7	1.1
Iraq	0.0	0.0	0.0	0.0	4.6	0.0	0.0	0.0	0.0	0.0
Pakistan	0.0	0.0	0.0	2.0	2.1	2.0	0.0	0.0	0.0	0.0
Philippines	0.0	0.0	0.0	0.0	0.0	0.0	0.6	1.2	0.0	0.0
Algeria	1.4	0.0	0.0	0.0	0.0	0.0	0.0	0.0	0.0	0.0
Others	1.8	0.0	0.8	0.0	1.3	0.0	0.0	4.1	0.7	1.7
TOTAL	720.4	672.0	698.6	767.4	886.5	918.7	945.6	1076.0	1105.9	1055.3
Av Price S$	34.93	37.39	32.68	33.65	39.86	34.64	30.31	26.79	23.85	24.44
Asia+										
Australasia	319.8	275.6	271.9	227.6	261.1	233.2	217.0	209.3	172.3	106.1
Middle East	398.9	396.4	425.9	539.8	624.1	685.5	728.6	861.9	872.5	885.4
Other	1.8	0.0	0.8	0.0	1.3	0.0	0.0	4.8	5.4	8.0
% Middle East	55.4	59.0	61.0	70.3	70.4	74.6	77.1	80.1	83.1	84.0

Sources: Own calculations from Singapore Trade Statistics, Indonesian and US Government Publications.

its supplies to Singapore by 460 thousand b/d between 1986 and 1994, while total imports have risen by 330 thousand b/d. When deliveries to partly Saudi Arabian owned refineries in the USA are excepted, Singapore emerges as a larger market for sales of Saudi Arabian crude oil than the USA. Among the other Middle East suppliers, Iran had managed some increase in sales over the early 1990s. However, the tightening of US sanctions on Iran in 1995 means that the Exxon and Mobil refineries can no longer run Iranian crude, nor can Caltex through its share of SRC, and Iranian exports have accordingly fallen back.

We noted in Chapter 4 the strong impact of exchange rate changes in Japan, overwhelming changes in dollar oil prices. Table 6.9 illustrates this process at work in the procurement of oil in Singapore. In nominal Singapore dollars, the average cost of crude oil imports had fallen by 1994 to just 70 per cent of its 1986 value, the low point for the dollar oil price. Indeed, even with the increase in crude runs, the total expenditure on crude oil in 1995 was only 2 per cent more in nominal Singapore dollars than it had been in 1986. The significance is of course far less for the Singaporean industry than for the Japanese, given its relatively low proportion of domestic currency denominated income flows. However, it is another demonstration that for many Asian economies oil is now extremely cheap in terms of resource cost.

This chapter has considered the development of the Singapore oil industry over a period of some 120 years. These developments have had a common thread, a series of natural advantages, and since independence an accommodatory government. This has resulted in the gradual addition of more sources of value added, culminating in the vision for the next century of the giant Jurong island refining and petrochemical complex.

The refining industry started in 1961, has been through distinct cycles. The oil price shocks of the 1970s and the reduction of the degree of vertical integration in the industry left Singapore with considerable excess capacity. A response was the growth of third party processing, which we have argued acted as a substitute for markets in the absence of a developed trading structure. The industry survived the problems of the 1980s, and the threat of competing capacity elsewhere, because of the surge in oil demand in the region as well as improvements in efficiency within the refineries. This has resulted in a flexible industry that has managed to maintain a comparative advantage.

Notes

1. R.J. Forbes and D.R. O'Beirne (1957), *The Technical Development of the Royal Dutch Shell*, E.J.Brill, Leiden.
2. F.C. Gerretson (1957), *History of the Royal Dutch*, (4 volumes), E.J. Brill, Leiden.
3. See Ralph W. Hidy and Muriel E. Hidy (1955), *Pioneering in Big Business 1882–*

1911: History of Standard Oil Company (New Jersey), Harper and Brothers, New York, for an account of Standard Oil's marketing operations in the Far East.

4. Herta E. Pauli (1947), *Alfred Nobel: Dynamite King – Architect of Peace*, Nicholson and Watson, London.

5. See Robert Henriques (1960), *Marcus Samuel: First Viscount Bearsted, Founder of 'Shell' Transport and Trading Company*, Barrie and Rockliff, London.

6. The *Murex*, together with many sister ships, was lost through German submarine action in the First World War.

7. Henriques (1960) op.cit. points out the utility of oil tins with a list of the items constructed from them. A truncated version of this account is also to be found in Daniel Yergin (1991), *The Prize: The Epic Quest for Oil Money and Power*, Simon and Schuster, New York.

8. For a slightly egotistical first person version of events, see Henri Deterding (1934), *An International Oilman*, Ivor Nicholson and Watson, London.

9. The time series for exports is taken from W.G. Huff (1994), *The Economic Growth of Singapore*, Cambridge University Press, Cambridge.

10. *Shell World*, October 1991, 'Bukom – and Beyond'.

11. Frank Owen (1960), *The Fall of Singapore*, Michael Joseph, London.

12. See 'Rebuilding Malaya', Petroleum Press Service, May 1950, p.130.

13. Amoco sold their share in 1990 to the Development Bank of Singapore (DBS), and 25 per cent of the company was publicly floated.

14. Lee Kuan Yew, 21 February 1967, quoted in Alex Josey (1968), *Lee Kuan Yew: The Crucial Years*, Times Book International, Singapore.

15. See Bennett H. Wall (1988), *Growth in a Changing Environment: A History of Standard Oil Company (New Jersey) 1950–1972 and Exxon Corporation 1972–1975*, McGraw-Hill, New York.

16. See Laura E. Hein (1990), *Fuelling Economic Growth: The Energy Revolution and Economic Policy in Postwar Japan*, Council on East Asian Studies, Harvard University.

17. For operational purposes the shares are equal, except for units other than crude distillation where the shares are proportionate to the investment put in.

18. Tilak Doshi (1989), *Houston of Asia: The Singapore Petroleum Industry*, Institute of Southeast Asian Studies, Singapore, gives the effective capacity in the late 1980s as between 800 and 850 thousand b/d, compared to a total nameplate capacity of 950 thousand b/d.

19. *Far East Economic Review*, 3 March 1983.

20. See Robert Mabro (1987), *Netback Pricing and the Oil Price Collapse of 1986*, Oxford Institute for Energy Studies, Oxford.

21. On the debit side, refiners did face a capital loss due to the fall in the value of inventory holdings. See Tilak Doshi (1989), *Houston of Asia*, op.cit.

22. The site is also used for blending of output from the Balongan refinery in Indonesia. The closure of the terminal is planned for early 1997.

23. The only other RFCC unit in the Asia-Pacific area as of 1996 is at Shell's Geelong refinery in Australia.

24. Third party processing still continues in some Mediterranean refineries, most notably the Saras plant in Sardinia. In the 1990s several East European refineries

have also begun third party processing on behalf of traders and major oil companies.

25. See *Far East Economic Review*, 20 January 1983.
26. Table 6.5 collates point estimate information from a variety of sources. In particular, the volumes shown are not period averages.
27. Texaco's independent involvement in the Far East grew with its crude oil production in the South China Sea.
28. The Okinawa refineries, operated by Idemitsu and Nansei Nishihara, have had a less restrictive regulatory environment compared to the mainland Japanese refineries.
29. Singapore does not publish data on trade with Indonesia for any commodity or manufactured good. A possible explanation arises from the desire to avoid contradicting official Indonesian data in the presence of factors that could lead to a discrepancy. See Tilak Doshi (1989), *Houston of Asia*, op.cit.
30. Tilak Doshi (1989), op.cit. estimates Singapore exports to Indonesia in 1981 as 197.0 thousand b/d, falling to 25 thousand b/d in 1987.
31. A proportion of Singapore exports to Hong Kong will be reexported into Guandong province. We use the phrase 'Chinese market' to cover flows to both Hong Kong and China.
32. The series for Indonesian exports is drawn from Indonesian official publications, and various issues of the annual US Embassy in Djakarta report on the Indonesian oil industry.
33. Tilak Doshi (1989), *Houston of Asia*, op.cit. provides a crude oil import series from 1980 to 1987. Saudi Arabian exports to Singapore in 1981 are given as 404 thousand b/d, a level that Table 6.9 shows was not reached again until 1992.

CHAPTER 7

THE PHYSICAL MARKETS

1. Introduction

In this and the following chapters of this part, we consider the oil trading activities that are associated with Singapore. We divide trading activity into physical, swaps and futures, this chapter being concerned with the physical markets in crude oil, and more extensively in oil products. A brief taxonomy of this (arbitrary) division runs as follows. We would define the physical market as being trades which will or can result in the transference of physical oil between the parties, and which are transacted outside of a formalized exchange.[1] This consists of a series of further subdivisions, shown in Figure 7.1.

We divide the physical market into two main subdivisions. The term market comprises of agreements made for the transfer of a series of cargoes over a time period, agreements which can be reached through bilateral discussions or through a tendering process. In contrast, the specific market consists of agreements for single cargoes, or for flows over a short

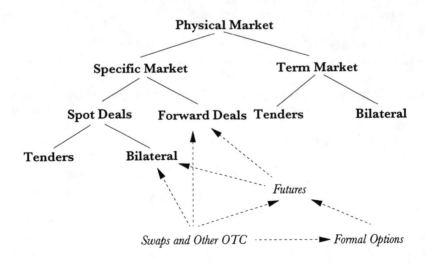

Figure 7.1: A Typology of the Physical Market and Price Linkages.

time period (the division we use is one month). The specific market divides into spot deals, and deals made through a market where forward delivery commitments are made. Deals for spot oil further divide into spot tenders, and spot deals made through bilateral bargaining.

The final division of the physical market we make then consists of five elements, spot tenders, spot markets, forward markets, term tenders and bilaterally bargained term deals. In reality, only spot and forward markets tend to have a role in the setting of overall price levels. Term deals tend to employ pricing formulae, whereby the oil moved under the deal is valued against the prices generated by spot or forward markets. Spot tenders also primarily use such formulae, although in some cases fixed price tenders are requested. Within spot and forward markets, a proportion of deals are also made with pricing formulae, i.e. they contain information on differentials, but not overall price levels. Therefore, in ascertaining exactly which part of the market generates price level information, we look to the subset of deals where an absolute price is agreed rather than a formula.

As the spot and forward markets are the elements of the physical market that tend to have a price setting role, they also normally represent the point of connection in price linkages with the other two forms of market we have distinguished. Formal futures markets are centralized exchanges that trade futures and options contracts. In some cases futures markets have physical delivery, and uncleared positions result in bilateral spot transfers of physical oil. In other cases, the contracts are cash settled against prices generated in other markets, and hence the linkage may be to forward or spot market prices. Oil futures markets in Asia are considered in Chapter 9.

The third form is informal swaps markets, and other OTC (over the counter) markets. The major interest in Asia is in swaps markets, which are in essence cash settled financial arrangements in which a fixed price is exchanged for a floating price. While, in the context of the Singapore market, the floating price involved is usually a spot market price, it could be any price generated in the system, and hence the form of the price linkages shown in Figure 7.1. We consider the swaps market in the next chapter.

Throughout this and the next chapter we make considerable use of the *Petroleum Argus Product Deals Database*. This is a record of confirmed deals, both physical and swaps, reported to Petroleum Argus in the course of their market reporting.[2] The remainder of this chapter is structured as follows. Section 2 covers the factors that have contributed to the growth of Singapore as the central location in the region for oil trading activity, with particular attention to the direct incentives offered by the Singapore government in the Approved Oil Trader scheme. Sections 3 and 4 consider components of the physical oil products market, namely spot markets and tenders. Section 5 gives a context to the role of the oil products trade into Asia from Middle East export refineries, and Section 6 considers the open specification naphtha forward market. Section 7 details the role of storage facilities and a final section offers some brief conclusions.

2.   The Singapore Oil Trade

In the mid 1980s the Far East market had two trading locations. Oil products trading, then almost exclusively physical, was centred on Singapore, while crude oil trading took place in Tokyo. By 1990 two structural changes occurred. First, crude trading migrated from Tokyo to Singapore, and secondly Singapore had a further infusion of new participants and a growing trade in oil swaps. Companies began to move trading operations out of Tokyo in about 1988. Led primarily by trading companies such as Marc Rich, Gotco and Transworld Oil (TWO), the exodus was joined by the major oil companies, and the oil trading components of both US financial institutions and some of the Japanese trading houses. By 1990 Singapore had become the centre for crude trading.

The demise of Tokyo trading was due to a series of interrelated factors. First, with yen appreciation and soaring land and therefore office space values, Tokyo was becoming an extremely expensive place to operate, particularly for foreign companies who had few yen based income streams. Secondly, by 1988 liquidity in the Tokyo market had fallen sharply. A series of losses had been incurred by Japanese traders, particularly in the Brent and Dubai markets through some highly speculative position taking. As a result, many had scaled back their overall trading activity, and had probably been encouraged to do so by government. The limited amount of trade that could be conducted in Tokyo made meeting the rising costs harder to justify. Thirdly, as liquidity fell it became harder to establish price levels and quickly close out positions. As a result, the implicit risk involved in the opening of the London market became greater. Finally, at a time when the Japanese government was signalling a disenchantment with oil trading, or at least a pronounced ambivalence towards it, the Singaporean government was signalling direct encouragement and providing incentives.

The Approved Oil Trader (AOT) scheme was introduced by the Singaporean government in January 1989. An AOT firm is granted a concessionary 10 per cent tax rate on trading activity, compared to the current normal Singapore corporation tax rate of 27 per cent. To qualify for AOT status an applicant company must fulfill four criteria. First, their turnover of trade in Singapore must exceed US$100 million. Secondly, they must employ at least three oil traders with more than five years experience. Thirdly, they must have a business expenditure per year in Singapore greater than US$500 thousand. Finally, they must be part of an international trading company that is deemed to be of good repute. All oil trades,[3] physical, paper or futures qualify, as long as they meet two criteria. They must not concern the supply of oil products to the domestic Singapore market (including the bunker market), or effect any physical change in the oil (hence refining itself or product blending activities do not qualify).

There were already enough push factors at work to lead to the demise of the Tokyo market, and enough reasons for trading to relocate. Singapore offered lower costs,[4] and a time zone position that allowed the opening of the London market to be covered within normal office hours. For companies who already had products traders there, the economies of scale involved in assimilating their Tokyo operations into their Singapore office were also attractive. Language was probably also a factor. While a non-Japanese speaking oil trader need not have been at any disadvantage in trading in Tokyo, they would still need English speaking support staff. For an international company, locating to Singapore to some degree meant that they could have more flexibility in moving trading staff between offices.

The AOT scheme provided a strong pull factor to Singapore. The physical location of trading does not have to equate to the location where taxation liabilities are incurred, as there has always been the opportunity to book out deals in offshore centres. The taxation concession involved in the AOT scheme is therefore not enough in itself to attract traders. It does however provide an incentive for Singapore traders to book out deals there rather than in an offshore centre. As such it provides spill-over benefits to the Singaporean service sector, and in particular to lawyers and accountants and other financial services. The AOT scheme can then be seen as providing for growth of the Singaporean service sector as a whole, rather than just oil trading. The scheme can also be seen as representing a signal that the government wished to ensure that the environment for oil trading was kept beneficial, and that there was a desire to maintain Singapore's competitive advantage over other regional locations.

No official list of AOT companies is currently released *in toto*. However, additions to the scheme are frequently announced in the Singapore press and oil trade journals, making it possible to create such a list from publicly available information. From such press reports (confirmed by interview evidence), and a previously published list[5] we have compiled Table 7.1. We believe that it is a reasonably close approximation to the companies in the AOT scheme. On the basis of this exercise we imply that there were probably 55 AOT firms in 1996, or 49 if subsidiaries of the same parent company are aggregated.

Table 7.1:   Approved Oil Traders (AOTs). 1996.

Japanese Companies (12)
C Itoh, Cosmo, Idemitsu, Japan Energy, Kanematsu, Marubeni, Mitsui, Nippon Oil, Nissho Iwai, Petro-Diamond (Mitsubishi), Sumitomo, Toyota

Korean Companies (4)
Hyundai, Hanwha, Yukong, Honam

Other Oil and Gas Companies (8)
Agip, BHP, Chevron, Elf, Enron, Neste, Statoil, Unocal

Petrochemicals (5)
China Resources Petrochem, Mitsubishi Petrochem, Mobil Petrochem, PCS, Tomen

Refiners and Affiliates (9)
BP, Caltex, Esso Singapore, Exxon, Mobil Singapore, Mobil Sales, Shell Eastern Petroleum, Shell Eastern Trading, SPC

Traders (14)
Astra, BB Energy, Cargill, Daxin, Hin Leong, Kuo, Mabanaft, Glencore, Petroleum Integration Trading, Sinochem, Sintra Oil, Stinnes, Vitol, Wickland

Wall Street Companies (3)
J Aron, Louis Dreyfus, Morgan Stanley

Source: *Straits Times, Business Times* and various other sources.

In the absence of any official list, we can not be certain of the current accuracy of this table. However, it does give an overview of the major companies active in the Singapore market, and indeed it is these companies that tend to be the most active in the individual markets that we consider in this and later chapters.

3.   Spot Oil Trading

We saw in Chapter 6 how, excepting naphtha, the balance of Singapore's oil exports had switched from the Japanese market and towards the emergent economies, and in particular China. To a great extent, the nature of the spot oil trading of flows within and to the Asia-Pacific region has also changed to reflect the new realities. An example of this is given by Table 7.2, in which we have collated all spot gasoil trades recorded by Petroleum Argus for 1988 to 1990 and 1994 to 1995. The deals are divided between those where a delivered price was agreed (i.e. c+f), and those where the pricing was at point of loading (i.e. fob), the former being an import centre price and the latter an export centre price.

Table 7.2:   Locations for Gasoil Spot Trade. 1988–90 and 1994–5. Per Cent.

	c+f 1988–90	1994–5		*fob* 1988–90	1994–5
	42.1 %	48.3 %		57.9 %	51.7 %
of which			*of which*		
Australia	2.4	3.8	China	1.3	-
China/Hong Kong	12.2	39.2	Japan (Okinawa)	0.4	1.8
India	-	10.8	Korea	-	7.7
Indonesia	0.6	4.4	Malaysia	0.4	6.5
Japan	61.0	4.4	Middle East	0.9	21.9
Korea	7.3	9.5	Philippines	-	0.6
Philippines	-	6.3	Singapore	96.9	61.5
Singapore	15.9	12.0			
Sri Lanka	-	0.6			
Taiwan	0.6	3.8			
Thailand	-	5.1			

Source: Own calculations from Petroleum Argus Product Deals database.

While the table shows some overall movement towards delivered cargoes in spot trade, the most important shift has been between delivery points. In the earlier time period 61 per cent of delivered price spot deals were c+f Japan, but by 1994–5 this had fallen to just 4.4 per cent. The major spot import market is now China. The number of spot import markets has also increased, with significant numbers of delivered price deals made into India, Korea, the Philippines and Thailand. Delivered cargo trades to locations other than Singapore and Japan have risen from less than 10 per cent of all trades to more than 40 per cent. Among export markets there has also been diversification, with the standard fob Singapore pricing of the early period now being accompanied by fob pricing of exports from Korea and Malaysia.[6] The proportion of spot Middle East material traded at point of loading has also increased very sharply.

Unlike the almost frictionless activity in paper markets, spot oil trading is largely governed by logistics and specification regulations. As such, spot oil trading in the Asia-Pacific region has simultaneously been moving towards greater regionalization and intra-regional differentiation, and also to greater globalization and connection to the European market in particular. We can illustrate these apparently contradictory tendencies best in the case of gasoil.

The first factor leading towards regionalism and fragmentation is the differences between the product specifications required in different countries. The permitted maximum sulphur content for gasoil varies considerably. It is 1 per cent in some of the countries of the Indian subcontinent and also Vietnam. Most of the rest of Asia uses 0.5 per cent, or at least has begun to cut from that level to 0.3, 0.25 or 0.2 per cent. A few countries, and most notably Japan, Korea and Thailand, are moving or have moved to ultra-low sulphur gasoil of a maximum of 0.05 per cent.

For the countries with requirements for low sulphur gasoil, this fractionation of the market has meant that they sometimes have to look outside the region to obtain the product. For example, in 1994 the largest exporter of gasoil (and also fuel oil) to Korea was the USA. The relative paucity of supplies of low sulphur gasoil within the region meant that Korea exported more (high sulphur) gasoil to Singapore than the (low sulphur) gasoil it imported from Singapore. However, sulphur is not the only major distinguishing feature. The typical gasoil specification for price assessment, shown in Appendix 1 in Table A1.1, has a pour point of 9 degrees, consistent with the bulk of Singapore production which tends to have a pour point of between 6 and 9 degrees. While perfectly adequate for most of Asia, such material would be useless as fuel for diesel engined vehicles in the cold winters of Korea, North China, Japan and the Russian Far East. Japan in fact requires a pour point of minus 20 degrees. North Asian gasoil is then in reality a totally different product to South East and South Asian gasoil.

The market was once split into two main parts, the low sulphur, low pour point countries, and the high sulphur, high pour point countries. With the tightening of specifications, but at different rates across countries, the divisions are increasing. The simple two way division has already been augmented by ultra-low sulphur with low pour, and low sulphur with high pour.[7] Gasoil is thus becoming a far less fungible product, splitting the market into a series of sub-markets across the region.

A second factor pushing towards greater heterogeneity is the purchasing strategy of many national oil companies. A general disinclination to act as traders means that tenders are still very important, as is discussed in the next section. With the volumes involved growing over time in many cases, a series of large, but essentially separate, sub-markets have grown to supply each regular tender. Even when supplies are purchased spot, there is a disinclination to buy fob, and a preference for delivered cargoes. As evidenced by Table 7.2, the proportion of c+f trades has been rising, and the number of pricing points increasing. There is thus no movement to any homogeneity, but the development of a series of delivered cargo markets. In each market trading is largely dictated by differing logistics.

While the above factors have pointed towards a growing regionalism, a further series of factors have pushed towards globalization. The first is the increasing ease of arbitrage between continents. Beginning in 1992 and 1993, some trading companies (and in particular Louis Dreyfus), some European oil companies (and in particular Finland's Neste and Norway's Statoil) and Singapore traders (notably Hin Leong) found it could be economic to move large cargoes of gasoil into Asia. Using either new or specially cleaned Very Large Crude Carriers (VLCCs), single cargo sizes of over one (and often two or more) million barrels could be brought in from Europe (and also Latin America), with considerable freight economies. The VLCCs could then be anchored, often between Indonesia and Singapore, and the cargo size broken down into smaller parcels loaded onto normal product tankers. Given the right alignment of prices, movements of gasoil from the Mediterranean on normal oil product tankers had been fairly common. However, the VLCC trade worked at lower differentials, and even from Scandinavia.

The gasoil VLCC trade was particularly economic because of the higher sulphur levels acceptable in Asia. With the trade of low quality gasoil into the Baltic greatly reduced by the collapse of the Soviet Union, material that was off specification for the western European market could be moved into Asia instead. Adding in the cost of making gasoil to fit low sulphur requirements meant that the VLCC arbitrage became even more attractive. We believe that the VLCC trade has had a major impact on the Asian market. The potential for such movements has tied Singapore prices far more closely to European, and the potential for VLCC transfers effectively overhangs the market even when no oil is actually moving.

One VLCC cargo is roughly equivalent to an entire month's gasoil output from an average 120 thousand b/d refinery. When that imaginary equivalent refinery is actually visible anchored off Singapore, its psychological impact on trading becomes even greater. The potential for several such cargoes to start moving if Singapore prices go up too far acts as a severe dampener. The extreme weakness of the Singapore gasoil market in 1994 could then be explained in both regional and global terms. The regional argument is that Chinese demand was weak and erratic, and that this depressed prices. The global is that Singapore prices had become very closely tied to European by the potential for VLCC movements. The European gasoil market was over supplied and weak, and therefore so was the Singapore market. While most attention has been given to the former explanation, we would tend to endorse the latter. Any likely further increase in Chinese demand in 1994 could have been easily filled by the European VLCCs. Singapore was effectively awash with gasoil, even if that gasoil was actually in Europe.

The second globalizing factor is the change in the composition of the companies involved with the spot gasoil trade. A series of new companies entered the market in the early 1990s, helped by the AOT scheme incentives. Many of these were global traders who thought naturally in terms of arbitrage possibilities, others were European companies who aimed to exploit the potential of physical links between their European operations and their trading in Singapore.

The most common gasoil cargo sizes traded spot are 100 and 150 thousand barrels, and 20 and 30 thousand tonnes (roughly 150 and 225 thousand barrels). The mean size for all spot cargoes is about 175 thousand barrels, and for those traded fob Singapore it is about 150 thousand barrels. As a result of this relatively large average cargo size, the potential number of spot gasoil trades is very limited. Table 6.9 showed the level of Singapore export of gasoil as being 187 thousand b/d in 1988, and 331 thousand b/d in 1995. At the average cargo size, this equates to about nine cargoes per week in 1988 and fifteen in 1994. A large proportion of gasoil does not move through the spot market, either going straight through the integrated channels of the refiners, being sold out in term contracts, or going directly from refiner to national oil companies through tenders. Given this, the number of spot gasoil deals recorded by Petroleum Argus of between four and five per week in 1994 and 1995 would seem to represent a very high proportion of total spot deals.

This limited potential pool of spot deals is the key difference between Singapore and other oil product trading centres. In the USA, products are traded at various points along pipelines. There are then no constraints on the volume that can be traded in a single deal, and the average size tends to be small. In Europe, while there is also a large cargo size market, the barge trade along the Rhine produces a core of small cargo size trades. In both areas the total volume traded is high, and the small cargo size markets mean that this translates into a large number of deals per day. By contrast, not only is the volume of oil spot traded in Singapore a small fraction of that in the USA and Rotterdam, but the large size means that the disparity in terms of deals done is even greater. In all, the necessity of large cargo sizes when combined with the volumes moving through other channels, produces a spot market that can be described as extremely thin.

The thinness of the spot market, in terms of the price information generated through physical deals, is increased when we consider how spot cargoes are priced. In 1988 all the gasoil spot trades recorded by Petroleum Argus were outright deals, that is to say the parties involved had agreed an absolute price for the deal. In 1989 the first formula deals were reported, where the parties set the price at an average over a number of days (usually five), of the gasoil assessments made by price assessment agencies.

By 1991 formula deals were as numerous as outright, and by 1996 they outnumbered outright deals by more than two to one. Outright trade in Middle East exports is all but non-existent, with deals normally arranged on the average of Platt's and Petroleum Argus price assessments for the Gulf or Singapore (minus freight). For both fob Singapore and delivered price trades in Asia, the dominant index is MOPS (Mean Of Platt's Singapore), with some trades done using a combination of Platt's and other assessment agencies. The price level setting information carried by spot market deals is then severely truncated. Of the four to five spot deals per week which we noted above are likely to be reported, only one to two are likely to be outright deals.

Table 7.3: Main Buyers and Sellers in Singapore Gasoil and Fuel Oil Spot Markets. Rank and Market Share. 1994–5. Per Cent.

Rank	Gasoil				Fuel Oil		
	Buyers		Sellers		Buyers		Sellers	
1	Louis Dreyfus	16.8	Hin Leong	38.3	Kuo Oil	25.0	Hin Leong	19.8
2	Hin Leong	9.2	Shell	17.5	J Aron	13.8	Mobil	16.2
3	Astra	5.9	Mobil	11.7	Hin Leong	11.8	J Aron	9.6
4	Shell	5.9	BP	5.0	Astra	8.6	SPC	9.0
5	Sinochem	5.9	Cargill	3.3	Vitol	8.6	BP	7.8
6	Vitol	5.9	Marc Rich	2.5	BP	7.9	Shell	4.8
7	Cargill	5.0	SPC	2.5	Mabanaft	5.3	Stinnes	4.8
8	Mobil	4.2	Cosmo	1.7	Elf	2.6	Vitol	4.8
9	Glencore	3.4	Louis Dreyfus	1.7	Glencore	2.0	Astra	3.6
10	Marubeni	3.4	Mitsubishi	1.7	Ipg	2.0	Elf	2.4

Source: Own calculations from Petroleum Argus database.

In Table 7.3 we have shown the shares of trade on both sides of the two most liquid Singapore spot markets, gasoil and fuel oil. The dominance of the trading company Hin Leong, among the sellers of gasoil stands out. It becomes even more of a dominant feature when we consider only trades in which an absolute price was agreed rather than just a formula. Among these trades 64 per cent have Hin Leong as the seller, with another 10 per cent in which it was the buyer. These reported outright trades are the key price setting trades in the most important Singapore spot market, which has a whole layer of regional physical trades and a high liquidity swaps market settled on its price.

The market is very highly geared in terms of the volumes priced off it. We noted above the low number of absolute price deals reported in Singapore per week. The potential problems associated with this are compounded by the fact that these deals predominantly involve one company, normally on the selling side. The concept of responsibility is of course not defined in an informal spot market. If it were, there would be a strong case for arguing that other market participants had abdicated responsibility for providing absolute price information through physical deals.

4.   The Market for Tenders

While tenders are now only a minor part of European and US markets, they remain central to the Asian market, particularly in physical deals involving national oil companies. While administratively extremely cumbersome, tenders do have advantages in some circumstances, as is discussed below. The general mechanics of a tender are as follows. The tendering company will contact a series of firms asking them to submit bids for the tender before a stated deadline. The tender will state a notional volume, product or crude oil specification, delivery (or loading for a sell tender) window and a designated port. Sealed bids are given, often with the lodging of a performance bond. Normally the bids will have a validity for another one or two days, i.e. the offer is considered to have lapsed after this time. The tenderer can evaluate the offers during this time, and then contact the winning bidders. The total number of cargoes accepted in the tender can be more or less than the amount stated in the original tender, and if none of the bids are deemed satisfactory a process of retendering is not unusual.

To provide a picture of how widespread tendering operations are in Asia, in Table 7.4 we have collated the tenders recorded in one oil industry journal over the course of 1995. In some cases the tenders are monthly (most importantly the regular Indian tenders for crude oil, gasoil and kerosene), in others quarterly, and in some cases the tenders are irregular depending on circumstances. The most important tendering country, in terms of both number of tenders and volumes accepted, is India followed by Taiwan. Not all tenders necessarily are reported in press journals, and hence Table 7.4 is intended as being illustrative rather than exhaustive. Bangladesh and Pakistan are also important tenderers, and large volumes of oil products from the Middle East are sold through tenders for term contracts. However, the table does demonstrate the extremely wide use of tenders, both geographically and in terms of the products covered. In all bar three of the cases shown, the tendering companies are state controlled. In most cases bids are made as a differential to an assessed market price, the major exception being Taiwanese tenders. In the very thin Singapore gasoline spot market, Taiwanese fixed price tenders serve as an important element in the price setting information set.

Table 7.4:   Selected 1995 Tenders.

	Country	Company	Type	Specification
Gasoil	Taiwan	CPC	Sell	0.3%S
	Thailand	PTT	Buy	0.25%S
	India	IOC	Buy	1%S HSD
	Sri Lanka	CPC (Sri Lanka)	Buy	1%S gasoil
	Philippines	Petron	Buy	0,2%S
Gasoline	Taiwan	CPC	Buy	95 RON Unleaded
			Buy	95 RON Leaded
	India	IOC	Buy	87 RON Leaded
	Philippines	Petron	Sell	95 RON reformate
	Vietnam	Petrovietnam	Buy	83 RON
Fuel Oil	Taiwan	CPC	Buy	1%+3%S 180 cst
			Sell	1%S 180 cst
	Thailand	PTT	Buy	1%, 2%+3%S 180 cst
	India	IOC	Buy	3.5%S 180 cst
Oil	India	IOC	Buy	
			Sell	3.5%S 180 cst
	Korea	Kepco	Buy	1.6%S 540 cst
	Sri Lanka	CPC (Sri Lanka)	Sell	3.5%S 180 cst
	Philippines	Caltex	Buy	3.5%S 180 cst
	Vietnam	Petrovietnam	Buy	3%S 170 cst
MTBE	Taiwan	CPC	Buy	
Naphtha	Taiwan	CPC	Buy	Heavy
			Sell	Open spec
	India	IOC	Sell	Light + heavy
	Sri Lanka	CPC (Sri Lanka)	Sell	Light
	Philippines	Petron	Sell	
	Malaysia	Petronas	Sell	
Jet	Taiwan	CPC	Buy	A-1
	Sri Lanka	CPC (Sri Lanka)	Buy	A-1
Kerosene	India	IOC	Buy	SKO
	Nepal	Nepal Oil Corp.	Buy	SKO
LSWR	Korea	Kepco	Buy	
	Taiwan	CPC	Buy	
	Malaysia	Petronas	Buy	
Crude Oil	India	IOC	Buy	
	Sri Lanka	CPC (Sri Lanka)	Buy	
	Philippines	Petron	Buy	
	Thailand	Bangchak	Buy	
		PTT	Buy	

Source: *Platt's Oilgram News.*

Indian tenders are the most influential for gasoil, kerosene and crude oil. Indian spot tenders for oil products are monthly, with offers made for delivery during the three thirds of each month. Tenders are specified for both the west and east coasts of India, with Middle East producers generally filling the west coast tenders, and greater competition on the east coast. Due to port constraints, cargo sizes for the east coast are normally 30 thousand tonnes, and for the west coast 45 thousand tonnes.

The offers made in an Indian oil products tender are differentials against a spot market assessment. In some tenders IOC has specified the geographical market they want differentials to be quoted against, but usually traders can choose. Mediterranean, Gulf and Singapore quotes are the three most commonly used. The choice of which to use can be a matter of the market view taken. If, say, a trader has the view that, by time of delivery, the Mediterranean market will be relatively weaker than the Singapore market, then they will put in a bid against Singapore prices and, if successful in the tender, aim to obtain the gasoil from the Mediterranean market.

Table 7.5 provides estimates of the volumes awarded in Indian crude oil tenders between January 1992 and August 1994. Over this period India increased its take from tenders from about 515 thousand b/d in 1992 to 575 thousand b/d in 1994. Roughly two-thirds of the total came from the large term tenders, with the remaining third from the monthly spot tenders. Among term tenders Middle East suppliers predominate, namely Aramco, KPC, NIOC and ADNOC. In Table 7.5, the companies that provided crude oil through spot tenders are shown in descending order of the total sold between 1992 and 1994. The number of months (out of a maximum of 32) in which each company had a tender accepted is also shown. A total of thirty companies were successful in spot tenders over this period, with the most successful being traders (Marc Rich, Gotco and Vitol) and Middle East national oil companies (Aramco, ADNOC and KPC).

The primary advantage of tenders is their transparency to the internal auditors of the tendering company. It is easier to defend a tender decision, with a mass of supporting information on paper, than it is to defend a spot market deal or a bilaterally negotiated term deal. Tenders are then primarily observed in organizations where the internal conditions for independent trading activity are not met. The primary disadvantages of tenders are their administrative cumbersomeness and their associated cost, together with the inflexibility they produce in the timing of market transactions.

The key question is whether the use of a buy (sell) tender results in the tenderer paying higher (lower) prices than they would through direct market transactions. The answer can be derived by revealed preference. Companies with the internal organization conducive to direct market operations rarely initiate tenders, while tenders are normally observed in

Table 7.5: Indian Crude Oil Tenders, 1992–4. Thousand b/d.

	1992	1993	1994	Number of Months
(a) Term				
Aramco	90.3	114.2	122.8	
KPC	72.3	90.9	94.8	
NIOC	65.4	42.6	66.5	
ADNOC	21.3	37.2	42.6	
BP	22.2	20.6	20.6	
Petronas	27.1	10.1	7.6	
Other	6.0	16.8	4.9	
TOTAL	304.6	332.3	359.8	
(b) Spot Tenders				
Marc Rich	44.6	42.4	33.3	22
Aramco	44.7	43.8	5.3	20
Gotco	15.2	44.3	17.6	15
ADNOC	21.1	21.0	15.9	26
Vitol	14.6	18.5	26.2	12
BP	9.2	18.2	31.6	16
Phibro	23.6	13.9	12.6	21
KPC	24.8	12.6	0.0	11
Shell	21.0	8.4	6.2	14
J Aron	13.8	8.5	0.0	8
Dreyfus	5.0	10.4	7.3	4
Morgan Stanley	4.2	6.9	12.5	2
Mobil	0.0	5.4	14.6	3
Conoco	7.9	0.0	8.0	8
CFP Total	6.3	3.8	4.2	8
Itochu	3.6	6.4	0.0	6
Coastal	4.6	1.3	5.5	11
NIOC	9.5	0.0	0.0	5
Elf	1.5	6.2	2.4	4
Neste	7.0	1.7	0.0	9
Mitsui	0.0	6.6	2.1	3
NOC	2.1	5.5	0.0	2
Caltex	5.3	0.0	0.0	1
Statoil	0.0	0.0	7.4	1
Repsol	1.5	0.0	4.2	3
Marubeni	3.0	0.0	0.0	2
Nissho Iwai	2.4	0.0	0.0	2
BHP	1.7	0.0	0.0	1
Sumitomo	1.4	0.0	0.0	1
Sinochem	1.2	0.0	0.0	1
TOTAL	211.3	285.7	216.9	

Sources: Various.

national oil companies that place constraints on direct market operations. Companies that can trade on open markets tend to do so; those that can not predominantly use tenders. The logical conclusion is that tendering produces outcomes inferior to those that arise from access to other markets. However, this raises the question of precisely what is the source of this disadvantage. We examine this in the framework of a simplified tender, and then introduce more realistic assumptions.

Consider the following in the context of a buy tender. The tenderer is atomistic, i.e. its operations do not affect prices. The constraint it faces is to satisfy internal procedures (set by an assessor) that it has achieved the lowest buying price. Its choice is between issuing tenders, or engaging in spot market trading. Assume to begin with that the potential respondents to the tender (tender participants) are the same companies it would contact in the spot market (market participants). The assessor will accept an observable market price as a yardstick, but the only evidence that it considers admissible in the spot market is the deal actually transacted. We also assume that the tender process is more costly to operate for both tenderer and tender participants.

If the market price is perfectly observable, then the tender is necessarily inferior. Market transactions will achieve the ruling market price which, with the above assumptions, will satisfy the assessor. For a tender, given the atomistic assumption, tender participants will only submit bids if their higher cost of tendering is covered. Their tender will be for that higher cost multiplied by the inverse of their perceived probability of getting the tender, in addition to the market price. In other words, as well as their own costs of the operation, through the price the tenderer will also have to pay, in effect, the increase in costs of all other participants. They will be unable to match the market price, and thus unable to satisfy the assessor.

The opposite case to this is where the market price is not observable, or at least not considered transparent by the assessor. The tenderer has to avoid the assessor's suspicion that a price in excess of the (unseen) market price has been paid, with the implied possibility that they have shared in the consequent rent. In this circumstance the tender is superior. In the intermediate case, a market price is observable, but this is not the relevant price for the tender. For example, an fob Singapore price for a grade is observable but not the basis between that and the tender's delivered price of a different specification of that grade. Again, to avoid doubt that the basis has been overpaid, the tender will be preferred as the only method of satisfying the assessor.

In these simple cases, the tender has not achieved a worse price than market operations, except to the extent that the price reflects the participants' higher cost of tendering. The three elements that can cause a differential to open up are, first, if the tenderer is large in the market and so can affect market prices, secondly the relative inflexibility of timing in the tender, and thirdly the case when the tender represents purchases more forward in time than spot market quotes.

Tenderers normally prefer to make their announcement of a regular tender at a preset time of the month, since flexibility in announcement introduces the question of timing as a variable in internal scrutiny. For the atomistic company considered above, this will over time make no difference to the average price they pay, provided there are no significant day of week or week of month patterns in the path of prices. If the company is large in the market, the set timing means a loss of flexibility. First, it means that their purchases are bunched, and secondly it means that their purchasing timing is predictable for other participants.

The only major variable left with the tenderer is the number of cargoes they will actually take as opposed to the nominal number specified in the tender. If they are perceived to be short of supply, prices will rise before the tender, and prices will fall if their requirements are thought to be less than usual. Further price adjustments will then occur after the tender, when the actual number of cargoes taken is compared with the expectation, and, if the tender is on market related rather than fixed price terms, these changes feed through into the actual absolute price paid. However, compared to a large purchaser with sufficient storage and discretion over timing, this bunching of purchases and predictability of timing will produce a higher average price paid. Note that the comparison between average tender price paid and average market price over the month is not a meaningful one, since the large purchaser, by definition, should be able to achieve a lower price than the monthly average through having an element of monopsonistic power.

If the timing of delivery of tender purchases is further out than that represented by the spot market quote, further effects arise. If the forward price is observable, i.e. risk management instruments exist, then the tender price is inflated by the cost of risk management. From the tenderer's viewpoint, given the audit constraints, these circumstances can provide a strong incentive to continue with a tender operation, i.e. in cases where spot prices are transparent but the forward curve of prices is not. We would argue however that this is not the case with the Singapore market, where the operation of swaps markets, as described in the next chapter, produces a very well defined forward curve.

5.   Oil Products from the Middle East

Exports of oil products from the Middle East (primarily from Saudi Arabia, Kuwait and Bahrain) play an important balancing role in the Asian market, primarily in supplies of LPG, naphtha, fuel oil and gasoil. While the balance of opinion in many Middle East national oil companies has now moved towards an emphasis on acquiring refining assets in consuming countries, relatively minor expansions and improvements of the Middle East export refineries continue.

An overview of downstream development in the six Middle East OPEC members is given in Table 7.6. Refining capacity has increased steadily in each decade, an expansion of 0.9 mb/d in the 1960s, 1 mb/d in the 1970s and 1.1 mb/d in the 1980s, reaching a total capacity of 4.46 mb/d by the mid 1990s. This pace of expansion is however rather distorted by the path of the Iranian industry, and in particular the destruction of the Abadan refinery (635 thousand b/d capacity in 1980) during the Iran–Iraq war. Among the Gulf OPEC members, one half of current capacity came on stream during the 1980s, primarily due to the large expansion of Saudi Arabian facilities.

Domestic consumption of oil products has also been rising, and has removed the potential for any significant oil product exportable surplus in Iran and Iraq. While countries have begun to take measures to cut the

Table 7.6: Refinery Capacity and Oil Product Consumption in Middle East OPEC Member States. 1960–95. Thousand b/d.

(i) Refining Capacity

	1960	1970	1980	1985	1989	1990	1991	1992	1993	1994	1995
Iran	467	594	1320	615	777	777	957	957	1092	1092	1092
Iraq	78	116	306	366	550	550	550	550	595	603	603
Kuwait	209	437	594	564	670	670	670	670	670	820	820
Qatar	1	1	11	63	63	63	63	63	63	63	63
Saudi Arabia	227	676	645	1440	1750	1750	1750	1550	1550	1670	1670
UAE	-	-	15	180	180	193	193	193	205	205	211
Total	982	1824	2891	3228	3990	4003	4183	3983	4175	4453	4459

(ii) Consumption of Oil Products

	1960	1970	1980	1985	1989	1990	1991	1992	1993	1994	1995
Iran	67	182	512	735	779	854	893	941	960	920	945
Iraq	34	62	189	256	303	325	309	440	564	624	613
Kuwait	7	14	37	102	72	65	66	101	95	106	118
Qatar	1	2	6	11	13	13	13	14	14	16	17
Saudi Arabia	13	43	407	632	701	680	676	729	768	792	751
UAE	0	2	58	101	114	116	119	125	133	132	126
Total	122	305	1209	1837	1982	2053	2076	2350	2534	2590	2570

Source: *OPEC Annual Statistical Bulletin*, various issues.

level of subsidy, domestic oil product prices in general remain below the level of their opportunity cost. This leaves the possibility of cuts in the real level of subsidies providing a dampener on demand growth, and indeed such cuts are responsible for the marked stagnation in demand seen from 1993 in Table 7.6. Saudi Arabian domestic demand absorbs most refinery production of gasoil and gasoline, leaving the exportable surplus concentrated on LPG, naphtha and fuel oil.

Table 7.7:   Exports of Oil Products from Middle East Countries. 1980–95. Thousand b/d.

	1980	1985	1986	1987	1988	1989	1990	1991	1992	1993	1994	1995
Bahrain	221	184	247	246	240	250	256	260	270	240	242	242
Iran	141	39	10	3	8	38	60	88	70	70	90	110
Iraq	35	12	8	110	110	145	80	11	17	18	20	19
Kuwait	343	493	525	553	635	690	380	40	256	419	622	794
Qatar	2	28	36	41	44	75	74	82	104	116	103	98
Saudi Arabia	413	574	712	1021	1141	1073	1306	1226	1288	1386	1360	1322
UAE	20	92	112	114	117	155	165	170	173	160	155	269
Others	77	110	107	124	123	124	128	129	126	151	140	128
TOTAL	1252	1533	1757	2213	2418	2549	2449	2005	2303	2560	2733	2982

Source: *OPEC Annual Statistical Bulletin*, various issues.

Table 7.7 shows the evolution of oil product exports, and indicates a total level in 1995 of around 3 mb/d, with 2.35 mb/d coming from the three main exporters, i.e. Saudi Arabia, Kuwait and Bahrain. Not all of the exports shown in Table 7.7 represent the output of refinery processes. In addition, the figures include exported LPG output from gas processing plants, which are particularly important in Saudi Arabia and Qatar. In terms of refinery output, Qatari exports stood at about 40 thousand b/d in 1995, and Saudi Arabian at about 700 thousand b/d.

By 1994, Kuwaiti oil product exports had begun to return to pre-Gulf War levels, and by 1995 they had reached a new high. However, the cumulative shortfall in Kuwaiti oil product exports had by 1994 amounted to some 850 million barrels since the Iraqi invasion. Crude oil is highly fungible, and the loss of Kuwaiti and Iraqi production was relatively easily made up by the utilization of spare capacity, in Saudi Arabia in particular. By contrast, the output of the Kuwaiti refineries, particularly the gasoil output of the hydrocracking units, proved much harder to make up. As we see in Chapter 13, the end of the Gulf crisis left the margins for complex units, based on Singapore prices, considerably higher, primarily due to the temporary loss of the Kuwaiti units.

The refineries in the Gulf area are shown in Table 7.8. They range from the high degree of sophistication of the Kuwaiti industry, to some, often very large, refineries that are little more than topping units. The major export refineries are those in Kuwait and Bahrain, plus the Saudi refineries in Ras Tanura, Al Jubail, and the larger of the two Yanbu plants. Plans for expansions of capacity have been greatly scaled back over the 1990s, with the policy emphasis having shifted decisively to obtaining refinery capacity in consuming areas, and away from the focus on large export orientated plants in the Middle East. In Asia, we have seen in previous chapters that Aramco has shares in refineries in the Philippines and in South Korea, and has been in negotiations for building capacity in China. Other Middle East national oil companies have also been in negotiations to build new capacity, *inter alia*, KPC in India, Thailand and Pakistan, and NIOC and ADNOC in Pakistan.

A major planned expansion of Saudi Arabian refineries was truncated after the merging of the state refiner, Samarec, into Aramco in 1993. Current expansions are primarily based on the upgrading of the Ras Tanura refinery, notably the addition of a visbreaker and a hydrocracker. After this is completed, on the basis of the structure shown in Table 7.8, the huge but extremely simple Rabigh refinery is first in the order of

Table 7.8: Refinery Capacity in Gulf Countries. Thousand b/d.

Country	Refiner	Location	Crude Capacity	Thermal Cracking	Catalytic Cracking	Hydro-cracking
Bahrain	Banoco/Caltex	Sitra	250	20	39	0
Kuwait	KPC	Mina Abdulla	242	59	0	36
	KPC	Mina Al-Ahmadi	415	0	28	36
	KPC	Shuaiba	144	0	0	82
Oman	Oman Refinery Co.	Mina Al Fahal	85	0	0	0
Qatar	National Oil Dist. Co.	Umm Said	58	0	0	0
Neutral Zone	AOC/Texaco	Ras Al Khafji	30	0	0	0
Saudi Arabia	Petromin/Shell	Al Jubail	310	32	0	0
	Aramco	Rabigh	325	0	0	0
	Aramco	Ras Tanura	300	0	0	0
	Aramco	Jeddah	42	0	13	10
	Aramco	Riyadh	140	0	0	34
	Aramco	Yanbu	190	0	0	0
	Aramco/Mobil	Yanbu	360	46	91	0
UAE	ADNOC	Ruwais	120	0	0	27
	ADNOC	Umm Al-Nar	85	0	0	0
Yemen	Aden Refinery Co.	Aden	110	0	0	0
	Yemen Hunt Co.	Marib	10	0	0	0

Sources: Various.

priority for upgrading. Kuwait's intention is to increase its domestic capacity in the longer term to 1 mb/d, but its primary focus in terms of downstream investment is now on the addition of assets in Asia.

The largest current expansion plans are in Abu Dhabi, with the intention of expanding the Ruwais refinery with new crude units and two condensate units, with the complex linked in to a petrochemical development. Elsewhere, plans remain for new condensate units in Qatar, and additional units are being built in Iran, primarily to cope with domestic demand rather than having any very significant export capability.

Virtually all Middle East oil product exports are sold on a term contract basis, normally priced against assessments of Gulf spot prices, or against Singapore prices net of freight. The one major exception is LPG, where the Saudi Arabian contract price serves as the major marker. The structure of LPG pricing has shown a marked change since Samarec was subsumed into Aramco. Previously, LPG terms sales were based on an explicit formula, whereby prices were primarily linked to the formula price of Arabian Light, with a residual (10 per cent) element based on the announced results of a monthly spot tender. After October 1994, the price has been declared as an absolute price, with no official explanation of its origin. However, the discretionary price appears to be based on the results (now undeclared) of three spot tenders per month, together with spot assessments and the position in both naphtha and crude oil markets.

For other products, Saudi term sales are priced on the basis of fixed differentials (renegotiated every six months) from the average of published assessments produced by Platt's and Petroleum Argus. For A-180 naphtha, for example, the premium (which, as an indication of its normal size, stood at $3.40 per tonne for the second half of 1996) is applied to the mean of Platt's and Petroleum Argus quotes for c+f Japan naphtha minus freight costs. For A-960 fuel oil exports, dual purpose kerosene and jet fuel, premia are applied to the mean of Platt's and Petroleum Argus Singapore quotes (i.e. MOPPAS).

The predominance of the term market creates some pricing difficulties. In the absence of any significant spot trade, there is no basis for any real local price formation. Instead, prices are primarily derived on the basis of Singapore prices, with an adjustment made relating to the freight cost to Singapore. In flat market conditions this causes few problems. However, normally there is a time structure to prices. Given that cargoes from the Middle East will take around ten days to reach Singapore and about fourteen to reach Japan or Korea, they will then carry either an implicit discount or premium.

To illustrate, consider market conditions where there is a contango, i.e. prompt oil sells at a discount to oil for more forward delivery. The Middle East cargo, for Korea or Japan effectively two weeks forward, is being priced relative to a prompt Singapore cargo with no allowance made for the time structure. A trader who wishes to hedge the absolute value of the cargo from changes in price in transit using the swaps market, buys prompt and sells forward, thus picking up the value of the contango as an effective discount to their Middle East cargo purchase. The profitability of these cargoes is then increased, and, were there a functioning Gulf spot market, prices would rise relative to Singapore. Likewise, in market conditions of backwardation, Middle East oil is overpriced compared to what would be its true spot market value. In total, whenever there is a time structure to prices, Middle East term cargoes are always incorrectly priced.

The tendency of the Singapore price minus freight cost method of calculating values in the Gulf, might not be seen as a problem for producers unless periods of contango dominated, and thus term prices averaged below the 'true' but invisible fair market spot price. In fact the problem exists under any relative frequency of backwardations and contangos. In backwardations, term lifters seek to minimize volumes, and seek alternative supply sources. In contangos, where term liftings carry an implicit discount, they seek to maximize liftings. The pricing formulae then create volatility in volumes, and the situation where Middle East term cargoes are also either the most or least desirable source of supply, but are only very rarely at their fair market value. In the absence of a spot market, the only solutions are either to price cargoes with a time delay to approximate delivered prices, as is used with crude oil pricing, or to incorporate an element of the time structure directly into the formulae.

6. Forward Markets – Open Specification Naphtha

The only forward oil market in Asia is the open specification (open spec) naphtha market.[8] The reason for its emergence lies in the development of petrochemical demand for naphtha in Japan, as shown in Table 7.9.[9] At the start of the 1980s, Japanese refineries were able to meet more than 75 per cent of domestic naphtha demand. However, while demand was stagnant until 1986, domestic production almost halved. Cuts in refinery runs reduced production, but, more importantly, increasing amounts were required for reforming into gasoline. As a result, imports of naphtha increased sharply.

The naphtha supply system in the mid 1980s was entirely composed of spot deals, tenders and term contracts. Petrochemical end-users obtained their imported naphtha through Japanese trading houses, the *sogo shosha*. In turn the *shosha* obtained naphtha primarily from term and spot supplies in the Middle East. As is shown in Table A5.5, while Japan is the most important outlet for Singapore naphtha, the volume was only just over 60 thousand b/d in 1995. In 1986 only about 15 per cent of Japan's supplies came from Singapore, leaving the balancing role to Middle East suppliers. The trade at this time was almost entirely in the hands of the *shosha*. To participate in this chain, a trader would not only need access to a secure supply of naphtha of suitable quality, but they would also need access to the end-users. Japanese petrochemical firms tended to prefer secure supply arrangements with *shosha*, and tended to be loath to enter spot markets on their own account to deal with traders.

Table 7.9:   Japanese Naphtha Supplies. 1981–95. Thousand b/d.

Year	Production	Imports	Supply	Exports
1981	314.7	124.1	411.0	0.0
1982	238.1	155.1	376.1	0.0
1983	198.6	225.7	394.5	0.0
1984	200.6	230.5	414.8	0.0
1985	178.3	256.7	419.7	0.2
1986	166.7	298.0	444.9	0.0
1987	150.5	345.4	473.3	0.8
1988	150.4	389.3	513.3	1.2
1989	151.7	408.0	534.7	5.5
1990	187.1	371.8	536.1	7.4
1991	243.1	344.0	561.2	7.2
1992	275.0	355.4	617.9	10.8
1993	299.1	332.3	622.0	3.8
1994	300.9	383.4	673.3	4.8
1995	306.6	473.1	765.4	5.7

Source: Petroleum Association of Japan.

By the middle of 1986, this system was showing signs of stress. There were four main factors at work. The first was the collapse of official pricing systems and the increase in the volatility of naphtha prices. Japan is about two to three weeks' sailing time from the Middle East, leaving the *shosha* with a price risk that was absent in the period of official (and relatively stable) pricing. Secondly, there was a strong perception that naphtha imports, and thus the volume on which there would be price exposure, were going to increase sharply. With a booming demand for ethylene, petrochemical demand for naphtha feedstock was already beginning to accelerate again, and there was no perception at that point that domestic refinery production would be allowed to increase. As shown in Table 7.9, imports did indeed increase until 1989 given the growth in end-user demand, and were then reined in by a doubling of domestic production, before increasing sharply after 1993.

The third factor was the start of a battle for market share among the *shosha*. Margins began to be cut as firms tried to gain customers. The fall in marketing profits made the *shosha* even more sensitive to price risk, i.e. for risk averse firms the same amount of price risk assumes greater significance the lower are expected profit levels. While *shosha* may have been averse to the risk involved in the physical transfer chain from the Middle East, at this time they showed a different attitude to pure trading activities. In 1986 there was a surge in mainly speculative trading by *shosha* in forward oil markets throughout the world, particularly in the forward Dubai crude oil market and the forward market for UK North Sea Brent. The fourth reason for the emergence of the open specification naphtha forward market was then a willingness and desire by the *shosha* to trade speculative instruments.

By the autumn of 1986, these factors combined to create the circumstances conducive to the start of the forward naphtha market. As with any forward market, there is no *a priori* need for absolutely all trades to be standardized. However, there is still a high degree of standardization in the trading contract used.[10] The basic structure was agreed by potential market participants in a meeting in Singapore in September 1986, and this general format remains much the same today.[11] Trades are conducted for two delivery periods each month, the first covering the first to the fifteenth, and the second the rest of the month. The standard cargo size is 25 thousand tonnes (25 kt equivalent to 225 thousand barrels), with a loading tolerance at the seller's option of plus or minus 10 per cent. The implications of the tolerance provision are detailed below. The location for the cargo is normally c+f Japan, i.e. a delivered cargo with freight costs already incurred.

The specifications for open spec naphtha are the subject of an annual conference. The major area of change since 1986 has been in the definition of acceptable origins for cargoes. Originally cargoes were only automatically acceptable if they had Middle East Gulf, Singapore or Malaysian origins. While the general definition was 'east of Suez and west of the international dateline', material from many sources, and in particular India and China, has been required to meet additional quality standards. While oil refinery distillation and cracking can normally cope with a wide range of inputs, petrochemical naphtha crackers tend to be more sensitive to a wide range of potential contaminants, and in particular to metals and chlorine. For this reason cargoes out of Singapore storage, which may be simply reblended Indian or Indonesian material, have been treated differently to cargoes from the Singapore refineries. Over time the origin clause has been widened to include Algerian and Egyptian material, and the 1995 conference accepted the need for a wider definition, in particular to allow easier sourcing from Europe. By the end of 1996, most European cargoes were included in the specification.

A major difference between forward and other markets is in the clearing mechanism, i.e. the translation from forward cargoes into physical cargoes. In the now dominant Singapore swaps markets detailed in the next chapter, trades are cash settled with no physical oil involved. In a futures market, which may or may not have physical delivery, positions are automatically cleared. Thus, if a trader has an outstanding buy contract for a delivery period, this can be cancelled by selling a contract for the same delivery period. This automatic offset feature is not shared by a forward market. Instead, given that the number of trades is a multiple of the number of physical cargoes, delivery chains are formed. The chains procedure for open spec naphtha works as follows.

A market participant who has a physical cargo they wish to sell forward, will inform one of the companies they have sold a forward cargo to, that they wish them to take delivery. That firm has two options, they can either take delivery in which case the chain stops there, or they can in turn pass the nomination to a company they have sold a forward cargo to, in which case the chain continues. Thus, a company that has both sold and bought forward cargoes can appear several times in the same chain. Likewise, there may be no direct trade between the actual supplier of the physical naphtha and the company that takes delivery. The nomination of the cargo will involve the naming of a five-day range for the timing of delivery, with nominations finishing fifteen days before the start of that window.

Consider a trader who has sold as many, say, second half April cargoes as they have bought. In a futures market, automatic offset would mean that the position had been terminated, but in a forward market each cargo is still in play and can enter the chains. Sometimes a trader will have a nomination passed to them and be unable to pass it on in time before the close of nominations. This is known is being 'five o'clocked', and is common in all markets cleared by chains. For the trader who was balanced, being five o'clocked removes that balance. They are now long one physical naphtha cargo, and short one forward cargo, as they will have one more forward cargo sold than bought. Another forward cargo will have to be bought to reachieve balance, and the physical cargo sold.

The possibility of being five o'clocked presents a powerful deterrent to traders who do not want to become involved with physical operations. This possibility can be reduced by a 'bookout', i.e. the identification of a possible closed chain of forward cargoes, which, subject to all parties involved agreeing, need not form part of the delivery chains and are cash settled on the standard 25 thousand tonnes. However, often a trader has an incentive not to agree to a bookout, primarily due to the 10 per cent tolerance clause noted above. This clause provides a method for maximizing gains or minimizing trading losses as in the following example.

Assume a trader has access to a Middle East naphtha source at $170 per tonne. In the open spec forward market they have made three sales at $180 per tonne, and two buys at $170 per tonne. If they put a standard 25 kt cargo into the chains, assuming all their forward trades stack up in the same chain, then their net profit for the series of deals is 3 x $10 x 25,000 = $0.75 million. However, invoking the tolerance clause, and maximizing the cargo, puts an additional 2.5 kt into the chain and results in an additional $75,000 profit. The trader then would have been disinclined to book out the deals, in the hope that at least one buy/sell pair of deals would be involved in the chain they were going to initiate. Likewise, a trading loss in the forward market could be reduced by minimizing the cargo and passing through only 22.5 kt.

The use of tolerance as a trading device is common in many forward markets, including the Brent market.[12] In Brent the tolerance is at the buyer's option, i.e the company at the end of the chain, and hence its use is attractive to any market participant that both buys and sells forward cargoes. In open spec naphtha however, it is at the option of the company at the start of the chain, and thus is only available to those who have a supply of physical naphtha to pass through the chains. For any trader without such a source, the tolerance clause will tend on average to work against them, as well as making bookouts harder and thus increasing the chances of five o'clocking. In all, the forward clearance mechanism means that the market is best suited to traders who have access both to physical naphtha, and also to the end-user market.

The development of the liquidity of the market is shown in Table 7.10 using the number of deals reported to Petroleum Argus. While the coverage of the Petroleum Argus database is less than total, we have no reason to believe that this coverage has varied widely over time. Thus, the data derived from this source should be representative of the overall market trends. In Table 7.10, liquidity is shown by quarter, and broken down into two categories. Outright deals are those in which an absolute price was negotiated and reported. Spread deals are those where the parties have negotiated the simultaneous sale and purchase of forward cargoes for different delivery periods. For example, a spread might involve company A selling a first half April cargo to B, while B simultaneously sells A a first half May cargo. These trades are then differential trades, with the major information they carry being the difference in price between the two periods. As two forward cargoes are traded, each spread deal in Table 7.10 has been recorded as two cargoes.

The market has been through several distinct periods. Until the Iraqi invasion of Kuwait at the start of August 1990, liquidity was almost entirely concentrated on outright deals, and, while there were some quarterly variations in the volume traded, there was no overall trend. The

Table 7.10: Liquidity and Forwardness of Open Spec Naphtha Market by Quarter. 1988–95.

Quarter	Year	Number of Cargoes			% Distribution of Deals by Half Month						Forwardness
		Total	Outright	Spreads	<3	3	4	5	6	7+	(Days)
1	88	55	55	0	33	64	4	0	0	0	39.6
2	88	56	56	0	13	48	36	4	0	0	48.4
3	88	56	56	0	29	61	11	0	0	0	39.7
4	88	43	43	0	19	67	14	0	0	0	45.8
1	89	31	31	0	10	39	35	16	0	0	51.4
2	89	42	40	2	10	40	38	13	0	0	51.5
3	89	53	53	0	2	15	57	21	6	0	60.7
4	89	60	52	8	4	19	33	38	6	0	63.4
1	90	45	43	2	2	2	44	37	9	5	67.9
2	90	66	66	0	5	8	9	23	52	5	78.4
3	90	40	36	4	0	0	3	40	49	9	82.1
4	90	25	9	16	0	0	0	56	44	0	81.5
1	91	21	13	8	0	8	23	62	8	0	69.7
2	91	102	36	66	0	17	25	44	14	0	67.2
3	91	201	57	144	0	2	16	30	45	7	77.2
4	91	174	64	110	0	2	2	5	34	58	92.4
1	92	134	66	68	0	0	5	35	44	17	78.7
2	92	141	37	104	0	3	37	40	17	3	67.4
3	92	97	59	38	2	2	5	38	38	16	78.1
4	92	13	9	4	0	0	33	44	22	0	77.4
1	93	121	57	64	7	11	30	45	7	0	63.8
2	93	39	39	0	8	28	59	3	3	0	54.3
3	93	72	72	0	3	15	25	44	13	0	67.6
4	93	30	24	6	25	21	29	17	4	4	52.3
1	94	26	20	6	21	16	32	16	16	0	55.3
2	94	15	13	2	15	8	54	23	0	0	55.9
3	94	15	15	0	27	7	7	47	13	0	57.4
4	94	10	10	0	10	10	50	30	0	0	57.7
1	95	48	26	22	8	12	50	27	4	0	60.8
2	95	80	52	28	2	4	27	60	8	0	68.7

Source: Own calculations from Petroleum Argus Product Deals database.

Gulf crisis had a profound effect, both in its immediate consequence and its longer-term impact on trading patterns. The immediate impact had two main aspects. First, liquidity in outright trades all but left the market in the fourth quarter of 1990 and the first quarter of 1991. The pressure was on obtaining prompt supplies, and traders rushed for the spot market. The risks of trading a large cargo size forward market also led to a decline in liquidity.

The Asian time zones were not best placed to deal with the flow of information from the Middle East, and most significant developments happened during the London trading day and outside of Singapore hours. In these conditions, leaving an uncovered position open at the end of Singapore trading left exposure to most of the London and all of the New York trading day.

As shown in Table 7.10, the average number of outright deals over the two full quarters of the Gulf crisis fell to below four a month, compared to the average of about seventeen per month in the first half of 1990. The second immediate aspect of the Gulf crisis was the growth in spread trades. As a trading device, spreads normally leave less price risk than outright trading, and this tendency was especially pronounced during the Gulf crisis. As a result, many of the traders left in the market preferred only to open positions using spreads, and more spreads were done during the Gulf crisis than in all the previous history of the market.

The end of the Gulf crisis brought the start of the zenith of the open spec market, which lasted until the second quarter of 1993. The total liquidity soared to levels three or four times those of 1988 and 1989. Upon closer inspection however, it is clear that the increase was due almost entirely to spread trading. The levels of outright deals were not very much higher than before, indeed the level before the Gulf crisis in the second quarter of 1990, while matched, was not exceeded before 1993, by which time total liquidity was falling.

For those mainly looking for a forum in which to trade, spreads did offer the advantage of lower risk. In a futures market, where the minimum size of trade is small, lower absolute risk can be achieved by simply trading smaller volumes. In the open spec market, where the units of trading are lumpy discrete 25 thousand tonne parcels, spread trading was attractive. Following a series of large losses in the 1980s through taking large speculative positions (mainly in the forward market for North Sea Brent), Japanese traders had become less attracted to large-scale outright position taking and were thus attracted by spreads. While this helps explain the growth of spread trading, it also explains its collapse in 1993.

With the growth of the Singapore swaps market, primarily trading in units four to five times smaller than open spec naphtha and without the same delivery and squeeze problems, there was less reason to trade on the forward market. In other words, for trading purposes the open spec market had been the only game available, and within that spread trading had become the preferred *modus operandi* for pure trading. With the development of better instruments, the pure trading motivation was better served elsewhere.

While the above reason considers traders who wanted less risk, spread trading was also useful for those who wanted more. Liquidity on the open spec market had never been sufficient for those who wanted to maintain a large open position for any considerable period of time. The growth of liquidity in spread trading provided a method for maintaining such a position. Once the position had been built up, it could be rolled over using spreads. For example, consider a trader who has a large position in first half April naphtha. By trading first half April to first half May spreads, they can balance their April position and move the position into May. What would appear in Table 7.10 to be large-scale spread trading would then actually be simply the roll-over of a large outright position.

The shift away from the open spec naphtha market, occasioned primarily by the movement of the spread trading liquidity towards the swaps market, resulted in the sharp decrease in liquidity in 1993 shown in Table 7.10, and was continued with the extremely low liquidity levels of 1994. The resurgence in 1995 is primarily a reflection of the shift in the nature and purpose of the market as reflected in the composition of its participants, as is discussed below. Table 7.10 also shows two metrics for the depth of the market, i.e. how far forward trades are conducted. We show first the percentage distribution of outright price deals by the number of half months forward. For instance, a deal conducted on, say, 20 April for first half June delivery is three half months forward. In 1988 trade was very heavily concentrated in the first three half months, with very little liquidity for any company wishing to trade further forward. Over the next years this liquidity appeared, and trade became deeper. By the end of 1991 most trades were being conducted six or more half months forward, with little volume for trades close to delivery. As the volume in the market fell away the mode forwardness became less again. While 92 per cent of trades were six or more half months forward in the last quarter of 1992, such trades have now become extremely rare.

By 1995, the majority of trade was being conducted in the fourth or fifth half month forward. Table 7.10 also shows an aggregate measure of forwardness measured in days. This measure is calculated as follows. A trade for first half June delivery made on 20 April has a forwardness of 48.5 days, i.e. 10 days in April, 31 in May, plus the 7.5 in June to the midpoint of first half June deliveries. Using this measure the pattern is confirmed of a considerable increase in forwardness until 1992, followed by a downwards trend with a partial recovery in 1995. Starting below 40 days, the measure peaked at the end of 1991 at above 92 days, before falling back to below 60 days and then recovering to 68.7 days by mid-1995. Since its zenith in 1991 and 1992, the market has then not only become less liquid, its trades have also become less forward.

While the volume of trade in the market has shown some sharp swings, the number of companies involved had until 1994 been fairly stable. Table 7.11 shows the number of companies recorded as making at least one forward open spec naphtha contract trade per year.[13] The number of participants fell gradually from 1998 to 1991, primarily due to Japanese end-users exiting the market and preferring to go back to relationships with *shosha*. The decline in liquidity in 1993 was not associated with a decrease in the number of participants, which increased to its highest level of thirty-seven. The start of the shift to the swaps market was then one where open spec volumes were scaled back, rather than one causing wholesale net exit. That happened in 1994, with a net exit rate of more than one-third, and the number of participants has remained low since.

Table 7.11: Number of Participants and Concentration in Open Spec Naphtha Market. 1988–96.

Year	Number of Participants	Concentration (Inverse Herfindahl)
1988	36	12.1
1989	34	16.6
1990	31	16.0
1991	31	9.8
1992	35	14.5
1993	37	14.1
1994	24	14.0
1995	26	12.1
1996	21	11.1

Source: Own calculations from Petroleum Argus database.

We have measured the concentration of market shares by calculating, on an annual basis, the inverse Herfindahl index as shown in Table 7.11.[14] The degree of concentration was in fact at its greatest (i.e. the inverse Herfindahl at its lowest) when the volume in the market was also at its peak. This is evidenced by the sharp fall in the inverse Herfindahl index from 16 in 1990 to 9.8 in 1991. This suggests that the increase in liquidity was heavily skewed towards a few companies. The level of concentration was reduced in 1992, and stayed roughly constant through the nadir of the market in terms of liquidity. The fall in value of the inverse Herfindahl in 1995 suggests that the upturn in market volume has not been evenly spread, and is associated with an increase in the concentration of the market.

The composition of the liquidity of the market by category of company is shown in Table 7.12. The table shows a major and profound change. What started as a primarily Japanese market has become less Japanese and more Korean. Indeed, in 1994 when Japanese participation was at its lowest ebb, Korean companies were more active than the Japanese, and overall the major trend in the 1990s has been the growth of Korean involvement. European companies (who have little access to physical

Table 7.12: Market Shares in Forward Naphtha Market by Category of Company. Per Cent. 1988–96.

	European	Japanese	Korean	Refiner	Wall Streeter	Western Trader	Other
1988	1.0	75.5	0.5	11.5	4.1	6.9	0.6
1989	4.2	65.8	0.0	9.3	6.8	12.7	1.1
1990	4.9	47.7	0.0	8.9	16.6	14.5	7.4
1991	2.1	42.9	0.8	26.3	15.4	10.8	1.6
1992	3.5	32.7	3.1	23.3	13.6	20.9	2.9
1993	1.5	28.7	11.3	23.7	13.0	20.0	1.7
1994	0.0	21.1	26.0	17.1	9.8	25.2	0.8
1995	4.7	33.0	19.5	11.1	12.7	18.7	0.3
1996	2.5	43.6	17.9	9.1	7.3	19.3	0.4

Source: Own calculations from Petroleum Argus Product Deals database.

naphtha or to end-users) have not been very active, although we will see in the context of other markets that they have elsewhere become a major force. Likewise, Wall Street companies and their affiliates never became as important as in other markets. Liquidity has tended to come from the Japanese, the Singapore refiners, western traders and now the Koreans. There has been no significant involvement by the Chinese, national oil companies or indigenous Asian based trading companies.

Disaggregating the data further, results in the market share by individual company shown in Table 7.13 for the largest fifteen traders in each year. The decline in Japanese involvement is best illustrated by the decline in Marubeni's trading. Using the forward market in conjunction with its dominant share of the Japanese end-use naphtha market, Marubeni was the largest trader until 1991. Its current involvement is limited, as is that of the other major trader in the early years of the market, Nissho Iwai. Other declines have been even more dramatic than Marubeni's. BP, while still in the market in 1996, has greatly reduced its activity compared to its leading position in 1991 and 1992.

We cover the naphtha swaps market in the next chapter. However, at this stage it is worth pointing out that the idea of a straightforward move of companies from trading in the forward market to trading swaps is too simplistic. Of the fifteen most active forward market traders in 1996, we record eleven among the most active fifteen in naphtha swaps. The three main naphtha swaps traders in 1996 were J Aron and Vitol, both very active in the forward market. The two markets have become intertwined, and trading is taking place involving positions in both. We believe that the upswing in open spec activity in 1995 is partly due to the increased use of trading optimization across the two markets. The swaps market may have

Table 7.13: Main Participants in Forward Naphtha Market. Rank and Market Shares. Per Cent. 1988–96.

Rank	1988		1989		1990		1991		1992	
1	Marubeni	18.9	Marubeni	12.2	Marubeni	11.3	BP	20.1	BP	15.0
2	Nissho Iwai	13.7	Nissho Iwai	9.8	Nissho Iwai	10.9	Nissho Iwai	17.1	Vitol	10.7
3	Shell	9.0	Mitsubishi	9.2	J Aron	10.3	J Aron	11.2	Shell	8.3
4	Mitsubishi	7.0	Mitsui	8.6	Phibro	7.1	Marubeni	6.7	J Aron	8.0
5	C Itoh	6.2	Shell	6.5	C Itoh	6.4	Shell	6.6	Nissho Iwai	7.1
6	Mitsui	5.2	Phibro	6.0	Vitol	5.8	Apex Intl	4.3	Marubeni	6.4
7	Nippon Oil	4.1	Sumitomo	4.8	BP	5.1	Mitsui	4.1	Phibro	5.0
8	Kanematsu	3.4	Nippon Oil	4.2	Mitsui	4.2	Mitsubishi	3.9	C Itoh	4.7
9	Phibro	3.1	BP	3.0	PRI	4.2	C Itoh	3.4	Mitsubishi	4.4
10	Sumitomo	3.1	Cargill	2.7	Shell	3.9	Phibro	3.4	Cargill	4.0
11	BP	2.3	Elf	2.7	Showa Shell	3.5	Vitol	2.7	Yukong	2.7
12	Asahi	2.1	Marc Rich	2.7	Coastal	3.2	Showa Shell	2.4	Showa Shell	2.5
13	Tonen	2.1	Tonen	2.7	Mitsubishi	3.2	Cargill	2.2	Kanematsu	2.4
14	Toyo Menka	2.1	Idemitsu	2.4	Neste	3.2	Toyota	1.8	Vanol	2.3
15	Idemitsu	1.8	Kanematsu	2.4	Cargill	2.6	Kanematsu	1.6	PRI	2.0

Rank	1993		1994		1995		1996	
1	Shell	16.8	Samsung	12.4	Samsung	15.3	Itochu	18.1
2	J Aron	11.4	Shell	10.7	J Aron	13.3	Samsung	14.1
3	Vitol	7.8	Cargill	9.9	Shell	9.4	Vitol	10.9
4	Mitsubishi	7.4	J Aron	9.9	Kanematsu	8.9	J Aron	7.5
5	BB Energy	6.9	Vitol	9.1	Itochu	7.8	Mitsui	6.2
6	BP	5.4	BP	5.8	Cargill	6.9	Glencore	5.4
7	Caltex	4.0	Itochu	5.8	Vitol	6.9	Kanematsu	5.2
8	Nissho Iwai	4.0	Daelim	5.0	Glencore	4.4	BP	4.4
9	Samsung	4.0	Marc Rich	4.1	Mitsubishi	4.4	Yukong	4.4
10	Cargill	3.1	Mitsubishi	3.3	Elf	3.3	Shell	4.0
11	Itochu	2.9	Mitsui	3.3	Mitsui	3.3	Cargill	3.6
12	Sumitomo	2.9	Hanyang	2.5	Marubeni	2.5	Honchu	3.2
13	Marubeni	2.7	Showa Shell	2.5	Sumitomo	2.5	Marubeni	3.0
14	Cosmo	1.8	Sumitomo	2.5	N. Iwai	2.2	Elf	2.6
15	Hyundai	1.6	BB Energy	1.7	BP	1.7	Showa Shell	2.6

Source: Own calculations from Petroleum Argus Product Deals database.

originally taken liquidity away, but with its development provided a boost to the forward market through the utilization of arbitrage possibilities.

Korean companies have become very active, as was noted above. The *chaebol* Samsung is a major trader, with a number of other Korean companies involved. Naphtha supplies were once primarily obtained from the *sogo shosha*, but now Korean companies have increased their own term contracts with suppliers and thus have material to put into the chains, as well as access to Korean end-users.

7.   The Role of Storage Facilities

To trade effectively in a physical oil product market, a trader normally needs to have competitive access to storage and blending facilities. The difference between product that is sourced in a deal and that which is sold on, can arise from location, timing, specification and quantity. Location is a matter of transportation, but the latter three can only be catered for in storage. Traders may wish to put oil in storage if market prices are such that there is a sufficient premium for deferred delivery. Likewise they will need storage if blending operations are necessary to meet the buyer's specifications. Alternatively storage may be needed to effectively break down large cargoes into smaller parcels, or to make up larger cargoes.

For a refiner this normally presents few problems, as all of the Singapore refineries have extensive storage facilities for both crude oil and oil products. For the trader without access to refining in Singapore, the absence of such facilities would effectively preclude entry to physical markets. Such traders would be confined then to any established forward or futures market in which they could either always close out positions so as not to take delivery, or in which taking delivery poses no special problems. The latter, while it holds in most crude oil markets, is generally not the case for oil products. With refiners tending to have storage geared for their own operations, the existence of physical oil product markets with many participants normally requires the presence of independent storage facilities.

Such independent operators started activities in Singapore during the 1980s, again with the encouragement of the Singaporean government. The government has been keen to promote a full range of oil industry infrastructure and services in Singapore, and to create the environment for an active trading centre. With those aims in mind, the lack of independent storage was seen in the early 1980s as an impediment to the development of the industry. The four current operators (three in oil and one in chemicals) are Van Ommeren, Tankstore, GATX and Oiltanking, with the location of their facilities having been shown in Figure 6.3.

Van Ommeren was the first to arrive, beginning operations on Pulau Sebarok, the same island that, as was noted in the last chapter, had been first developed for storage by Standard Oil in the 1890s. The Dutch parent company owns the majority share, but shares are also held by the Development Bank of Singapore and the Port of Singapore Authority (PSA), showing the active government interest in and promotion of the venture. Two years later Tankstore began operations, originally from floating storage using a Very Large Crude Carrier (VLCC), but now from Pulau Busing close to Bukom. The GATX facility is for petrochemical storage, and is another joint venture with the PSA. The final entrant was

Oiltanking who commenced operations on Pulau Seraya in 1989.

The majority of independent storage is leased on annual terms, leaving perhaps one-quarter to one-third for short-term storage. With utilization rates high, there can often be a shortage of spare capacity. This has of course meant that the longer-term arrangements have become more attractive to traders, who are prepared to buy space on an extended contract, which effectively carries an insurance value. The refiners own most of the storage facilities. They have about 40 million barrels of dirty storage (i.e. crude oil, grades of fuel oil and bunker oil), and 25 million barrels of clean storage (i.e. lighter oil products). The independent operators have 6 million barrels and 9 million barrels of dirty and clean storage respectively.

During the early 1990s a trend emerged for traders to obtain their own storage facilities outside of Singapore, for strategic purposes as well as to cut costs. For example, following the US military withdrawal from the Philippines, Coastal acquired a lease on the Subic Bay storage facility.[15] Chemoil in collaboration with Itochu has also acquired a lease for storage in the Philippines. Some storage has been started in China,[16] while others have seen floating storage as VLCCs as a viable proposition, particularly at strategic points. For example, the trading company Louis Dreyfus with a Chinese partner company once positioned a VLCC as floating storage off Southern China for a period when operations were viable. Traders with their own tankers, most notably Hin Leong, have also routinely used them as floating storage off Singapore and elsewhere.

The expansion of independent storage facilities in other areas is a reflection of the general increase in trading possibilities across the region, rather than indicating any decline in Singapore. Coastal bought a ready-made installation, and floating storage is essentially used to fill gaps in areas where there is a trading niche, such as breaking down large gasoil cargoes into smaller barge lots for the Chinese market. Singapore's main advantages of an already developed infrastructure and communications, as well as its location, continue. Generally an expansion of trading implies the expansion of independent storage. As we have argued above, physical oil trading needs independent storage, and there is a symbiosis between the two activities which means that their growth is interlinked.

8. Conclusions

Spot trade in Singapore is constrained by the logistical factor of the standard cargo size for a long distance export market. Because of this feature, it can not support a base, with a large number of spot transactions, such as that which is found in the US pipeline or Rotterdam barge market. The price level information generated by the physical base, is fairly limited. As we have seen above, there is a marked reluctance to trade in absolute prices in the spot market, leaving one company (Hin Leong) to act in the role of key market maker in the major product. We return to this reluctance to trade in Chapter 14, where we term it the 'wallflower complex'.

We would express no particular concern about the dominant price setting role played by this company, as we believe that this position has not been maintained by the presence of barriers to entry. In particular, the company has no real inherent advantages over other market participants. In global terms it is a small trading company, it is non-integrated, and indeed it has no oil production or refining assets. Beyond its trading assets, its only market infrastructure is its tanker fleet. In short, while it is in a position to influence price formation behaviour, it has not achieved that position by the exercise of market power. It has taken on a leading role mainly by the default of others to do so, and the role is in no way unchallengeable.

In considering spot trade, we noted the coexistence of pressures both towards regionalization and also towards globalization. The question of which effect is stronger is to a large extent a non-question, as they affect different prices. Globalization will set overall price level ranges, for instance the binding together of European and Singapore prices through the potential for VLCC arbitrage. The range of independent movement of Singapore prices is greatly reduced by this blurring of the edges between Asian and European markets. We return to this impact in Chapter 13. On the other hand, regionalization primarily affects price differentials. The global factors can coexist with considerable dislocations in what become regional niche markets, due to logistical or product specification factors. A greater diversity of delivered price markets is still consistent with there being overall less independence of fob Singapore prices from global trends.

The above raises a major issue for the organization of trading. Simply trading spot fob Singapore is probably no longer viable. Such trading leaves one at a disadvantage to the global players who are trading between regions, and misses out on the niche intra-regional markets. In short there are two (non-exclusive) options. Either trade on both relatively frictionless spot and paper oil globally, or concentrate on the niche delivered markets where the friction from logistics, and also trading experience, play a role.

We have seen that direct encouragement by the Singapore government, and in particular the incentives offered by the AOT scheme, have had an important role in expanding the number of oil trading companies in the market. While we believe that factors were already in play that would have resulted in the demise of Tokyo as a regional trading centre for crude oil, the AOT scheme did accelerate and concrete that process. The scheme did far more than simply move trade from Tokyo, it also led to an influx of trading companies wishing to achieve an overall expansion and greater globalization in their activities. Given the small potential physical base in terms of cargoes, most of the increased activity has not been felt in spot markets, but has instead been associated with the expansion of informal swap markets built on top of the physical base. From being an almost purely physical based market in 1990, in contrast to the greater depth of paper and futures markets in Europe and the USA, the early 1990s have seen a change in emphasis towards more informal non-physical oil trading. This development provides the focus of the next chapter.

Notes

1. The latter condition is made simply so as not to create a distinction between futures markets with physical delivery, e.g. IPE gasoil, and those which are cash settled, e.g. IPE and SIMEX Brent blend crude oil.

2. Price assessment is covered in Appendix 1, which also gives details of the markets covered by the assessors.

3. Oil trades in the context of the AOT scheme, cover crude oil and all of the major oil products with the addition of sulphur, gasoline components such as MTBE, and even liquefied natural gas.

4. A 1996 report by the Corporate Resources Group found Tokyo to be the most expensive city in the world for foreign workers, 99 per cent more expensive than New York. Singapore was tenth, at 27 per cent above New York, and cheaper than Beijing (62 per cent above New York), Shanghai (52 per cent), and Hong Kong (50 per cent).

5. See Petroleum Intelligence Weekly (1993), *Petroleum in Singapore 1993/94*, Petroleum Intelligence Weekly, New York.

6. As noted in Chapter 5, Korea is both an import and an export pricing centre due to quality differences. A mismatch between refinery configurations and product quality regulations means that Korea imports low sulphur material and exports high sulphur.

7. North China, North India and the Russian Far East can be counted as high sulphur low pour regions.

8. The forward market in Dubai crude oil, which strongly affects Asia, is considered in Chapter 11. The market for paper Tapis is really a swaps market, and is considered in Chapter 10.

9. In Table 7.9, domestic supply of naphtha is less than the sum of domestic production and net imports. This is because some of the imported naphtha is used for reforming into gasoline and thus is not part of the total supply to petrochemical end-users.

10. As a matter of terminology, while *contracts* are traded in a futures market, in a forward market, while each trade is of course bound by a contract, it is *cargoes* that are traded.

11. The meeting was organized by the broker Libra, which remains the dominant broker in the open spec naphtha market.
12. See Paul Horsnell and Robert Mabro (1993), *Oil Markets and Prices*, op.cit. for an account of the Brent tolerance game.
13. Data for 1996 in this and in following tables in this section, cover the period up to August.
14. The inverse Herfindahl index is computed as the inverse of the sum of squared company market shares. The lower this index, the greater is the degree of market concentration. The number can be thought of as the equivalent number of equal sized firms that would result in the same degree of market concentration.
15. Storage in the Philippines was considered in Chapter 5.
16. See *Singapore Oil Report*, July 1994.

CHAPTER 8

THE SWAP MARKETS

1. Introduction

Swaps represent just one form of oil derivative. As the word 'derivative' is sometimes used by different people in different contexts, it is as well to begin with a definition. A derivative is any form of trading instrument whose value is based in some way on (or derived from) outcomes in another market. Following this definition, all informal forward markets trade derivatives, as do all the futures markets considered in the next chapter. The term derivatives then covers forwards, futures, swaps of all varieties, and options.[1] A further distinction can be made between exchange based derivatives where trade takes place in a highly regulated environment, and the more informal structures known as 'over the counter' (OTC) markets. Thus futures trades are exchange based derivatives, and forwards and swaps are OTC derivatives. Options are available in both OTC and exchange based forms.

In the Singapore market the major growth area in trading over the course of the 1990s has been in OTC derivatives, and in particular swaps. Swaps are now at the leading edge of the market in terms of both volumes and influence. As we detail in this chapter, their advantages have enabled these instruments to grow in a market which had not been conducive to the growth of forward trades. Swaps have become such a central component of the Singapore market that it is sometimes easy to forget how recent an innovation they are. The first swap trades were made in financial markets towards the end of the 1970s, and over the next few years their use as an instrument to hedge interest rate and exchange rate risk grew sharply. Their use in oil began around 1986, when the same financial institutions that had pioneered swaps and other instruments in financial markets began to diversify into oil and become active oil market participants. The first oil swap was probably provided by the Chase Manhattan Bank in a deal to manage the risks attached to the purchase of jet fuel by Cathay Pacific Airlines.[2] The next section of this chapter details both the structure and mechanics of typical swap deals, and also the advantages of the short-term swaps prevalent in Singapore over forward markets. The current Singapore oil product swaps market is examined in detail in Section 3, and the limited market in Tapis crude oil swaps is considered in Section 4. A final section provides some conclusions on the role of swaps in the Singapore market.

2.　The Principle of Oil Swaps

A swap in its simplest form is essentially an exchange of floating price risk for a fixed price. The swap can be disaggregated into two linked and putatively physical transactions. Imagine that an agent wishes to purchase a swap so as to achieve a fixed buying price for a given volume of oil. The swap involves the swap provider accepting a commitment to sell that volume to the swap buyer at a fixed price, while buying back the same volume from the swap buyer at a floating price. The two putatively physical transactions can then be cancelled out, leaving a purely financial settlement based on the difference between the fixed and floating prices.

To illustrate, consider the following example. Company A, the swaps buyer, wishes to lock in a fixed buying price over the next year for the 1 million barrels of gasoil they buy each month. Company B, the swaps provider, offers a price swap at $19 per barrel. This swap involves a volume of 1 million barrels per month, with settlement made each month and the floating price being that of a designated grade as published by a price assessment agency. The workings of the swap arrangement are shown in Table 8.1.

The first column of Table 8.1 shows a series of floating prices for gasoil.[3] When the floating prices are higher than the fixed price specified in the swap, the value of the putative oil that A, the swaps buyer, has sold to B, the swaps provider, is higher than the offsetting sale from B to A. The buyer of the swap will then receive a cash transfer equal to the volume of oil swapped multiplied by the difference between fixed and floating prices. For example in January, with the floating price 96 cents higher than the fixed, company A will receive $0.96 million from B. Likewise, if the floating price is lower than the fixed, the transfer will be from the swaps buyer to the provider.

From the point of view of the swaps buyer, if they are also buying 1 million barrels of gasoil per month in the physical market at monthly average prices, the swap gives them an effective fixed price. Any divergence of their monthly expenditure on physical oil from $24 million will be offset by the flow of funds arising from the swap. For example, in May in Table 8.1, company A will spend $26.11 million buying 1 million barrels of gasoil in the physical market, but will receive a transfer of $2.11 million from the swaps provider, thus maintaining an effective buying price of $24 per barrel. The mechanics of the swap if its buyer had wished to fix a selling rather than a buying price are similar. In this case the putative

Table 8.1: Mechanics of a Swap Deal.

Month	Floating Price $/b (A sells to B)	Fixed Price $/b (B sells to A)	Fixed - Floating $/b	Transfer B to A $million	Transfer A to B $million
January	24.96	24.00	-0.96	0.96	-
February	25.14	24.00	-1.14	1.14	-
March	24.89	24.00	-0.89	0.89	-
April	25.68	24.00	-1.68	1.68	-
May	26.11	24.00	-2.11	2.11	-
June	23.86	24.00	0.14	-	0.14
July	22.81	24.00	1.19	-	1.19
August	22.77	24.00	1.23	-	1.23
September	23.09	24.00	0.91	-	0.91
October	24.28	24.00	-0.28	0.28	-
November	23.39	24.00	0.61	-	0.61
December	21.07	24.00	2.93	-	2.93

deal at floating prices is a sale from swaps provider to swaps buyer, and *vice versa* for the fixed price component. Hence, in this case transfers would flow to the buyer of the swap whenever floating prices are lower than the fixed price.

In Table 8.1 the average floating price for the year turns out to be less than half of one cent above the fixed price, and the swap would involve a total net transfer from provider to buyer of just $50,000 over the year. Despite the small net transfer over the year as a whole, note that the swap does have cash flow implications for the provider, with a transfer of $6.78 million to the buyer over the first five months. By contrast, the buyer can budget for a fixed price of $24 throughout the year, and hence the offsetting flows to the swaps provider in the latter half of the year require no additional precautionary cash balances. Indeed, one of the main advantages for the swaps buyer is the streamlining of their cash flow and the minimization of the amount of assets that need to be held in low yield liquid forms for precautionary purposes.

The swap has transferred price risk to the swaps provider. However, it does not follow that the swap provider is risk neutral, or even that it is less risk averse than the buyer. The parties are likely to be differentiated not only by their attitude to risk, but also in their ability to handle risk. The swap provider could have a greater facility for coping with risk due to a series of factors. They might have a larger and more robust asset base. They could have economies of scale in trading or a comparative advantage in trading. Finally, they may have access to other markets so that diversification across trading instruments can reduce their risk. For example, compared to a petrochemical plant primarily engaged in purchases in naphtha, a swaps provider may also be offering swaps in other petroleum products, in interest rates, exchange rates and other commodities. Overall, even if the petrochemical plant owners were less risk averse than the swaps provider, the greater diversification and hence ability to diffuse risk of the provider, means that an *a priori* mutually beneficial swap could still be transacted. When one adds to the above the consideration that swaps can also be bought for speculative purposes, it becomes clear that a generalization about the relative attitude to risk of swap providers and buyers is impossible.

The swaps provider is normally not risk neutral, and so will try to reduce the risk they have incurred. The simplest and most effective way would be to provide an offsetting swap. If, using the example shown in Table 8.1, they could also find a seller of physical gasoil who wanted to fix the price of the same volume of gasoil over the same period, then the provider's risk could be completely eliminated. Having provided the gasoil buyer's swap at $24, they could use a lower fixed price in the gasoil seller's swap, and thus lock in a guaranteed margin. Having received payment for taking risk from both parties, the risks can be offset.

The above example raises the question as to why the two physical gasoil market participants do not simply deal with each other on a one year fixed price, and cut out the role of the swaps provider. There are four main reasons why such a transaction may not take place. First, there may be less than perfect information in the market, and buyer and seller can not identify each other quickly or costlessly. The margin taken by the swaps provider is then in part a share of the reduction in information costs, and they have in fact acted in a similar way to a broker. Secondly, there may be transport costs and other logistical factors. For instance, if the buyer is in Boston but the seller is in Singapore, transport costs will probably exceed the benefits of the direct match up. Likewise, factors such as differences between buyer and seller product specifications may rule out the transaction. The third factor is credit risk. The gasoil buyer and seller may be unwilling to trade with each other, on the grounds that they are concerned about the default risk involved in a longer-term high volume contract. Both may prefer to deal through a counterparty they consider to be financially robust. Finally, there could be bargaining failures, and in particular while spot prices may be visible, lack of information on longer-term or year average prices may mean that the agents simply can not come to terms. The presence of a swaps provider may mean that at least one of the agents might be able to make an arrangement. It is perhaps possible that both will be able to deal if they give more credence to the provider's view of the market than they do that of another purely physically based firm.

While the example of the perfectly offsetting swap provides a theoretical idealization, in reality such a fit is extremely rare. The swaps provider will find that customers want different volumes, different length of swaps, or different indices for the evaluation of the floating price reflecting locational or quality differences. In reality the degree of risk reduction that can be achieved by providing other swaps is likely to be limited, except in the case of short-term swaps as explained later in this section.

If perfectly offsetting swaps are not possible, the provider may need other methods to handle risk. However, in some cases this will not be necessary, simply because providing a swap can in itself be a method for reducing the risk exposure of a provider who also has a position in the physical oil market. For example, consider a refiner with surplus gasoil above the requirements of their own marketing system. Providing a swap to a gasoil consumer reduces their risk exposure, and allows them to deal with a wider set of agents than they could with a physical deal, as the swap does not involve incurring transport costs. If the refiner is in Singapore but can not find an Asian party wishing to buy on fixed price terms, providing a swap allows them to deal with any consumer throughout the world (while accepting some basis risk) but still selling out their physical production cargo by cargo on the local market.

When other methods for reducing the swap provider's risk are necessary, this is known as 'warehousing' risk. Warehousing involves the use of other instruments, such as taking positions in forward, futures or options markets. In the situation shown in Table 8.1, the swaps provider is exposed to increases in the gasoil price. One solution would be to buy 1 million barrels of gasoil futures contracts in each of the delivery months over the year, and hence offset changes in the financial flows in the swaps position with changes in the value of the futures position. However, often warehousing is not as easy, and tends to leave one or more of three main types of basis risk for the swap provider.

The first form of basis risk is exposure to differential price movements between commodities. If the swap is for, say, jet fuel, then there is currently no jet fuel futures contract to use. Instead the provider will tend to use a gasoil contract, still leaving them exposed to the basis risk between the commodities. Secondly, a geographical basis may be left. In Singapore, as we see in the next chapter, oil product futures contracts have been unsuccessful, so if futures were to be used for warehousing, positions would have to be opened in New York or London. The risk of differential movements in prices between Singapore and the other centres would then be incurred. The third form of basis risk is probably the most important, especially in the case of longer-term swaps, and consists of exposure to movements in the time structure of prices. For a long-term swap, there may not be sufficient liquidity in futures markets so far out. As a result, they may decide to hedge a long-term position with positions in the more liquid nearby months. As we see in the next chapter, such a strategy can be fraught if the risks are not fully understood.

The forms of swaps we have considered to this point have been simple, what are known as 'plain vanilla' swaps. In addition there are more exotic varieties. Since each swap is a bilateral deal between the buyer and the provider, it can be tailored to the specific risk management needs of the buyer. For instance, a swap can be used to set either a maximum price for a physical market buyer or a minimum price for a seller. A cap sets a maximum price, so that the fixed and floating prices are the same under the swap unless prices go higher than the maximum. Similarly, a floor defines a minimum price. A collar is a combination of a cap and a floor, leaving the fixed price equal to the floating only for a defined range. A plain vanilla swap is nothing other than a special case of this, i.e. it can be thought of as a collar with zero spread between cap and floor and with a zero premium.

The provider can also offer the swaps user, in exchange for the payment of a premium, the option, valid for an agreed length of time, to activate a swap. For example, the deal could offer the option, valid over three years, to activate a one year swap to sell gasoil at $20 per barrel. Should prices fall below that level during the three years, the user, depending on their view of the market, may decide to activate the swap. This is a hybrid instrument involving elements of both swaps and options, and is known as a 'swaption'. Hybrid instruments can also be created by combining several swaps into one. For example, a user may be concerned not only by oil price risk, but also with the exchange rate risk involved in its oil transactions. A hybrid swap can then offer the user an oil price fixed in domestic currency terms through the use of only one instrument.

Swaps may also be offered to cover long-run oil prices, for instance as a long-term tradable oil index.[4] However, the performance of such instruments has been to date (from the point of view of their providers) disappointing. Long-term indices and 'synthetic oil fields' such as the Salomon Phibro Oil Trust did not achieve any real sustained liquidity.

The structure of the example we have used in Table 8.1 of a swap deal with payments made over a series of periods, is typical of most swaps involving end-users or small producers, or more generally if the swaps buyer is not an active trader. However, it is not typical of the bulk of the swaps trade in terms of deals done. The greatest volume of trade in Singapore and elsewhere is in short-term swaps, primarily covering a period of a month or less and mainly confined to nearby months. We cover the Singapore swaps market in the next section, but at this point we note that the most common swap involves the average prices for a month that is usually within six or less months of the time of trade. In these deals there tends to be just one transfer of funds to conclude the arrangement, rather than a series. These markets have evolved in Singapore in preference to forward markets in most oil products, and this is due to a series of advantages that swaps have over forward markets.

There are several main differences between short-term swaps and forward markets. The most obvious is the impossibility of physical delivery with a swap. To a great extent this is the reason why swap markets have developed in Singapore to a degree that forward markets never managed. In the USA products are traded along a pipeline, and so there are no constraints on volumes and small parcels can be traded. In Europe, the trade in small parcels of oil in Rhine barges generates a large volume of trade. There is no equivalent low parcel size, high volume market in Singapore. Physical cargoes are large enough to deter many users from entering a forward market where there is even a possibility of taking delivery. This also means that a standardized cargo size would be so large as to greatly reduce the number of deals that could be done, and so liquidity would be limited. In a forward market there is no guarantee that a participant will not have to take delivery. A bookout, i.e. identification of a closed ring of deals, solves the problem, but a bookout is not always possible. Deals that enter the delivery chains can be subject to five o'clocking.[5] As a result, even a participant who has made as many buys as sales can find themselves having to take delivery and also having to rebalance their forward market position. By contrast, the swaps market can deal in small volumes of oil, increasing the number of possible trades, and provides no risk of taking physical delivery.

A further difference arises from the method of closing a position. A swap requires no further attention, the gain or loss on the swap is determined by the absolute price agreed in the deal and the floating price average that will be determined in the specified time period. A trader may want to cover the position by making other swap trades, but the original deal is in effect closed out the moment the deal is made. In a forward market the position will require some further attention, either being closed out by taking delivery or by making an offsetting transaction.

A final major difference concerns the treatment of time spreads, and in particular the spread which is locked in. A swap will be settled as the difference between the market price relating to the expectation, at the time of the deal, of what the monthly average spot price will be, and the actual monthly average spot price. By contrast, a balanced forward market position produces a net settlement of the difference between the market expectations at the times when the deals were made. As a result, attempting to hedge a physical position using forward prices often leaves a basis risk between forward and spot prices. To take an example from the North Sea, the basis risk between prompt Brent cargoes (i.e. dated Brent) and forward Brent has spawned a swap market to cover that risk (the Brent CFD market). Because a swap trade is normally settled against the spot price, no similar basis risk arises.

We would contend that the Singapore swaps market represents a very particular form of swap, and that the market can be termed a quasi-forward market. In its operations, we would define a quasi-forward market as being one that completely mimics the operation of a forward market, except for the physical delivery provisions and the necessity of delivery chains. Swaps outside quasi-forward markets tend to be highly diverse, and the market can usually be split into swaps providers (usually finance houses or major oil companies) and swaps buyers (usually those with a risk to hedge). Completely identical deals are rare, and the swaps are primarily used as hedging rather than speculative vehicles. As such, there are four main features that distinguish a quasi-forward market from other swaps markets. The first is a very high degree of standardization in the swaps. The second is a very narrow spread between bid and offer prices, indicative of a transparent market with a high degree of competition. The third, closely related to the second, is a high level of liquidity per trading day. Finally, in such a market, a significant number of participants would be expected to trade frequently on both sides of the market on a regular basis.

Unlike the longer-term swaps, a trader is likely to hold a portfolio of short-term swaps at any one time, and through further swaps trading change the characteristics of the overall portfolio. If the swaps are being used to take positions on the price level, then a diversity of different grades of the same basic product, or different pricing locations, or methods of floating price determination, make the portfolio harder to manage. Hence, while each deal is a bilateral contract that can be tailor made, the use of short-term swaps, particularly as an actively traded speculative instrument, tends to lead to some form of standardization. This motive for standardization is also present in forward markets, but the standardization is primarily forced by the need to have a smoothly operating clearance mechanism in the delivery chains. Trading a volume of swaps in Singapore that is not a multiple of 50 thousand barrels is possible, but it is harder to accomplish than a standardized trade, and the bid-offer spread tends to be unfavourable in comparison. Hence there is also a strong tendency towards standardization in volumes.

The bid-offer spread is normally very low in the trading of the standardized swaps. There is a significant level of trade on a daily basis, and there is no significant difference overall in the companies that buy and those that sell swaps. It is possible to use a portfolio of swaps for speculation rather than pure hedging. For a participant, the market then performs exactly as a forward market in which they could book out all their deals and face no risk of taking unwanted physical delivery. In total, the market for short-term swaps in Singapore fulfils the conditions for a quasi-forward market.

3.   Oil Product Swaps

As noted in the previous section, while there is no necessary reason for any standardization in swaps, using a portfolio of swaps as a position taking instrument tends to lead to standardization for ease of trading. The Singapore swaps market shows a high degree of homogeneity, consistent with its use as a substitute for forward market trading or, as termed above, a quasi-forward market. Both crude oil and oil product swaps are traded, and we concentrate on oil products in this section. Crude oil swaps are centred on Malaysian Tapis, and this market is considered in the next section. In oil products, swaps are made for gasoil, fuel oil, jet fuel, naphtha and gasoline. While swaps for LSWR have been made, they remain extremely rare given the narrow participant base and specialized nature of the physical LSWR trade.

Oil product swaps take three main forms in Singapore. First, there is the swap of a monthly average price (usually the average of Platt's assessments for the defined grade), normally for near months, i.e. within four or five months of the time of the deal. The mechanics are exactly the same as for the swap used as an example in the previous section, with there being one cash settlement of the difference between the fixed swap price and the realized monthly average. From the analysis of deals reported to Petroleum Argus, we conclude that in over 90 per cent of cases the swaps are for a volume of 50 thousand barrels for all products except fuel oil, where the standard size is 5 thousand tonnes (approximately 33 thousand barrels). In virtually all other cases, the volume is an exact multiple of the standard volume, or occasionally half that volume. Deals for non-standard volumes are extremely rare. For example, if a company wished to hedge, say, exactly 38 thousand barrels, then brokers will tend to have extreme difficulty in finding a counterparty except at a relatively large bid-offer spread. This follows directly from the features of a quasi-forward market. If a company is using a portfolio of swaps, having one for a non-standard volume can be awkward in terms of the management of that portfolio.

The second main form of swap is for a quarterly rather than a monthly average. While the monthly average trades tend to be confined to near months, it is these quarterly swaps that provide the depth to the market. With the swap being cash settled against each monthly average, a quarterly deal is then exactly the same as making three monthly swaps all at the same price, making those swaps completely fungible within a portfolio with monthly average trades. They are then subject to the same tendency towards standardization, and the usual volume is either three or one and a half times the standard monthly average volume. While the next full quarter is the most often traded, deals can usually be done four or five quarters out. For the liquid swaps markets, the forward curve of prices is then usually very well defined, and observable to market participants mainly through contact with brokers.

The third form of swap can be termed as a 'roll over swap'. Consider a trader who has a swap to buy 50 thousand barrels of gasoil against the, say, October average. With October approaching, the trader wants to maintain their position in the market, but to hold it against the November average. The equivalent operation in a forward market would involve selling an October cargo to balance the October position, and buying a November cargo. Doing these deals with the same counterparty means selling the October to November spread, and this is the exact analogy of a roll over swap. It is then the swap of the difference between two monthly averages, and has the effect of rolling a position into another month.

We document in the next chapter the failure to date of oil product contracts on the Singapore futures market, and note the reasons for traders preferring to use swaps. However, at this point we can show one factor that is *not* a reason for the lack of futures liquidity. It has often been claimed that trading went into swaps rather than futures because of the inherent flexibility of swaps. This is a spurious argument. It is certainly true that for a one-off deal, particularly as a risk management tool for the end-user, swaps are highly flexible. They can indeed be transacted for almost any volume or product quality. However, to use swaps as a trading instrument in a quasi-forward market, that flexibility is deliberately foregone, and hence the very high degree of standardization we have noted.

The difference is between having a single swap for a particular purpose, and holding a position made up of a portfolio of swaps. For the former flexibility is an advantage, for the latter it is a disadvantage. Indeed, in some respects the futures market is *more* flexible than the Singapore swaps market. For example, consider the company used above as an example who would have difficulty in trading a swap for 38 thousand barrels. They can trade futures for that amount (subject to the market being liquid), as futures contracts are for smaller volumes (i.e. 1 thousand barrels) than the standard swap volume. Flexibility is then not the key to the predominance of swaps. Indeed, the high volume of the Singapore swaps market has arisen because of the suppression of flexibility.

Where the flexibility does matter is in an aspect that is common to all informal markets and not just to swaps. The development of a formal market requires design and regulatory approval, whereas an informal market can spring up in a short period of time, with many of its structural features evolving with use. Hence, the flexibility argument can only be used to explain why informal markets arise before formal. The feature of the Singapore market is that the formal markets have not yet followed, and indeed in the case of fuel oil, the informal market arose after the formal market.

The role of brokers is key to the swap market. The brokers effectively take on the information functions of a centralized exchange, and are the major source of transparency to market participants. Beyond a role equating to that of an exchange in price discovery, brokers also have a role in providing opinion and advice, and in actively matching trades. There is some specialization between brokers on the products covered, but generally the main brokers for Singapore product swaps are Star Supply, Shogun, TFS, Fearnoil and Libra. The market has enough niches for a process of comparative advantage to work. To give but one example, Intercapital Commodity Swaps is a major broker and promoter of swaps in Europe. Its Singapore operation thus has a comparative advantage, and has actively developed the niche, for swaps that seek to work an arbitrage, for instance, between the Singapore swaps and European gasoil markets, or for trades further down the time curve.

We first attempt to derive a series for the volume of trade in swaps in Singapore. As with any informal market there are, by definition, no centrally collated records of all trading activity. There are records for trades reported to price assessment agencies, such as Petroleum Argus whose database we use extensively in this section. However, the problem is of course that not all trades are reported, and hence the critical measure is the proportion of deals that price reporters hear of.

To answer this we have used three methods. The first was to ask a sample of market participants for their own estimates of a range for how many trades were made per trading day in swaps for each oil product, at the time of interview, and these were compared with the volumes reported to Petroleum Argus for the relevant month. Secondly, we asked other participants for an estimate of their own total swaps activity. This number could then be scaled up according to the proportion of reported deals that involved that company. The third approach was to ask a different sample of market participants and price reporters what proportion of swap deals they believed were reported. On the basis of all three exercises, we believe that a fair measure for the coverage rate for recorded deals is about 25 per cent. In Table 8.2 we have taken the reported volumes, and scaled them up according to this coverage rate (rounding to the nearest 5 thousand b/d). This of course involves the assumption that reporting rates have remained constant, but we have no evidence from either market participants or from price reporters that there has been a change on a yearly average basis. Table 8.2 shows yearly averages per trading day, but it should be noted that there is considerable volatility in the volume of trade, depending on market conditions. On a monthly basis, the market has sometimes traded above 4 mb/d. The volumes shown should be treated with some caution, but the trends within the table are clear enough to produce a series of propositions about the development of the market.

Table 8.2:   Estimated Swap Volumes Traded. 1991–5. Thousand b/d.

	Total	Gasoil	Fuel Oil	Naphtha	Jet	Gasoline
1991	320	125	30	85	80	-
1992	630	195	95	265	75	-
1993	1765	850	215	400	270	30
1994	1855	975	315	270	145	150
1995	1280	570	415	225	35	35

Source: Own calculations from Petroleum Argus Product Deals database.

Until 1994, the market was in expansion. A few gasoil trades were made in 1989, and then liquidity began to build in 1990 and 1991. Growth was undoubtedly spurred by the AOT scheme, for reasons described in Chapter 7, and particularly by an expansion of the scheme in 1990 that brought in a series of major Wall Street firms and other traders. The invasion of Kuwait put a severe brake on the market, for the same reasons we described in Chapter 7 in the context of the open specification naphtha market. However, the liberation of Kuwait heralded a period of high, but not extreme, levels of volatility that were highly conducive to trading, and removed the time zone disadvantage that had held back Singapore trading during the crisis.

Prices in Singapore had been left relatively higher, compared to other areas, than before the crisis, primarily due to the removal of the Kuwaiti refineries from the market. As a result, arbitrage was encouraged between Singapore swaps and European forward and futures markets. Liquidity also tends to have its own momentum, i.e. a 'virtuous circle' whereby an increase in liquidity encourages more trading, and to a large extent the growth of the market fuelled itself after the initial impetus.

It should also be noted that the AOT scheme encouraged a considerable number of companies either to enter the Singapore market, or, probably more significantly, for the already active companies to expand their activities. The number of oil traders employed in Singapore rose sharply over a fairly condensed time period. One could argue that physical trading had already reached a limit on the number of traders it could viably support. With an effective excess supply of traders, there were only the paper markets in which to channel a company's expansion of trading. Without pushing the argument too far, we believe that in the cases of a significant number of companies the causality ran from a perception that Asia was the major growth area, to a desire to increase their trading presence in the region, and then to an increase in their paper market activity. In other companies the causality ran the other way, i.e. the growth of the paper markets led to a desire to expand their trading activity. However, while the reverse also held, to a certain extent the influx of traders brought liquidity to the market.

The market reached a peak in terms of yearly liquidity in 1994. There has now been some retrenchment in volumes, with the growth of the market stalled. As Table 8.2 shows, liquidity in the naphtha and jet fuel markets declined in 1994, and continued to fall in 1995. In the case of jet fuel this has been almost to the point of the complete demise of the market. Naphtha swaps have been used to hedge or take a position on gasoline, and some of the decline in naphtha in 1994 was due to the sharp growth in gasoline swaps. However, the collapse of the gasoline market in 1995 has not stemmed the fall in naphtha liquidity. We consider the naphtha market, and its relationship with the open specification forward market, later in this section.

The gasoil market has also fallen back. While Table 8.2 shows that it is still the leading market in terms of the total number of barrels involved in swaps, in 1995 it was no longer the largest in terms of deals done. As the standard fuel oil swap is for a smaller volume than gasoil (5 thousand tonnes, roughly 33 thousand as compared to 50 thousand barrels for gasoil), the most liquid market in terms of deals done in 1995 was fuel oil. This is the one product that has shown constant growth, with an increase that represents the reverse of the constant decline over the 1990s in the liquidity of fuel oil futures. With the last remaining volumes moving from fuel oil futures to fuel oil swaps over the course of 1995, fuel oil liquidity has been able to move against the more general decline.

Just as liquidity can create more liquidity, illiquidity can create a vicious spiral downwards. The further liquidity falls, the more the bid-offer spread in the market widens, and the harder it becomes to make trades. This process has certainly affected the jet fuel and gasoline swaps markets. As noted above, naphtha can be used as a proxy for gasoline positions, and gasoil can be used for jet fuel positions. This suggests that a demise of the jet fuel and gasoline swaps markets would make a rebirth of these markets more difficult. The low levels of liquidity in 1995 are enough to mean that neither market now meets the condition for a quasi forward market. In other words, while there will still be swaps, the trading of those swaps no longer has enough liquidity to be considered a market in aggregation.

Effectively, gasoline and jet fuel swaps have reverted to being individual swap deals, rather than the more specific form of quasi forward market swaps where it is the portfolio of swaps rather than the individual deal that is central. For jet fuel, longer-term tailor-made swap arrangements are common. Risk management in the Asian airline industry has increased dramatically over the course of the 1990s. Compared to the European industry, acceptance of the use of jet fuel swaps was slower, but now most regional airlines are heavily involved. To give two examples,[6] Singapore Airlines tends to hedge at least 30 per cent of its needs for over a year in advance, and when prices appear attractive the proportion may increase to 60 per cent. Qantas hedges half its needs for up to eighteen months. Other airlines cover a smaller proportion of their requirements, but the demand for swaps is steadily increasing.

There are currently only three standardized swaps markets of importance, and we now consider the structure of participation in each. The major participants in the gasoil swaps market are shown in Table 8.3, together with their market shares.[7] The table also shows the number of firms that made at least one recorded deal in the year, and the inverse Herfindahl index measure of concentration. There has been some fluidity in the list of major participants. Of the ten major traders in 1992, only four remained in the first ten in 1996. The number of firms in the market, and the relatively high inverse Herfindahl index (i.e. a low level of market concentration), are both indicative of an active and competitive market. By way of comparison, we estimate that the forward Brent market (the most active informal oil market in the world in terms of barrels traded), has since 1993 had both a higher level of concentration and a lower number of participants than Singapore gasoil swaps. The comparison is not totally fair (since there are further companies who trade Brent on the formal but not the informal market), but it does emphasize the size of the participant base in Singapore swaps. However, after the peak of 1994, the fall in volumes in gasoil swaps has been associated with the withdrawal of 20 per cent of the companies who had been active in the market.

Juxtaposing the firms shown in Table 8.3 with those most active in spot gasoil, as shown in Table 7.3, there is a broad correspondence between participants, albeit a wide variance between relative involvement in physical and swap trades. The most notable absentees in the swaps market from the physical participants are the Middle East producers, Korean companies and national oil companies of importing nations.

Compared to the relative decline of gasoil swaps after 1994, Table 8.4 shows that the participant base in fuel oil swaps has continued to expand in 1995, before falling back in 1996. As noted above, the period of fuel oil swaps growth coincided with the decline to extinction of the futures market in fuel oil. We consider this in more detail in the next chapter, but

Table 8.3: Main Participants in Singapore Gasoil Swaps Market. Rank and Market Share. Per Cent. 1992–6.

Rank	1992		1993		1994		1995		1996	
1	J Aron	19.3	J Aron	13.1	J Aron	10.3	Elf	12.6	Elf	12.3
2	Morgan Stanley	12.0	Louis Dreyfus	10.8	Morgan Stanley	8.6	Sinochem	11.1	J Aron	8.1
3	BP	8.4	Cargill	10.5	Cargill	7.9	Vitol	7.8	BP	6.9
4	CFP Total	7.2	Mobil	5.9	Sinochem	7.3	Hin Leong	6.5	Sinochem	6.4
5	Astra	6.0	Morgan Stanley	5.0	Elf	6.4	Mobil	5.3	Shell	5.7
6	Apex Intl	5.4	BP	4.4	BP	6.1	Cargill	5.0	IPG	4.9
7	Phibro	4.8	Elf	4.3	Louis Dreyfus	5.2	J Aron	5.0	Vitol	4.7
8	Louis Dreyfus	4.2	Shell	4.3	Mobil	5.0	Morgan Stanley	4.8	Hin Leong	4.2
9	Cargill	3.6	Vitol	4.1	Hin Leong	4.7	Louis Dreyfus	4.4	Morgan Stanley	4.2
10	Coastal	3.0	BB Energy	4.0	Vitol	4.5	BP	4.2	Phibro	3.7
11	BB Energy	2.4	Sumitomo	2.9	Cred Lyonnais	3.2	Phibro	4.0	Cargill	2.9
12	Mobil	2.4	Mabanaft	2.4	Mabanaft	3.0	Shell	4.0	Banker's Trust	2.7
13	Vitol	2.4	Total	2.4	BB Energy	3.0	Mabanaft	2.7	Mabanaft	2.7
14	Daxin	1.8	Marc Rich	2.0	Neste	2.7	Neste	2.3	Mobil	2.7
15	Caltex	1.2	Neste	2.0	Shell	2.7	Itochu	1.7	Hyundai	2.2
16	Chevron	1.2	Tosco	1.8	Galaxy	2.0	IPG	1.5	Idemitsu	2.2
17	Elf	1.2	Marubeni	1.8	Caltex	1.7	Caltex	1.5	Itochu	2.2
18	IPG	1.2	Astra	1.6	Itochu	1.7	Glencore	1.5	Conoco	2.0
19	Mabanaft	1.2	Phibro	1.3	Astra	1.5	BB Energy	1.3	IOT	1.7
20	Mitsubishi	1.2	Coastal	1.2	Statoil	1.2	Marubeni	1.1	Galaxy	1.5
Number of Firms	33		55		60		52		48	
Concentration	12.4		16.2		18.4		17.5		20.1	

Source: Own calculations from Petroleum Argus Product Deals database.

Table 8.4: Main Participants in Singapore Fuel Oil Swaps Market. Rank and Market Share. Per Cent. 1992–6.

Rank	1992		1993		1994		1995		1996	
1	Cargill	15.8	Cargill	13.1	BP	10.3	BP	12.6	BP	12.3
2	BP	12.9	Louis Dreyfus	10.8	Louis Dreyfus	8.6	Elf	11.1	Vitol	8.1
3	Phibro	7.2	J Aron	10.5	Cargill	7.9	Vitol	7.8	Mobil	6.9
4	Mobil	5.8	BP	5.9	J Aron	7.3	Morgan Stanley	6.5	Shell	6.4
5	J Aron	5.0	Mabanaft	5.0	Vitol	6.4	J Aron	5.3	Hin Leong	5.7
6	Shell	5.0	Banker's Trust	5.0	Morgan Stanley	6.1	Mobil	5.0	Elf	4.9
7	Vitol	5.0	Mobil	5.0	Mobil	5.2	Phibro	5.0	Sinochem	4.7
8	Chevron	4.3	Morgan Stanley	4.3	Phibro	5.0	Cargill	4.8	IPG	4.2
9	BB Energy	3.6	BB Energy	4.1	Sinochem	4.7	Hin Leong	4.4	Morgan Stanley	4.2
10	Kuo Oil	2.9	Vitol	4.0	Elf	4.5	Shell	4.2	BB Energy	3.7
11	SPC	2.9	Caltex	2.9	Shell	3.2	Sinochem	4.0	Caltex	2.9
12	Sinochem	2.9	Itochu	2.4	Hin Leong	3.0	Kuo Oil	4.0	Stinnes	2.7
13	Astra	2.2	Shell	2.2	Astra	3.0	Texaco	2.7	Idemitsu	2.7
14	Itochu	2.2	Marc Rich	2.2	Banker's Trust	2.7	Banker's Trust	2.3	Neste	2.7
15	Mabanaft	2.2	Astra	2.2	Daxin	2.7	Glencore	1.7	Mabanaft	2.2
Number of Firms	33		42		51		62		46	
Concentration	12.4		10.0		14.9		18.2		18.3	

Source: Own Calculations from Petroleum Argus Product Deals database.

at this point we note that daily trade volume in futures in 1993 was the equivalent of about 800 thousand b/d. The reduction in this volume (to an average of about 475 thousand b/d over 1994, and to zero by the end of 1995) is large enough compared to the increase in swaps volume shown in Table 8.2 to suggest that the decline of futures was a major factor behind the rise in swaps.

For an alternative use of the two markets, consider the following trading play. A trader buys a position in Singapore swaps, but sells forward cargoes in the open specification market. They now buy aggressively in the physical Singapore naphtha market so as to push up the Singapore price, and use the cargoes they source in Singapore to fulfill their obligations in the forward market. Their cost is to have bought physical naphtha at a high price, added to the cost of the freight into Japan or Korea, but they profit on their Singapore swaps position. If that position is large enough, then overall the joint use of Singapore physical, Singapore swaps and the open specification market will be profitable.

The above example rests crucially on whether aggressive spot buying in Singapore will force prices up. With the level of spot naphtha trade in Singapore being slight, prices can normally only be assessed on the basis of the netback from Japan. If Singapore prices were only ever assessed as a netback, then the example would not lead to a dislocation between the two markets and thus would not be a profitable strategy. In fact the example is representative of the trading pattern of one company during February and March 1995. Swaps were done against a price whose assessment was pushed up on the basis of the brief flourishing of the Singapore spot trade. The differential between Japan and Singapore was assessed by the major price assessment agency about $1 per barrel lower as a result, resulting in the discontinuity shown in Figure 8.1.[8]

The naphtha swaps market coexists with the open specification forward market described in the last chapter. The swaps have no physical delivery, a small (usually 50 thousand barrels) lot size, and are settled on a monthly average fob Singapore price. The physical delivery forward market trades 225 thousand barrel lots, with prices that represent the c+f Japan value over a half month. This lack of fungibility between the two markets creates opportunities in trading them both. For example, a long (i.e. net buy) position in one and a short in the other is a way to trade or hedge the freight cost between Singapore and Japan.

The major participants in naphtha swaps are shown in Table 8.5. The inverse Herfindahl index of concentration shows this market to be markedly more concentrated, and also to have fewer participants, than the gasoil and fuel oil markets considered above. In 1995 in particular, one company, J Aron, had a dominant market share of 20.8 per cent (i.e. they were involved in 41.6 per cent of all naphtha swap trades). Among

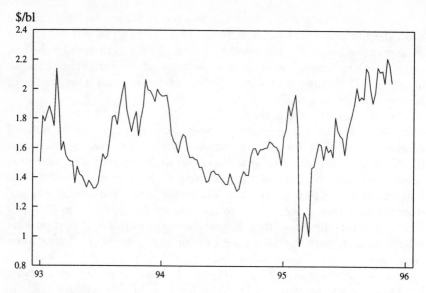

$/bl

Figure 8.1:   Price Differential Between c+f Japan and fob Singapore Naphtha. 1993–5.
Dollars Per Barrel.

the other companies, note the representation of Japanese companies that
was not evident in the other swaps markets. Comparing Table 8.5 with
the composition of forward naphtha market participants shown in Table
7.13, reveals a very strong correspondence. The major participants in
both are broadly the same. The non-fungibility between the markets we
noted above is such that to a large extent they can act as complementary
markets. However, note the large fall in the number of participants after
1994, primarily due to the trading manipulation in the first quarter of
1995 detailed above.

There have occasionally been trades of LSWR swaps. However,
liquidity has always remained low in what is a specialist niche market with
a very limited number of potential physical market participants. This is
compounded by the nature of that physical market. LSWR primarily goes
into Japan, where, as we have seen, to date the use of pure risk management instruments has been rare. At other times, LSWR goes into the
USA as a refinery feedstock, and could usually be fairly effectively hedged
using WTI, rather than using an instrument with such a large bid-offer
spread as LSWR swaps. It could then be argued that the demand for
LSWR swaps has been strictly limited.

A major feature of the existing swaps market is the relative lack of
involvement to date by the Asian industry outside of Singapore. To
demonstrate this, we have categorized the sources of liquidity in the three

Table 8.5: Main Participants in Singapore Naphtha Swaps Market. Rank and Market Share. Per Cent. 1992–6.

Rank	1992		1993		1994		1995		1996	
1	Vitol	21.8	Vitol	16.7	J Aron	13.5	J Aron	20.8	J Aron	16.2
2	BP	19.1	Shell	13.3	BP	11.3	Shell	7.5	Vitol	13.1
3	J Aron	8.0	J Aron	12.6	Vitol	10	Vitol	6.5	Marubeni	8.4
4	Cargill	6.9	BB Energy	7.6	Itochu	9.6	Itochu	6.3	Idemitsu	7.3
5	Apex Intl	5.4	BP	7.5	Cargill	7	Elf	6.1	Elf	6.8
6	Mobil	4.3	Cargill	6.7	Shell	6.7	Idemitsu	6.1	Sumitomo	6.3
7	Shell	3.8	Itochu	4.1	Samsung	4.4	BP	4.8	Itochu	5.8
8	Daxin	2.9	Morgan Stanley	3.8	BB Energy	3.5	Mitsui	4.6	Cargill	4.7
9	Morgan Stanley	2.8	Sumitomo	3.7	Mobil	3.2	Marubeni	4.3	Shell	4.2
10	Sumitomo	2.8	Nicor	2.2	Marubeni	2.6	Cargill	4.1	Total	4.2
11	Caltex	2.4	Caltex	2.1	Caltex	2.5	Glencore	3.4	Interlink	3.1
12	Itochu	2.1	Idemitsu	2.1	Sumitomo	2.5	Mobil	2.8	Samsung	3.1
13	Nissho Iwai	1.6	Toyo Menka	2	Unocal	2	Samsung	2.8	Toyo Menka	3.1
14	CFP Total	1.4	Mobil	1.9	Elf	1.9	Mitsubishi	2.6	BP	2.1
15	Vanol	1.4	Nissho Iwai	1.8	Subic Bay	1.8	Total	2.3	Glencore	2.1
Number of Firms	34		44		42		35		21	
Concentration	9.4		11.7		14.6		12.7		12.7	

Source: Own calculations from Petroleum Argus Product Deals database.

main swaps markets as is shown in Table 8.6. We have used nine categories. Refiner covers the companies with Singapore refining assets. The rest of the Asian industry has been divided into Chinese and Japanese companies, Asian trading companies (predominantly located in Singapore), with a residual other category comprising Korean firms and Asian national oil companies. Western firms are divided into Wall Streeters (i.e. financial entities and their offshoots), western oil trading firms, and US refiners and European oil companies not falling into any other category.

Table 8.6:   Market Shares in Swap Markets by Category of Company. Per Cent. 1992–5.

	1992	1993	1994	1995
(a) Fuel Oil Swaps				
Asian Trader	3.5	1.1	5.8	5.4
Chinese	2.8	1.1	3.2	3.4
European	0.7	2.7	5.3	13.1
Japanese	13.3	6.4	3.5	8.6
Other	1.4	0.8	1.2	0.9
Refiner	27.3	14.4	21.0	23.8
US Refiner	7.0	1.9	2.2	3.6
Wall Streeter	12.6	34.5	29.5	21.4
Western Trader	31.5	37.2	28.5	19.8
(b) Gasoil Swaps				
Asian Trader	2.9	1.6	6.1	7.7
Chinese	8.2	0.9	7.2	11.3
European	5.8	9.3	11.1	16.1
Japanese	0.6	8.5	4.4	5.6
Other	0.0	1.3	0.4	0.6
Refiner	13.5	15.9	15.7	18.4
US Refiner	5.3	4.0	2.7	1.5
Wall Streeter	39.8	29.0	27.6	20.3
Western Trader	24.0	29.6	24.8	18.4
(c) Naphtha Swaps				
Asian Trader	2.9	2.2	5.3	5.3
Chinese	0.7	0.3	0.6	0.3
European	3.3	1.5	2.8	6.7
Japanese	11.4	16.8	21.7	27.8
Other	1.7	1.3	4.7	5.9
Refiner	29.6	24.9	23.7	17.4
US Refiner	1.0	1.2	3.3	0.8
Wall Streeter	11.9	17.1	13.5	21.3
Western Trader	37.4	34.6	24.2	14.3

Source: Own calculations from Petroleum Argus Product Deals database.

In the gasoil and fuel oil markets there has been a sharp increase in the trading of European companies, emphasizing the importance of arbitrage from Europe, and the move by these companies (primarily Elf and Statoil) towards global trading. Trading in these two markets is primarily made up of the Singapore refiners, Wall Streeters, western traders and the Europeans. In 1995 these categories accounted between them for 78 per cent of fuel oil swaps and 73 per cent of gasoil swaps. The involvement of Asian companies in these markets is generally limited, most markedly in the case of Japanese and Korean companies.

The potential for greater participation by Asian companies was shown in the naphtha swaps market, with the increasing share of Japanese companies between 1992 and 1995. Naphtha is certainly a special case, given the size of physical flows into Japan and the coexistence with the open specification forward market. It is also an unfortunate case, given the highly deleterious impact of trading manipulation on the robustness of that market. However, it is still perhaps significant that, in the one product in which to date there has been a strong demand for risk management in Japan, the involvement of Japanese companies was significant. We would then conclude that there is considerable scope for greater use of the swaps market (either direct or indirect) by both Asian companies and western companies involved with the trade into and within liberalizing oil markets.

4.   The Tapis Swaps Market

There are no forward markets for any Asian crude oil. The only market with any depth of trading is the informal Tapis swaps market. The reason why no forward market for Tapis, i.e. one that trades deliverable forward cargoes, has arisen is primarily a function of the structure of the physical market. Tapis is a blend from a series of fields off the east coast of Malaysia, pipelined ashore to the terminal at Kerteh. The sole equity holder for the fields is Exxon, who operate under a production-sharing agreement with the state company Petronas. Current output is about 350 thousand b/d, with around 200 thousand b/d accruing to Petronas.

The amount available for trading is extremely limited. Petronas runs more than half of its entitlement through its own refineries, and sells most of the rest through term contracts. Exxon similarly uses most of its share as an internal transfer to its affiliates. Virtually all the term contracts go to regional refiners (in particular from Japan, Korea, India and Thailand) who have tended only very rarely to trade cargoes further. We consider a set of desirable characteristics for a marker crude oil in Chapter 12. However, in advance we can note that Tapis is ruled out as a suitable marker (and also as the basis for a deliverable forward market) *inter alia* by the combination of two companies acting as both exclusive first sellers and also as the dominant users, together with the highly limited number of cargoes available for trading.

Conditions have never been conducive to the development of a liquid forward Tapis market. However, it should be noted the extremely low level of spot availabilities is a relatively recent phenomenon. It is primarily associated with the start-up of the Melaka refinery (a Conoco and Petronas joint venture) at the start of 1994, which has on average absorbed around 80 thousand b/d of Tapis. Previously, a proportion of this volume found its way onto the spot market, primarily through Petronas term contracts with traders (notably Phibro, Astra and Glencore).[9]

While the creation of any active forward Tapis market is impossible, a Tapis swaps market started to grow in the 1990s. Originally this was composed entirely of single Tapis price swaps, that did not fulfill the conditions for what we termed above as a quasi forward market.[10] Over time, the conditions were met, i.e. liquidity in volumes arose in standardized short-term swaps, with participants trading on the basis of a portfolio of swaps and being prepared to take both sides of the market.

The standardization that emerged was for swaps of multiples of 100 thousand barrels, cash settled against the APPI (Asian Petroleum Price Index) assessments for Tapis averaged over a month. As with oil product swaps, the quasi forward market suppresses the natural flexibility of swaps in order to gain the advantages of standardization. Quarterly swaps are also traded, which *de facto* break into three separate monthly swaps following the mechanism described in the previous section.

The details of APPI assessment are covered in Appendix 1. Some features of APPI methodology impact on the question of what the value of, say, a March Tapis swap actually represents. In particular, it does not represent the expectation of the average value of Tapis in March. As detailed in Appendix 1, APPI quotes (produced weekly on Thursdays) for Asian and Australasian crude oils, during the first half of a month represent oil loading in that month. During the second half, they represent next month delivery. Hence, if March has four Thursdays, the swap is settled against the average of two quotes for March, and two for April. If March happens to have five Thursdays, the swap is settled against the average of three quotes for one month and two for the other. In terms of value, a March Tapis swap is then in effect a proxy for March/April Tapis, and indeed, depending on when the Thursdays fall, sometimes it can be more April than March.

The level of reported trade in Tapis swaps is shown in Table 8.7. Few clear trends emerge. The level of liquidity is highly variable, ranging from more than one reported deal per trading day down to less than one per week. Even allowing for reporting rates, we would put the average level of liquidity in 1995 at no more than 250 thousand barrels per trading day, i.e. very limited compared to the liquidity of oil product swaps.

Liquidity is provided by a limited number of market makers in what is a highly concentrated market. Table 8.8 shows the market shares of the ten major traders of Tapis swaps by year. J Aron, BP, Morgan Stanley, Phibro and Vitol have been the dominant players, with the recent addition to the set of Koch. As shown in Table 8.8, in 1995 there was little significant participation beyond this core group of six companies.

Table 8.7: Composition of Liquidity in Tapis Swaps Market by Quarter. 1993–5. Reported Deals.

Year	Quarter	Total Deals	Outright Deals	Spread Deals	Brent Spreads
1993	1	17	17	0	0
	2	26	22	4	4
	3	47	41	6	5
	4	81	74	7	7
1994	1	26	23	3	1
	2	66	58	8	5
	3	19	13	6	3
	4	41	36	5	4
1995	1	36	23	13	11
	2	45	37	8	5
	3	22	22	0	0
	4	10	9	1	1

Source: Own calculations from Petroleum Argus database.

Table 8.8: Participants in Tapis Swap Market and Market Shares. 1993–5. Per Cent.

	1993		1994		1995	
1	J Aron	18.4	BP	21.7	J Aron	18.3
2	Morgan Stanley	14.4	J Aron	20.2	Vitol	17.0
3	Louis Dreyfus	14.1	Morgan Stanley	15.7	BP	15.7
4	BP	9.2	Vitol	8.7	Koch	11.4
5	Mobil	8.9	Phibro	7.5	Phibro	10.5
6	Vitol	7.1	JP Morgan	3.9	Morgan Stanley	7.9
7	Shell	5.2	Bankers Trust	2.7	Glencore	3.9
8	Phibro	4.6	Marubeni	2.7	UBS	3.5
9	Caltex	3.7	UBS	2.7	Mobil	3.1
10	Marc Rich	2.1	Marc Rich	2.1	Elf	2.2

Source: Own calculations from Petroleum Argus database.

5.   The Growth of the Swaps Market

We have seen that oil product swaps in Singapore have become the central part of Singapore oil trading. The two main products, fuel oil and gasoil have grown into what we have termed quasi-forward markets, and all exhibit a high level of liquidity and standardization of trades. When the minor products are added, we have in Table 8.2 estimated the average volume of trade to have been about 1.8 mb/d in 1993, 1.9 mb/d in 1994 and 1.3 mb/d in 1995. There are also a number of tailor-made individual swaps and swap based risk management packages transacted, (which tend never to be reported to price assessment agencies due to commercial confidentiality), sometimes for large volumes and long time periods. The total volume of swaps in both quasi-forward and tailor-made deals is therefore likely to be significantly higher than Table 8.2 suggests.

The main question that Table 8.2 raises is why the growth of the swaps market is stalling, and indeed going into a decline. Part of the explanation lies in the nature of the participants, as well as the non-participants, as we will demonstrate below when considering the categorization of the companies active in each market. The other reasons are all in some way connected with the bottom line profitability of trading.

Market conditions provide one argument for a temporary decline. In the latter part of 1994 and throughout 1995, the level of market volatility was low, and thus the potential gains from and the motives for trading have both been adversely impacted. Further, the relative weakness of the Far East product market has closed physical arbitrage possibilities, and thus reduced the use in trading of playing Singapore swaps positions against European market positions. Other profitability considerations may be longer lasting. First, the cost of trading operations has been rising, and for foreign based companies this has been compounded by the continuing appreciation of the Singapore dollar. The result has been some retrenchment in Singapore, reinforced by restructuring of trading operations at a more global level by some companies.

Arguments that suggest something is a victim of its own success are often very unconvincing. However, we believe that there are some elements of this in operation in the Singapore swaps market. The market at its peak had become extremely efficient, and extremely transparent. A completely efficient (in economic terms) and completely transparent market is in fact not very conducive to trading activity. For example, a completely efficient market will change prices in response to new information faster than a company can trade off any inefficiency. This is reflected in the observation made by several companies in interviews, that the trading opportunities that they utilized during the early growth of the market, are now of a lower magnitude and are also traded away far faster. This is not to argue that the market has declined because it expanded, but rather that it may have expanded too fast to sustain a prolonged period of growth. This view would then imply that the decline is merely an adjustment back to a more sustainable level from which a slower rate of growth can be maintained.

Volumes may have expanded faster than the demand for risk management and trading instruments within Asia warranted, and given the paucity of this base, reached a maximum level of sustainable liquidity. Having reached that maximum level, liquidity then becomes a function of market conditions such as volatility, arbitrage possibilities, and the global profitability of trading activity. For the sustainable maximum to rise would then require an increase in the use of risk management in the Asian oil industry. This would generate either an influx of new participants, or market activity that would require ancillary operations on the swaps market.

The purpose of Part II of this study was to demonstrate that the elements for this structural change are already in place. Throughout Asia we found an increasing exposure to the price signals generated in Singapore. Risk exposure has been increased by this process, and the conditions necessary for a more proactive response to risk are being met, albeit on very different timetables across countries. This does provide an underlying motive force for further development of the risk management markets.

Notes

1. In some cases derivatives are based on instruments that are themselves derivatives, for example an option to buy or sell futures. Higher level derivatives are possible, for instance in the next chapter we see that North Sea Brent futures are derived from Brent forwards which are derived from spot Brent. Therefore an option to buy or sell Brent futures is a third level derivative.
2. See Intercapital Brokers and Petroleum Intelligence Weekly (1990), *The Complete Guide to Oil Price Swaps*, Petroleum Intelligence Weekly, New York.
3. We have used the 1993 figures for 0.5 per cent sulphur gasoil, average monthly mean of Platt's Singapore (MOPS) as the floating prices in this example.
4. See Xavier Trabia (1992), *Financial Oil Derivatives: From Options to Oil Warrants and Synthetic Oilfields*, Oxford Institute for Energy Studies, Oxford.
5. i.e. when a participant is unable to pass on a cargo nomination before a deadline, and hence must take delivery.
6. These examples are drawn from *Jet Fuel Intelligence*, 17 July 1995.
7. Data for 1996 in this and in following tables in this chapter, reflect the first eight months of the year.
8. That agency changed its methodology to one of netbacks only as a result of the obvious distortion. Other agencies were unaffected as they maintained a netback approach to the assessment of Singapore naphtha.
9. See *Weekly Petroleum Argus*, 15 April 1996.

10. We defined a quasi forward market as a swaps market that mimics the functions of a forward market, but without physical delivery and thus without the need for delivery chains.

CHAPTER 9

OIL FUTURES MARKETS

1. Introduction

Europe has a thriving oil futures market in the International Petroleum Exchange of London (IPE). The USA has the successful energy complex of the New York Mercantile Exchange (NYMEX). In contrast, oil futures have been extremely slow to take root in Asia. In the previous chapter we considered the forms of informal markets in oil derivatives that now constitute the major part of the Singapore market. Compared to other major oil trading centres, Singapore is different because of this reliance on informal mechanisms, rather than the use of formal futures markets which is intrinsic to trading in London and New York. As yet, the formal markets have not had the same impact in Asia as they have had in the West. This raises a series of questions about the nature of the Asian market, of which the most fundamental is whether Singapore is merely evolving slower, or whether the development of the market has taken a different track from that in the West? Fundamental to this is the nature of the relationship between market structure, the behaviour of market participants and the growth of formal markets.

We advance the hypothesis that futures markets can arise in two circumstances. First, centralized exchanges can be a result of a poor information and communication infrastructure, with their major function being market participant identification and mitigation of third party credit risk. We would term these exchanges undeveloped futures markets, and they are likely to be characterized by less than perfect regulation and an initial proliferation of exchanges, followed by potential decline through the growth of integrated marketing and improvements in information flows. They arise as a substitute for other market forms, given market failures that prevent the dominance of more informal trading relationships.

Secondly, exchanges can arise in the event of declines in integration. In conditions of good communications and information flows, their primary purpose switches to a desire for risk management or speculative trading instruments. These are developed futures markets, which will tend to be few in number (since information and communications reduce spatial advantages), and also more tightly regulated. Instead of beginning as a substitute for the absence of other market forms, they tend to grow after informal mechanisms have arisen.

Oil futures markets in the West have taken both forms of development. A profusion of exchanges arose in the early US industry in the nineteenth century, to decline primarily due to the increasing integration of the industry. In the late 1970s and beyond, developed versions have arisen following the growth of spot and informal forward markets. The industry then showed a pattern over time of undeveloped futures markets, vertical integration, informal markets and then developed futures markets.

Both forms of futures markets have been manifested in the Far East. The tradition of developed futures markets is reflected in the continuing effort to form an oil futures complex at the Singapore International Monetary Exchange (SIMEX). The growth of oil futures in China in the 1990s is however closer to the tradition of undeveloped futures markets, followed by a transitionary phase to a developed form. Our proposition is then that SIMEX follows in the development of modern western exchanges, while the Chinese markets, and the reasons for their growth, have a closer parallel in the nineteenth century.

Given that we wish to draw the parallels between early undeveloped futures markets and the Chinese exchanges, in Section 2 we have included a brief discussion of nineteenth-century oil markets. Section 3 details the operation of developed oil futures in the Asia-Pacific region, represented by open outcry trading in Singapore and electronic trading in Australia. Section 4 presents a brief history of the modern Chinese oil exchanges.

Derivatives trading, of which futures markets are just one form, has acquired a high profile following a series of large losses incurred by a variety of organizations. The most spectacular was the collapse of Barings in 1995, but similar and often far larger losses have been incurred by others. For example, within the oil industry, in Japan both Showa Shell and Kashima Oil have sustained large losses in the use of exchange rate derivatives. The fear of derivatives failure has been a factor used as a reason for avoiding their use by a large number of Asian oil companies. In the conclusion (Section 5) we shall argue that the derivatives failures have more to do with management than markets.

2.   The Early Oil Exchanges

The existence or not of markets, and their role and importance, is a function of the structure and regulatory environment of an industry. Their growth is facilitated by a series of factors. The degree of vertical integration is important, since integration results in potential market trades becoming simply internal transfers. Likewise the presence of a dominant buyer or seller at any stage of an industry tends to reduce the possibility of a market, as does the presence of any price setting regulator. The presence of and sensitivity to risk is another factor. If agents are not risk averse or have little exposure to risk, perhaps by being able to achieve full pass through of prices to final users, the market has no hedgers and is unlikely to persist.

These conditions have rarely held in the oil industry, and the modern history of markets of any form, let alone organized futures markets, has been a brief one. Before the modern era there were oil futures markets in the US industry. Following in the wake of Drake's strike at Oil Creek in 1859, a profusion of exchanges grew up. To begin with they acted as a way of reducing transport and time costs and providing a centralization of information flows. According to an account of the setting up of the Titusville exchange in a hotel in the 1860s, 'Buying agents who had formerly ridden like Lone Rangers from well to well, and purchased oil at wildly fluctuating prices, now gravitated to the hotel.'[1] In a market where buyers and sellers had poor information on current market prices, and where travel between sellers was difficult, the exchanges had a useful centralizing function. When the Titusville exchange moved to its own premises in 1870, its importance and perceived permanence is shown by its construction, expensive bricks rather than the cheap, flimsy and flammable wood used in the rest of the town.[2]

High transport and communication costs meant that the industry needed more than one exchange. At their peak there were more than twenty. As information was less than perfect, control of news was an extremely effective trading weapon. 'Prices were manipulated up or down; rumours were manufactured to confuse rivals and make a fast killing on the market.'[3] There was also what can only be described as insider dealing. Companies would fence off drilling as it neared completion and employ armed guards. A gusher could be rigged to make others believe it was dry. Knowing the price was about to fall, the company, through agents, sold heavily on the market, ready to buy back when news of their gusher became known.

Histories of the oil industry tend to concentrate on the drillers, those who made their fortunes, and those who lost it all when they did not strike oil. In reality more fortunes were made and lost on oil futures trading than through the physical oil business. The exchanges attracted gamblers and syndicates who had never seen an oil lease, or in modern futures market parlance, a high proportion of liquidity was provided by 'locals'. While we have categorized them as undeveloped futures markets, this does not imply that they were not complex in their operation, indeed in some aspects they were at least as sophisticated as modern exchanges.[4] The volume of trade was a large multiple of world production. For instance, to take a typical day in 1889 (after the peak of the exchanges when their number had fallen to just four), on 1 November the market volume for the Bradford exchange was 1.0 million barrels (mb), at Oil City it was 0.6 mb, at Pittsburgh 1.2 mb, and at New York 1.6 mb. As world production in 1889 was of the order of 0.2 mb/d, the total daily volume on the four exchanges was some 22 times world production. Modern crude oil futures volumes are normally between two and three times world production. At the peak of the markets in the early 1880s, Weiner reports that the multiple was often several hundred.[5] The early exchanges traded six days a week; unlike modern markets Saturday was a full trading day. Whereas today the dollar is the universal unit of account for transactions, traders in Bradford and the other exchanges traded the delivered price of kerosene into Antwerp, London and Bremen in foreign currency terms.

The exchanges had highly effective arbitrage mechanisms. The telegraphic communication of price changes was used by some traders to enable them to trade off any differential movement in prices, and as a result the price was very swiftly equalized. As today, they reacted to official inventory figures, and to production estimates. Similarly, they traded several delivery months, and some traders played the spreads between delivery months. Regulation was completely internal to the exchanges, and trading tended to be what to modern eyes might be considered a little on the rumbustious side.

The demise of the exchanges was due to the appearance of vertical integration, involving the control of both railroad carriage of oil, and of the refineries, moving the industry closer to downstream monopoly. From the point of view of the producers there had been a move towards monopsony. A monopsonist has the power to set buying prices, and the refinery posted price was born, essentially a 'take it or leave it' price as far as producers were concerned. While some trading could continue for oil moving outside of integrated channels, the market structure of the industry had changed in such a way as to remove the conditions necessary for a market mechanism to operate. The first oil futures markets had become defunct by the mid 1890s, and the next generation of futures markets proved to be very different in form.

3.   Developed Oil Futures Markets in the Asia-Pacific Region

Oil futures arose in the nineteenth century because of lack of integration in the oil industry, and died out because of the growth of integration. For more than eighty years there was no possibility of futures markets reappearing. The industry was a fairly closed oligopoly, with oil moving through integrated channels. At no point was there an interface where many buyers were in the same market as many sellers, all facing unregulated prices without dominant buyers or sellers. Depending on the production stage, there was always some significant asymmetry of power towards one or other side of the market. Prices were primarily set rather than being produced by a market mechanism, and they were often also heavily regulated by governments. With there being no viable high liquidity spot market, there was then no basis for the effective development of a futures market.

Conditions began to change in the 1970s. In the USA and elsewhere, price controls began to be removed. In addition, the wave of OPEC nationalizations led to a reduction in the level of vertical integration in the industry. Independent refining grew in Europe, and new sources of supply were developed, particularly in the North Sea, where large amounts of oil were produced outside integrated channels. Trading became an important function within the industry, and spot and then informal forward markets began to grow. While many attempts have been made to launch oil futures contracts in the USA and in Europe, only two exchanges have succeeded, the New York Mercantile Exchange (NYMEX) and the International Petroleum Exchange of London (IPE).

Oil futures contracts have been launched on the Singapore Monetary Exchange (SIMEX) which grew out of a restructuring of the Gold Exchange of Singapore, in line with government policy. The government wished to build up Singapore as a regional finance centre, and to internationalize operations. The result was SIMEX, opened on 7 September 1984 with an immediately outward looking stance. Its major innovation at the start was a mutual offset system with the Chicago Mercantile Exchange (CME) on the eurodollar futures contract. By mutual offset, clearing members of either exchange can clear positions set up on one exchange by trading on the other, expanding the hours of trading to cover both the US and Pacific time zones. The idea was highly innovative at the time, but the benefits were obvious and helped launch the eurodollar contract as SIMEX's flagship contract.

As of 1997 SIMEX listed thirteen futures contracts and six options contracts, shown in Table 9.1 with their dates of launch.[6] As shown, the contracts are predominantly financial, with a range of currencies covered as well as stock market indices for both the Japanese and Hong Kong market. The legal framework for activity on SIMEX is the 1987 Futures Trading Act, which makes the Monetary Authority of Singapore the regulator. Regulation is consistent across Singapore's financial markets given that the same regulatory authority has control over banking and other securities markets. The regulatory climate is by international standards tough, with tight regulation explicitly aimed at promoting the reputation and future growth of the Singapore financial sector. Regulation is designed also to have a developmental role.[7]

Table 9.1:   SIMEX Contracts as of 1997.

Contract	Futures Launch	Options Launch
Eurodollar	7 Sep 1984	25 Sep 1987
Euromark	20 Sep 1990	-
Euroyen	27 Oct 1989	19 Jun 1990
Nikkei 225 Average	3 Sep 1986	19 Mar 1992
Nikkei 300 Average	3 Feb 1995	3 Feb 1995
MSCI Hong Kong Index	31 Mar 1993	
Japanese Bonds	1 Oct 1993	11 May 1994
Dollar/Yen	1 Nov 1993	-
Dollar/Deutschmark	1 Nov 1993	-
Gold	7 Sep 1984	-
Fuel Oil	22 Feb 1989	
Brent Crude Oil	9 Jun 1995	-
MSCI Taiwan Index	9 Jan 1997	9 Jan 1997

Source: SIMEX.

The regulatory structure of SIMEX came under close international scrutiny in February 1995, with the collapse of Barings Bank. Losses were made that were greater than the bank's total asset value, through trading of SIMEX's Nikkei 225 options contract. While the incident was clearly highly embarrassing to Singapore as a financial sector, and to SIMEX in particular, the problem was one of a lack of internal company controls, rather than any direct deficiencies in what is one of the tighter financial regulatory regimes in the world. All losses made by Barings were covered, and all contracts honoured. We explore the issue of control in derivative trading failures in the context of oil derivatives later in this chapter.

Table 9.2 shows the total growth by sector of the Singaporean economy from 1983 to 1993, and their shares of GDP in 1993. As shown in this table, finance has been the fastest growing sector of the economy, and represents a similar share of GDP to that of the manufacturing industry. SIMEX shared in this fast growth, but its liquidity has been heavily concentrated in the Eurodollar and Nikkei 225 contracts. The exchange has thus always sought to diversify the base of its liquidity, and energy futures have been seen as a major potential source of development. However, progress in this area has been slow. In launching its energy contracts, SIMEX faced one problem that the IPE and NYMEX did not. The IPE has only ever traded energy futures, and, while NYMEX is an old exchange that used to trade dairy and agricultural products and still trades metals, the launch of energy futures was essentially a relaunch for the exchange. By contrast SIMEX is a thriving exchange with its financial contracts. On the one hand, existing members of SIMEX might have no

Table 9.2: Growth Rates and Share of Singapore GDP by Sector. Per Cent.

Sector	Growth 1983–93	Share of GDP
Manufacturing	113.2	27.6
Finance	127.9	26.9
Commerce	100.1	17.9
Transport/Communications	122.9	14.6
Other Services	68.3	9.8
Construction	12.7	6.7
Utilities	104.9	2.0
Agriculture/Fishing	-49.0	0.2
Quarrying	-55.2	0.1

Source: Singapore Yearbook of Statistics.

interest in energy futures. On the other, those interested in the large-scale trading of energy contracts might have no interest in financial futures. The cost of membership of SIMEX, i.e. the cost of buying a seat, reflected the success of financial futures, and could act as a deterrent to potential newcomers to SIMEX interested in energy trading. By 1997, the value of a SIMEX seat was close to S$300,000 (i.e. over US$200,000).

To combat this problem, SIMEX issued trading permits to companies only wishing to trade energy, as well as setting up a new category of membership, commercial associate members. As of 1996 there were twelve such members, including oil companies BP, Caltex Trading, Caltex Asia, Mobil, Shell Eastern Trading, and Singapore Petroleum Company. All of the Singapore refiners bar one are thus represented. Commercial associate membership is also held by Hin Leong Trading, Itochu, Marubeni, Mitsui, Petrodiamond (an oil trading arm of Mitsubishi) and Yutaka Futures. To this pool can be added full clearing and non-clearing members of SIMEX who have substantial energy trading interests, such as Phibro and Morgan Stanley. Other SIMEX member firms have also been active in energy trading.

The first, and most successful oil product contract to date in terms of longevity, has been the High Sulphur Fuel Oil (HSFO) contract, which was launched in February 1989. The contract trades lots of 100 tons of HSFO, with trading taking place on the SIMEX floor between 9.30 a.m and 12.30 p.m., and 2.30 p.m. to 7 p.m. Singapore time. Nine consecutive months are listed, with the expiry date for the first month contract being the last business day of the previous month. For the expiring contract, trade finishes at 12.30 p.m. on the expiring day. SIMEX managed to launch a contract in an oil product where success has eluded both NYMEX and the IPE. Given how far the development of futures lagged behind in Singapore, floating at the first attempt a successful contract that had proved impossible elsewhere, might imply that the formation of a full energy futures complex was going to be a relatively easy task. However, the initial success of the contract lay in the one feature that NYMEX and the IPE could not call on, namely the largest market for bunker fuel for shipping in the world. The SIMEX grade is almost fungible with bunker fuel,[8] and hence there was a large underlying market.

However, the concept of success in the HSFO fuel oil contract needs to be clarified. It was a success to the extent that it survived a considerable time. The successful contracts in the IPE and NYMEX, all went through a period of exponential growth. The HSFO contract did not; indeed the total volume traded in 1989, its first year, is only slightly less than the combined total volumes over the next five years. There was growth between 1991 and 1993, but average 1993 volumes were still only 1266 contracts per day, and liquidity then fell through 1994 to about 722 contracts per day. By the autumn of 1995 the contract traded only very rarely, and the level of open interest was negligible, and at the start of 1996 liquidity had disappeared. A handful of contracts were traded in the last quarter of 1996, after a year's absence of trade, but at the time of writing, the original contract was effectively dead.

Success in terms of longevity is not the same as success in terms of liquidity. Even in 1993, its peak year during the 1990s, the HSFO contract represented less than one half of one per cent of total world liquidity in oil futures and options. There are two main sources for the failure of the contract to achieve rising volumes. The first is the cut-off from the market of what should have been a natural constituency. While the demand for power generation using fuel oil was greatly contracted in the USA and Europe by the oil price shocks, there remains a demand there from utilities for hedging fuel oil. Indeed a substantial part of the volumes on NYMEX's natural gas contract comes from users looking for a vehicle for fuel oil risk management. In Asia there is far greater reliance on oil-fired power generation, yet the national power utilities were never drawn to SIMEX, due primarily to governmental regulation of both prices and utility behaviour.

The second reason concerns the asymmetric development of markets. The informal paper fuel oil market was thin in 1989, and provided little volume that could potentially move to SIMEX. In fact, the movement of liquidity went the other way, features of the flexibility of the paper market, and in particular the lack of any need for physical delivery, started drawing users to the informal market. NYMEX and IPE contracts have tended to supplant the informal markets, with the exception of IPE Brent which grew on top of the informal market and then attracted volumes across. By contrast the SIMEX contract grew before the informal market. There is no example in oil futures of any contract both surviving and growing without the existence of an already high liquidity informal market, and numerous examples where they have failed even when the informal markets did exist.

On the criteria of building on the base of an existing informal market, SIMEX should have been on firmer ground with its next contract launch, a contract in Dubai crude oil which was launched in June 1990. This on paper had two main merits – there was and still is an existing informal Dubai forward market, and Dubai is used by most Middle East producers as an index for the pricing of their crude oil sales to the Far East.[9] The contract began well, but a few weeks after the launch Iraq invaded Kuwait. Conditions of high price volatility certainly encourage the development of futures markets, but extreme price volatility is not the ideal environment in which to launch a new contract. However, even the two merits identified above did not really act in the contract's favour. The informal Dubai market was primarily based in London, and London traders' volume did not move to Singapore. Further, while Dubai was used in pricing formulae, it was the monthly average that was used, and so it was harder to hedge using a daily Dubai price. The SIMEX Dubai contract had a brief existence, and stopped trading in 1991. A third contract, for gasoil, was launched in June 1991, and after initial good liquidity, also failed to survive. The gasoil contract was relaunched in June 1992, but failed again, being delisted in November 1993.

NYMEX and IPE contracts often grew by supplanting, or, in the case of IPE Brent, acting as an auxiliary market to existing market structures. Because of the difficulties in establishing informal forward markets in Singapore noted in Chapter 6, SIMEX futures tended to have no high liquidity underlying informal market to build on. Indeed, swaps markets have tended to grow after SIMEX has attempted futures launches.

As with the swaps markets, futures have also been held back by the nature of market participants. In the USA after price control, there were large numbers of producers, refiners and distributors facing price risk, who were prepared to trade as a defence. The development of the North Sea brought the same conditions to European crude oil markets, and the development of the Rotterdam market and independent refining brought them to European product markets. In Asia, the major national oil companies are often naturally hedged by the ability to pass price rises on to the consumer or government, and have preferred to continue to use tenders rather than directly trade. Markets have also been subject to heavy price regulation. The awareness of trading and of the role of risk management that helped NYMEX and the IPE to develop, has not developed so quickly in Asia. Fiscal regimes have not acted as a brake on western markets, indeed in the early history of the North Sea they were a positive incentive to start trading. In Asia, fiscal regimes for oil production have tended to be retrospective, and structured in such a way as to be virtually unhedgable using derivatives.

The failures of its gasoil and Dubai contracts, as well as the gradual death of the fuel oil contract as trade migrated to the paper markets, convinced SIMEX that a change of direction was needed. SIMEX's own history started with a strategic alliance, that with the CME, which enabled the trading of an established contract from the first day of the exchange. SIMEX's energy complex had tried to go it alone, except in the alliance with the IPE in trying to launch a new contract (i.e. Dubai) jointly. The independent route had failed, as had the idea of launching new contracts in alliance with others. The only remaining option was to enter into strategic alliances with the IPE or NYMEX, to allow the trading of some of their already successful contracts in Singapore. SIMEX in fact entered into negotiations for alliances with both these exchanges. A letter of intent was signed with NYMEX in November 1994 for the placement of terminals for NYMEX's electronic trading system in Singapore. A similar scheme has gone into effect in Sydney and Hong Kong, as discussed below, and at time of writing there are no plans to implement the content of the letter of intent in Singapore.

The deal SIMEX made with IPE was for a mutual offset arrangement on the IPE Brent crude oil contract, and this was launched on 9 June 1995. This involves trading Brent in open outcry in a SIMEX futures pit. Positions opened in Singapore can be carried over to London and closed there, and vice versa. The deal is attractive for both sides. For the IPE it increases the hours of trading of its flagship contract, and gives it the globalization it will need for the longer-term development of the exchange. For SIMEX, it offers the opportunity to finally list a liquid crude oil contract. It is also in some senses a holding operation. With a belief that an indigenous contract can eventually succeed, the IPE link is aimed to keep SIMEX in oil futures, while building up liquidity for later expansion of contracts.

When London is using British Summer Time (roughly from the end of March to the end of October), SIMEX trades Brent in a morning session between 9.25 a.m. and 12.30 p.m. local time, and in an afternoon session between 2.30 p.m. and 4.58 p.m.. Four minutes after the SIMEX close (i.e. 10.02 a.m. London time), the contract begins trading on the IPE. During the winter period when the UK reverts to Greenwich Mean Time, trading on SIMEX continues for one extra hour, (i.e. until 5.58 p.m. local time) leaving the same four minute gap before IPE Brent trading commences. Close of trading in the IPE equates to 4.15 a.m. Singapore time (or 3.15 a.m. during summer), and thus the link-up produces a total trading day spanning eighteen or nineteen hours.

The contract traded on SIMEX is essentially the same as that on the IPE. Each contract is for one thousand barrels of Brent blend of current quality, with twelve consecutive months tradable at any one time. Cash settlement of open positions is made at contract expiry against the same Brent index compiled by the IPE to settle open IPE Brent positions. The Brent index is computed on the basis of reported forward Brent trades, and the price assessments made by five price assessment agencies.[10] The expiry day for any contract month is the 15th (or the first London trading day thereafter) of the prior month. While the same final settlement price is used at expiry, for the purposes of daily marking to market and making margin calls, SIMEX uses its own price (equivalent to trading values at the end of the SIMEX session).

The performance of Brent on SIMEX in its first year can be described as creditable if not spectacular. Daily volumes have tended to be about 500 contracts (i.e. 0.5 million barrels). While of course dwarfed by the 30 to 40 million barrels regularly traded in Brent on the IPE, it does represent enough liquidity to initiate a position. It should also be noted that the common fear in low liquidity markets of not being able to close out a position is removed by the ability to utilize the mutual offset and close out positions in London. To the extent that the offset is used to close out a Singapore position, the official volume figures could be an underestimate of the volume that has arisen because of the link. In Chapter 12 we will note the growing influence of Brent related crude oil in Asian markets. With this evolution, SIMEX Brent has the possibility of spawning a series of sub-markets, such as the development of informal swaps of the basis risk between Brent and regional crude oils, i.e. CFD (contract for differences) markets.

While SIMEX linked up with the IPE, NYMEX went into alliance with the Sydney Futures Exchange (SFE), and the Hong Kong Futures Exchange (HKFE). The NYMEX–SFE relationship involves the positioning of workstations for NYMEX ACCESS, an electronic after-hours trading mechanism, in the SFE, linked through the SFE's own after-hours trading system (SYCOM). The link was established on 8 September 1995. ACCESS is available for trading of the light sweet crude oil contract (colloquially the WTI contract) from 4 p.m. to 8 a.m. New York time, i.e. 8 a.m. to midnight Sydney time. Trading of heating oil, New York harbour gasoline, natural gas and propane is also possible through the link. By the end of 1995, average trading volume per ACCESS session was around 4000 contracts. No official breakdown of ACCESS liquidity by geographical location has yet been released. However, given the distribution of terminals, the bulk of it is conducted in the USA in the US late afternoon and early evening. According to NYMEX figures in late 1995, 52 per cent of trade was conducted in the first four hours of each session, i.e. up to 8 p.m. New York time. The best available estimate for the trades through the SFE in the first six months of the link was an average of between 100 to 200 trades (i.e. 100 to 200 thousand barrels) per session.[11] NYMEX has also linked up with the Hong Kong Futures Exchange (HKFE) for trading of NYMEX contracts, with the link due to become operative in the first quarter of 1997.[12]

The gap remains for an Asian oil product futures contract. The options available are limited by the nature of the physical market. Singapore (and indeed Asia in general), has no equivalent of the small volume markets represented by pipeline markets in the USA and the Rhine barge trade in Europe. These markets generated a large number of spot trades, and a complex of first informal forward and then futures markets formed on this physical base. Singapore is a cargo market, with large standard volume sizes. Added to this there is growing proliferation of different specifications for oil products across Asia. Changes in and multiplication of gasoline grades due to environmental legislation in the USA proved to be a problem, albeit eventually surmountable, even for the established high volume New York futures market. In this context, the splintering of the Asian market does suggest that the launching of a new physically delivered futures contract is made more difficult. The problems inherent with physical delivery were a strong element behind the demise of gasoil futures in Singapore. Likewise, physical delivery (of a grade that matched neither the common cargo nor bunker market specification) in the fuel oil contract was one factor that led hedgers to prefer to use swaps. Physical delivery in the HSFO fuel oil contract worked with what could potentially be very short notice for the buyer. Buyers then sometimes had problems with arranging tanker fixtures, and theoretically could face the freight costs involved with taking delivery of fuel oil at several different terminals. In short, the logistical features of the Singapore market are such that the design of a workable oil product contract with physical delivery poses some additional problems that were not experienced to the same degree in London or New York.

A relaunch of SIMEX's HSFO contract would perhaps represent the last attempt to get physical delivery to work in the context of a Singapore oil product futures contract. Modifications to the contract, discussed by the exchange in 1996 but not as of 1997 implemented, remove two key weaknesses of the original. In particular, they involve the standardization of the specification rather than keeping the cargo and bunker hybrid, and tightening the delivery procedures by removing the potential for multiple deliveries of small parcels at different terminals for the same physical clearance.

If physical delivery for oil product contracts has proved to be difficult, another option is cash settlement. This after all is the method used in the Brent futures market to bridge the difference between a large parcel size physical market and a small parcel size futures market. The next problem is what to settle against. One option would be an index for the day of expiry, similar in design to the Brent index which is essentially an average across a day's trading. The alternative, which would more closely mirror terms in the physical markets (including term, tender and spot deals) is to settle against a monthly average.

Settling on a monthly average of physical prices, produces a futures contract that effectively mimics the swap market. In the last chapter we termed the main Singapore swaps markets as quasi-forward markets, in that they approximated to a forward market in every detail except the possibility of physical delivery. The monthly average futures contract would then be a true hybrid, a futures contract mimicking a swaps market mimicking a forward market. However, in our view this would be the best structural design for any Asian oil products futures contract. The possibility of a monthly average cash settled contract has been raised by SIMEX in the context of a gasoil contract. The major problem encountered has been the specification of precisely which measure of the physical market to cash settle against.[13] However, we would note that the Brent futures market functions well with a settlement price which is a proxy to a daily average based on a fairly complicated calculation. In comparison, devising an index for a monthly average price that does leave a significant basis risk from whatever index is used in the physical market, would appear to be a far simpler undertaking.

The question remains as to whether there is actually a role for a futures based contract designed on the lines above, given the existence of a liquid swaps market. Consideration of the differences between such a futures contract and the existing swaps market provides a series of tradeoffs. In the description below, we consider solely the hypothetical monthly average cash settled futures market, and the highly standardized quasi-forward form of swaps that make up the main Singapore swaps market. The importance of this distinction is that, as pointed out in the last chapter, the frequently used argument that swaps are more flexible than futures does not hold. To achieve a narrow bid-offer spread and high trading volumes, the mainstream Singapore swaps market foregoes that flexibility. In fact futures are more flexible than standardized swaps, given the smaller parcel size. A hedge can be constructed for, say, 38 thousand barrels using futures, it can not using swaps without attracting a widening of the bid-offer spread.

There are differences in credit risk and cash flow between the two forms. Swaps markets leave counterparty risk, which is removed in futures markets. For example, had the Barings collapse occurred through informal market trading, there would have been implications for other companies that did not occur in the context of SIMEX trading. Counterparty risk tends to be a dominant factor inhibiting the growth of longer-term informal market swaps, but it is still a concern in shorter dated swaps. The tradeoff for the removal of counterparty risk is the cash flow implications. Trading futures requires the positioning of cash margins, and also the holding of precautionary balances should a position's margin requirements be increased. Swaps are less onerous on cash flow, but that brings with it the aspect of counterparty risk.

A further difference arises from hours of trading. Informal swaps have no set hours, which can be an advantage if arbitrage positions with other regions are being used. For instance, one factor that pushed liquidity into fuel oil swaps and away from futures was the trade in crack spreads. The most popular of these is the spread between Dubai crude oil and heavy fuel oil. Given that the Dubai market is primarily traded in London, the time period when a trader could lock in the spread by dealing with a London Dubai trader while opening a SIMEX position would be extremely limited. With swaps not being bounded by exchange hours, the room for manoeuvre in arbitrage spreads is greater. This difference should however perhaps not be overstated. The vast majority of all Singapore swaps are transacted between the hours of 2 p.m. and 8 p.m. Singapore time. It is not a market with a high level of liquidity spread out over an inordinately long trading day.

It is harder to build up a large open position without moving the market in a futures pit than it is in the swaps market with its greater information lag. In other markets, and most particularly Brent, this has been one reason for the coexistence of formal and informal markets. A company may often do its routine hedging on the formal market, and the same company uses the informal market when it wants to take a market view and build an open position quickly. In other words, whether this feature is a merit or a demerit for futures is entirely dependent on the aims (and size) of the company involved.

SIMEX is a highly regulated exchange, compared to the looser regulation of the informal market. We make this distinction not because we believe it is necessarily relevant to current market participants, but because it might be relevant to potential market participants. In particular, the marketization of the Asian oil industry has led to a gradual move towards interest in direct involvement in trading.

The form in which any new entrants make at least their initial forays is likely to be governed by either internal company or government guidelines. In some cases, trade on organized exchanges will be mandated, while trade on informal derivative markets will not. For example, financial liberalization in Taiwan has involved explicitly specifying the foreign derivative markets in which Taiwanese capital can enter. The list includes SIMEX, but not informal swaps. For some potential market participants in liberalizing economies, the major difference between futures and standardized swaps is that they may be excluded from the latter.

Ultimately, while the above factors will be of import, we suspect that for most companies the main consideration is relative trading cost and liquidity. The relative weaknesses of physical delivery based futures compared to swaps in Singapore would appear to be removed by a cash settled contract that mimics a standardized swaps market. Our conclusion is that while we see little possibility of a complex of oil product contracts with physical delivery being built up, given the structural features of the Singapore market, there also appears to be no *a priori* reason why a well designed cash settled contract should not succeed.

One further potential new contract of note is that planned by the Tokyo Commodity Exchange (TOCOM). TOCOM are planning a gasoline contract, provisionally with physical delivery in Japan with prices denominated in yen, subject to government approval. The development of a gasoline contract fits in well with the increased exposure of the Japanese industry to international prices. However, it should be noted that, if approved and launched, such a contract is unlikely to become an international market. We noted in Chapter 7 the exodus of oil traders from Tokyo in the late 1980s, and the relative costs of Tokyo trading would not appear to have been reduced enough to occasion a reflux. Denominating a contract in yen also greatly reduces the appeal for international traders.

The one other major question mark is about domestic market price volatility. In Chapter 4 we noted that Japanese deregulation has led to a tendency towards import price convergence. However, the cost and logistics of trading are such that Japanese prices need not converge on the import substitute price at all times. Any severe and persistent misalignment can be corrected, but there is a range in which arbitrage is not profitable. The implication of this is that large short-term volatility in international gasoline prices (particularly if this is volatility around a trendless average level) will not necessarily translate into the same degree of volatility in Japanese wholesale prices. The reduced degree of domestic volatility, and therefore of risk, would reduce some of the motivation to trade. Further, the yen denominated Japanese delivery price may then have a significant short-term basis risk from international prices, and thus be less effective in the hedging of the short-term commercial risk involved with the transfer of international cargoes.

4. The Chinese Exchanges

The history of futures trading in China seems to equate more closely to that of the nineteenth-century exchanges considered in Section 2, than to that of SIMEX. The reasons for the growth of the Chinese exchanges follow those listed for their earlier counterparts. Trade had become deintegrated, and the liberalization within the Special Economic Zones had created price volatility and therefore price risk.[14] Further, the government saw the development of futures trading as one element in the modernization of the financial sector, and actively promoted the foundation of exchanges. The first commodity exchange was the Shanghai Metals Exchange, which opened in May 1992. The announcement of the first oil exchange was made in December 1992, with the state trading company Sinochem a major backer of the Nanjing gasoil contract.

Gasoil trading began in Nanjing in March 1993, followed by gasoline in May. Trading in Shanghai began in May 1993, with contracts for crude oil, gasoline, gasoil, and fuel oil. Initial liquidity was low, hampered by unfamiliarity and the lack of any foreign participation. The prohibition on external capital came from a direct government regulation. In fact, even without state restrictions it is highly unlikely any significant foreign liquidity would have entered the markets. The exchanges traded in local currency rather than dollars, with all the attendant currency conversion problems. The lack of regulation, and, as noted below, the lack of definition in the role of government in the exchanges would also have acted as deterrents. However, liquidity on the exchanges soon increased, as local traders and state corporations began to see their use as both hedging and speculative vehicles. New contracts were added, and new exchanges opened up.

By the spring of 1994 China had more than thirty futures markets, between them trading an almost full range of commodities. Of these exchanges, crude oil or oil product contracts were listed at three main exchanges at Beijing, Nanjing and Shanghai, as well as at Daqing, Guangzhou and Huanan.[15] At this time there were nine futures exchanges listing oil contracts in the world, and China had six of them, although within a fragile system. The major Chinese oil exchange was the Shanghai Petroleum Exchange (SPEX), which had over 80 per cent of the total volume, and traded about 1.7 billion barrels of oil between May 1993 and June 1994.

The nineteenth-century exchanges were destroyed by the imposition of a fixed price regime. The same happened to the first incarnation of the Chinese exchanges. A reverse in the policy of price liberalization led to the reimposition of price controls. The obvious disagreements among central policy makers, and the consequent uncertainty about implementation, increased volatility in the run-up to announcements. In April 1994 the Shanghai gasoline contract achieved daily volumes of above 1000 lots, equivalent to over 100 thousand tonnes of product. However, trading ceased completely when the State Council implemented a directive (known as the 'rectification programme'). The directive banned futures trading in a series of key commodities such as coal, rice, steel and petroleum products. The only commodities where futures trading was permitted to continue tended to be marginal and non-strategic, for instance plywood. As price controls for the strategic commodities came into effect, the motives for trading were also removed regardless of the direct ban. The petroleum exchanges were effectively placed in limbo.

Despite attempts to launch contracts in products not affected by the ban, such as lubricating oils, the policy changes brought an end to the first phase of the exchanges' development. There was however some progress after the directive. Most importantly a primary regulatory body emerged, somewhat equivalent in function to the Commodities and Futures Trading Commission in the USA, the China Securities Regulatory Committee (CSRC). Trading continued in some commodities, and the central bureaucracy began to consider an overall framework for commodity trading. The groundwork is therefore being laid for the movement to a smaller number of more regulated exchanges, in other words a move closer to the framework of developed futures exchanges.

The need for tighter regulation of futures trading was reinforced by events in the bond futures market at the Shanghai Stock Exchange. On 23 February 1995 traders caught in short positions collapsed prices in a heavy bout of selling in the last ten minutes of the day's trading. The government reaction was to tighten regulations, and aim to cut the number of exchanges trading bonds, consistent with the line taken in the review of petroleum and other commodity exchanges. Within a year of the new government policy, the number of recognized exchanges had fallen to fifteen and could be expected to fall further. As part of this rationalization of exchanges, the Shanghai Petroleum Exchange has reemerged as part of the Shanghai Commodity Exchange, a merger of four exchanges. The other features of the changes being made before petroleum futures resume trading are also suggestive of a move from undeveloped to developed futures markets. The first Chinese exchanges, like the early US markets, differed from more developed structures in a series of key areas. Membership of the exchanges was loose and sometimes undefined, the independence of the exchange and clearing house was often unguaranteed, speculative limits on the size of position that could be taken were rare, margin systems to prevent defaults were ineffective, and the exchanges operated outside of the ambit of any specific national law or statutory regulatory body.

In addition, the role played by government in the Chinese exchanges was somewhat less than transparent. Government departments and parastatals played a key role in establishing exchanges, and were also important traders, often trading against other government organizations, while still having a hands-on and often very highly interventionist role in the physical market. Just to compound this, the supervisory administrative boards of the exchanges tended to be made up of officials drawn from both central and local governments. The principal-agent problems involved in the multiple roles of the state, and the multiple forms in which the state participated, are only too apparent.

The abandonment of trading in economically strategic commodities such as oil represents a hiatus, during which the above problems can be addressed. In particular, the exchanges that are emerging from the process are closer in organization to those outside China, with a transition from undeveloped to developed futures trading. The drafting of an overall law, and the powers of the CSRC, represent the addition of a more structured regulatory layer. In addition, controls have been added in cases where communications systems are currently insufficient for smooth and transparent functioning. For instance, in 1996 the CSRC banned intra-city trading, whereby some exchanges had set up and accepted members in other cities. Exchanges have moved towards a more formal membership system, with administrations electable by members rather than by central and local government. More recognizable margin, position limits and position reporting requirements are being imposed. In the case of oil, the Shanghai Commodity Exchange appears to be in the position to emerge as the dominant exchange when trading resumes.

The early growth of the Chinese exchanges also demonstrates one circumstance in which trading can go straight to futures without passing through a stage of informal markets, i.e. where market transparency and information flows are poor. In the development of western informal markets in the 1970s and 1980s, identification of potential counterparties was not a problem, information flows were good, and market transparency was provided by brokers and price reporting agencies. In addition communications were good enough to obviate the need for centralized trading. At the start of Chinese liberalization (just as in the early Pennsylvanian industry), communications were sometimes difficult, identification of counterparties (and in particular assessment of their default risk) was a problem, flows of information poor, and there were no sources of market transparency for the internal market. Centralized futures markets provided a solution that informal markets could not. Without these problems in the West, when futures evolved there was no need to replicate contracts in the same region. In China (again in parallel with the nineteenth-century exchanges) futures markets could proliferate in such a loosely regulated environment as the problems would persist in any market location without an exchange.

The contrast between the first Chinese exchanges, and the organizational forms that have been evolving after their forced demise, is a striking one. While still likely to be exclusively Chinese and not international, at least at the beginning, the second incarnation of the exchanges owes far more to the structure of the international exchanges.[16] While perhaps in the immediate future something of a curiosity in international terms, the short history of the Chinese exchanges demonstrates clearly the relationship between regulatory environment, industrial structure, and the development and organizational form of futures trading.

5.   Conclusion

Large-scale derivative trading failures have had a psychological impact considerably greater than their true importance. Fear of such failures is often cited by oil companies, predominantly state owned, as a reason for resistance to engaging in forms of risk management. In questions of regulation, the impression is sometimes given that there is a suspicion that derivatives are inherently unstable, and catastrophic failures are an inevitability.

Asian companies, and Japanese in particular, have been prone to derivatives failure. In addition, within the context of SIMEX, there was the Barings collapse. Few of the high profile cases have involved oil trading, but substantial losses have been made in oil. The major case was the collapse of the US operations of the German firm Metallgesellschaft (MG). Fear of such catastrophic failure has been cited by many Asian oil companies as a reason not to trade on SIMEX. In addition, the MG case and other major failures related to derivatives trading raise a series of issues on the nature of derivatives and their correct use. We would suggest that the major lessons all concern the internal operation of a firm, and in particular information systems and management structure. While we contend that the perceived problems of derivatives use and their solutions are internal, the MG case and others have been followed by calls for greater external regulation of derivatives. We now consider whether there is any case for a tighter regulatory regime.

Greater intervention by governments and monetary authorities if motivated by a desire to reduce the incidence of large derivative losses, would represent a relatively rare basis for regulation. Normally government intervention in a company's activities is designed to avoid outcomes which are considered not to be in the public interest, but which might be in the company's interest. The regulations aim to put bounds on some outcomes that might arise from what would be optimizing behaviour by firms.

Regulation of derivatives in order to reduce company failures would be different. It would be regulation designed to prevent something that a rational firm would not want to happen. Consider two cases. In the first there has been no information or control failures, and the management of the firm fully understands the risks their derivatives trading is exposing them to. In an idealized case imagine that sufficient information on the company's use of derivatives is also available to shareholders, who are aware of the risk profile of the asset they hold. The company and the holders of its shares are essentially gambling, and it so happens that the gamble is incorrect and large derivative losses are incurred, perhaps leading to the company being liquidated. Should regulation have been in place to stop this failure? We would suggest not. There is in this case no difference between the risk incurred in derivatives and that involved in other commercial decisions. The derivative portfolio under these circumstances has the same status as choices about speed of expansion, entry into new markets or development of new product lines. For each of the latter choices there are multiple examples of companies incurring losses or collapsing due to an *ex post* incorrect strategy. Likewise under such a high level of information if the losses were greater than the asset value of the firm and defaults occur, these are still the correct market solutions as agents dealing with the company also had knowledge of its risk profile and were thus in effect speculating in dealing with it.

This first case is of course extreme, but at that extreme there is no justification for regulation designed to stop large derivative losses. Now consider the other extreme, the nature of the derivative risk was unknown to the top management, shareholders and other agents dealing with the firm. A large derivatives loss ensues, and the company collapses with debts greater than its assets. There are thus both internal and external information failures, and the first-best solution would involve correcting these. The internal failure is a product of the managerial organization of the firm, and a firm which is optimizing should be able to prevent these failures. Regulation is then a second-best solution, applicable in the case where government feels that companies are incapable of producing sufficient efficiency of internal information flows.

The one area in relation to derivatives where intervention might be justified is in ensuring the external transmission of information on the company's use of derivatives. Here there is a justification more in line with the general motive for regulation. The more risk a company is exposed to, the lower its current value will be if the equity market is risk averse. In such a case, there would be an incentive for management to disguise those risks. The reverse also holds; if the use of derivatives has, as it should, reduced the riskiness of the company then it will be in the interests of management to make full disclosure. In the case of oil companies however, the situation is less clear. A significant number of equity investors buy oil shares because they do want exposure to commodity risk. Hedging by the company would then in fact reduce its value to such investors. Overall it is ambiguous whether there is an incentive for managers to reveal the nature of their derivative portfolio, and hence there could in some circumstances be a case for a requirement to reveal the information.

In fact in the wake of the MG débâcle and other failures, there has been a move by regulatory bodies towards the encouragement by companies in their published accounts of the scale of their derivative business. Table 9.3 shows the value of contracts held by a selection of companies at the end of 1994 as collated by *Petroleum Intelligence Weekly*. The figures are not really comparable across companies as the definitions used have varied widely. Some have taken the full definition of derivatives, and have included the value of futures and forward market positions. Others have included only swaps. However, there are three main points to be taken from Table 9.3. First, the energy derivative business has become extremely important. Secondly, as evidenced by Shell's $751 million in swaps alone, swaps constitute a major part of companies' total portfolio of derivative instruments. Finally, even with the above disclaimer on the difficulty of comparison, there is clearly a considerable divergence between companies in their use of derivatives.

Table 9.3: Value of Energy Derivative Contracts. Selected Companies. End 1994. $ Million.

Company	Value
BP	2,173
Elf	1,506
Mobil	1,294
Amerada Hess	1,223
Shell	751
Texaco	511
Total	411
Exxon	19

Note: Amerada Hess covers crude oil contracts only. Shell covers swap trades only.

Source: *Petroleum Intelligence Weekly*.

In fact Table 9.3 only shows a part of the companies' use of derivatives, since it includes only commodity derivatives. Swaps and other derivatives are also heavily used for interest rates and exchange rates. For example, at the end of 1994 Mobil held positions worth $3600 million in financial derivatives, and Exxon held interest rate contracts worth $604 million.[17] While of some interest in itself, disclosure of the figures shown in Table 9.3 does not carry very much information. There is no real relationship between the value of derivatives held and risk exposure. The problem with MG was not so much that it held a large number of derivatives, but rather how it held them in relation to its physical positions.

A large involvement can simply be indicative of a well hedged company. Disclosure in the form currently used might even be a disincentive to enter into perfectly sound derivative programmes that would increase the value of the company. A snapshot of the end of year position may not be representative of a normal position. Overall, imposing disclosure of simply the total value of derivatives may be worse than no disclosure at all. On the other hand, disclosure of too much detail runs the risk of compromising commercially sensitive information. We believe the correct solution is independent risk auditing.

In simple terms this would involve a third party inspecting a company's portfolio of derivatives, its evolution and the hedging and trading strategy employed. Most importantly, this should occur at a random point of time. This has two uses. Internally it provides a check on the company's own risk evaluation, and reduces the chance of internal misreporting which has been central to nearly all derivative failures. Externally, it provides an evaluation of the company's trading activity. In our view random independent risk audits would have greatly reduced the dangers in cases such as MG, or at least identified them at a point when remedial action was still possible.

The case for regulation, viewing each company in isolation, is then rather weak. However, one further possible justification concerns externality effects, i.e. the possibility that the collapse of one firm has a domino effect and collapses others. This is the 'meltdown' scenario feared in financial markets. It should be noted that this is extremely unlikely on a formal market, where the system of margins makes the development of a large uncovered loss a purely intra-day possibility, and even then the clearing house guarantees performance. While technically more possible in informal markets, it should be noted that the flow of information in informal oil markets on counterparty exposure and credit tends to be good. Even in extreme market moves, such as those occasioned by the invasion of Kuwait, there have been no externality effects.

Thus, regulation in response to derivative failures does not appear to be either necessary or desirable. Further, as Bellew notes,[18] there is no inverse correlation between the degree of regulation and derivative failures. Given that derivatives trading takes place in both highly regulated futures exchanges and in often very lightly regulated OTC markets, one would expect that failures would be concentrated in OTC trades if a lack of regulation was the problem. In fact, as Bellew points out, the most spectacular failures, and in particular Barings and MG, have involved derivatives on formal exchanges.

The future development of derivatives in Asia depends on an understanding of the true nature of the risks involved and of the management structure required to deal with these risks. Nothing is gained by demonizing derivatives, but great losses can be incurred when trading without a full appreciation of the nature of the beast.

Notes

1. Hildegarde Dolson (1959), *The Great Oildorado*, Random House, New York.
2. W.E. Stanton Hope (1958), *The Battle for Oil*, Robert Hale, London.
3. Hildegarde Dolson (1959), op.cit.
4. The following description is drawn from a series of *Daily Era* petroleum market commentaries for the Bradford Exchange, for 1889 and 1890.
5. Robert J. Weiner (1992), 'The Origins of Futures Trading: The Oil Exchanges in the 19th Century', Université Laval, Quebec.
6. For a description of SIMEX's financial contracts see Keith K.H. Park and Steven A. Schoenfeld (1992), *The Pacific Rim Futures and Options Markets*, Heinemann Asia, Singapore.
7. See George Teo (1991), 'Regulation of the Securities Market in Singapore', in Charmaine Lye and Rosalind Lazar, *The Regulation of Financial and Capital Markets*, Singapore Academy of Law, Singapore.
8. SIMEX HSFO specifies a maximum sulphur content of 4 per cent and a viscosity of 180 centistokes. This is a hybrid of the most usual bunker specification (4 per cent sulphur and 380 centistokes) and the most traded cargo market specification (3.5 per cent sulphur and 180 centistokes).
9. The Dubai forward market is discussed in Chapter 11.
10. Platt's, Petroleum Argus, London Oil Reports, Telerate and Reuters.
11. Figures given in *Petroleum Argus Energy Trader*, 8 March 1996.
12. NYMEX press release, 12 September 1996.
13. SIMEX suggested cash settlement against the mean of Platt's and FEOP monthly averages (for a description of price reporting agencies see Appendix 1). Platt's however refused permission for its gasoil price to be used for settlement purposes (see *Platt's Oilgram News*, 17 January 1996). Platt's prices are employed in conjunction with other agencies in the settlement price used to settle the SIMEX Brent contract.
14. The Chinese reform process is considered in more detail in Chapter 3.
15. The Huanan and Guangzhou exchanges merged in August 1994 to form the Guangdong United Commodities Futures Exchange.
16. One major difference, at least visually, is the tendency of futures trading in China to be computer based, using automated trading systems rather than direct open outcry in futures pits. While making the Chinese exchanges appear, at least on the surface, rather more restrained, the practical implications of the difference are minor.
17. These figures are taken from *Energy Risk*, June 1995.
18. Christopher Bellew (1995), 'The Future of Derivatives Markets', *Oxford Energy Forum*, no. 22.

PART IV

Crude Oil Markets

CHAPTER 10

THE ASIAN PRODUCERS

1. Introduction

The high rates of oil demand growth we evidenced in Chapter 2, have led to Asia becoming a region of ever greater crude oil deficit. The deficit in terms of total flows, raises the issue of the severity of the deficit at the margin. While falling far short of current total demand, the extent to which Asian and Australasian production can provide some contribution in meeting incremental demand is a matter of some import. A fall in regional production means that increases in the absolute import gap will be larger than those in demand, while incremental regional production helps at least to moderate the rise in the dependence on longer haul sources.

Physical deficits have also been accompanied by a market deficit in terms of price formation. Europe has its active Brent complex, the USA has the WTI market, and (while of a lesser scale and more of a façade than a real centre of price formation), there is the Dubai market in the Middle East. There has been no equivalent in Asia, mainly due to structural factors relating to the imperfections of possible candidates, but also partly due to the same historical lack of marketization across the region's oil industry that we found in previous chapters in the context of oil product markets. These three deficits, two in physical flows and one in markets, are the major focus of this and the remaining two chapters in this part.

The next two sections of this chapter detail oil production in Asia and Australasia, and seek to provide a discussion of the underlying determinants of the future path of production, i.e. a structure for addressing the question of the potential physical deficit at the margin. We draw on the lessons of the North Sea experience in the 1990s, and in particular focus on the central role of the fiscal terms offered in the Asian upstream oil industry. New technology combined with cost savings offer promise, but fiscal issues hold the key to realizing that promise. In common with Part II of this book, the determining feature of the dynamics of the situation is an area of government policy which is largely endogenous to the underlying trends.

After having set the context in terms of the physical base, the following three sections are concerned with the deficit in markets. Futures markets for crude oil were considered in the previous chapter, and there are no forward markets for any regional crude in the manner of Brent and Dubai. There is however a swaps market in Malaysian Tapis, albeit a fairly thin one, which was described in Chapter 8.

The structure of official prices, and in particular their retroactive nature, has been a major bar to the development of any price discovering spot market. These prices are discussed in Section 4, and the nature of spot market trading for Asian and Australasian crude oil is considered in Section 5. The final section provides a summary of the chapter's implications in regard to the three forms of crude oil deficit we have identified.

2. Crude Oil Production

The level of crude oil production in Asia and Australasia is shown in Figure 10.1, with Indonesia and China shown separately. Total production reached 7.3 mb/d in 1995, an increase since 1975 of 3.6 mb/d. It should be noted that, except in 1980 and 1982, the total has increased in every year over the period shown. Within this overall total, Indonesian

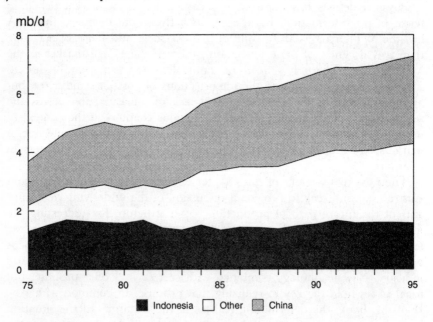

Source: *BP Statistical Review of World Energy*, various years.

Figure 10.1: Crude Oil Production in Asia and Australasia. 1975–95. Million b/d.

production has been stagnant, with the increases confined to other areas. As detailed in Chapter 3, the major single source of production growth has been China.

The composition of total crude oil production is further disaggregated in Table 10.1. A comparison with the oil demand figures in Table 2.8, provides a quantification of the first of our three regional oil deficits, i.e. the net amount of crude oil and oil product imports which is required from outside the region. In 1995, the residual of total demand over Asian and Australasian crude oil production was about 10.5 mb/d. The major part of this figure consists of the Japanese and Korean deficits, and outside these countries the Asian deficit in 1995 was about 3 mb/d. Our second oil deficit was that at the margin in the future, and we discuss this further below. However, looking at the pattern of marginal increments in the 1990s, we note that while demand increased by 4.2 mb/d between 1990 and 1995, regional crude oil production increased by 0.57 mb/d, i.e. just 13 per cent of the demand increment.

Of the countries shown in Table 10.1, only five (i.e. Indonesia, Malaysia, Brunei, Vietnam, and Papua New Guinea) are currently net exporters in terms of crude oil and oil products combined.[1] Australia and China, while a level of exports remain, have both moved from being overall exporters to being overall importers. We consider the position of Indonesia below, but it should be noted that Indonesia's overall surplus only fell slightly (from 0.9 mb/d to 0.7 mb/d) between 1985 and 1995. This change has been relatively small compared to those in some producer countries. India's deficit has increased from 0.25 mb/d in 1985 to 0.7 mb/d in 1995, and over the same period, China has moved from a 0.7 mb/d surplus to a 0.4 mb/d deficit.

Of the 3.6 mb/d increase in output in Asia and Australasia between 1975 and 1995, 2 mb/d occurred outside China. This figure is put in context by noting that the equivalent rise in total UK and Norwegian production over the same period equates to 5.5 mb/d. We would contend that the North Sea[2] experience does contain pointers for the future of Asian production. At its heart, the question is whether the elements of the surge in North Sea might be reproducible in Asia and Australasia.

A starting point for a comparison is the reserve base and the evolution of estimates of its size. While the UK and Norway are dwarfed in terms of geographical size by Asia, were they simply lucky in the distribution of oil reserves? Further, consider whether the superior output performance of the North Sea is simply a matter of geology. In other words, has the performance of UK and Norwegian output, relative to Asia, been driven primarily by the discovery and exploitation of new fields?

The levels of remaining economically recoverable reserves, given as contemporaneous estimates as of the end of 1975, 1985 and 1995, are

Table 10.1: Crude Oil Production in Asia and Australasia by Country. 1960–95. Million b/d.

	1960	1970	1975	1980	1985	1986	1987	1988	1989	1990	1991	1992	1993	1994	1995
China	0.10	0.50	1.49	2.12	2.51	2.62	2.69	2.74	2.76	2.78	2.83	2.84	2.89	2.93	2.99
Indonesia	0.41	0.85	1.31	1.58	1.34	1.43	1.42	1.38	1.48	1.54	1.67	1.58	1.59	1.59	1.58
India	0.01	0.14	0.17	0.18	0.63	0.66	0.65	0.68	0.73	0.73	0.70	0.64	0.62	0.71	0.79
Malaysia	-	-	0.10	0.29	0.45	0.50	0.50	0.54	0.59	0.62	0.65	0.66	0.65	0.66	0.74
Australia	-	0.18	0.41	0.38	0.65	0.58	0.62	0.59	0.56	0.64	0.61	0.60	0.57	0.61	0.58
Brunei	0.09	0.15	0.19	0.23	0.17	0.17	0.16	0.15	0.15	0.15	0.16	0.18	0.18	0.18	0.18
Vietnam	-	-	-	-	-	-	0.01	0.02	0.03	0.06	0.08	0.11	0.13	0.14	0.15
Papua New Guinea	-	-	-	-	-	-	-	-	-	-	-	0.06	0.13	0.12	0.10
Pakistan	0.01	0.01	0.01	0.01	0.04	0.04	0.04	0.05	0.05	0.06	0.07	0.08	0.06	0.06	0.06
Other	0.03	0.03	0.03	0.06	0.12	0.12	0.12	0.11	0.12	0.13	0.13	0.13	0.15	0.14	0.13
TOTAL	0.65	1.86	3.71	4.85	5.89	6.11	6.18	6.25	6.47	6.70	6.89	6.87	6.94	7.13	7.27

Source: *BP Statistical Review of World Energy* and *International Petroleum Encyclopaedia*.

shown in Table 10.2, together with the cumulative production in the two decades. Consider first the comparison between the combined position of the UK and Norway on the one hand, and of Asia and Australasia excluding China (for brevity referred to as simply the East in this section) on the other. From these figures, the answer to the first question posed above is clearly yes – for small countries the UK and Norway have certainly been extremely lucky in the relative size of their reserves. However, the evidence suggests that the answer to the second question is no, the process has not been driven primarily by geology. Indeed, in some respects, the North Sea's performance has been worse than Asia's.

Below we trace what would be the case for any defence of the relative performance of Eastern producers.

Table 10.2: Reserves and Cumulative Production for the North Sea Compared to Asia and Australasia. 1975, 1985 and 1995. Billion Barrels.

		Norway and UK	*Asia+Australasia (excluding China)*	*India*	*of which Malaysia*	*Indonesia*	*China*
Reserves	end 1975	23.0	21.4	0.9	2.5	14.0	20.0
	end 1985	23.9	18.9	3.7	3.1	8.5	18.4
	end 1995	12.7	20.1	5.8	4.3	5.2	24.0
Cumulative	1976–95	23.3	24.7	3.9	3.3	11.2	17.9
Production	1976–85	8.0	10.6	1.4	1.1	5.6	7.6
	1986–95	15.3	14.1	2.5	2.2	5.6	10.3
Addition to	1975–95	13.0	23.4	8.8	5.1	2.4	21.9
Reserves	1975–85	8.9	8.1	4.2	1.7	0.1	6.0
	1985–95	4.1	15.3	4.6	3.4	2.3	15.9

Source: *Oil and Gas Journal* and own calculations from various sources.

In Table 10.2, note that the two regions had similar estimated reserves in 1975, figures that proved to be considerable underestimates. In both cases, cumulative production in the next two decades was greater than those estimates. Imagine that the reserve estimators in 1975 had been able to see what would be the true cumulative production over the next two decades. They might have been more surprised at the performance in the East, where cumulative production was 3.3 billion more than their figure for total reserves, compared to 0.3 billion in the UK and Norway. If the estimators in 1985 had also been allowed to see the future, they would have noted that 65 per cent of their figure for UK and Norway had been produced in the next decade, compared to 75 per cent in the East.

The relative performance of Eastern producers does not in these terms appear inferior to the North Sea. In terms of additions to reserves (resulting from new discoveries and the upgrading of known fields), it is in fact superior. As shown in Table 10.2, additions in the East amounted to 23.4 billion barrels between 1975 and 1995, only 1.3 billion barrels less than cumulative production over that period. While North Sea additions were greater in the first decade, in the second they lagged far behind those in the East. Including China, the remaining reserve estimates for Asia and Australia in 1995 were higher than those in 1975, despite the total of 42.6 billion barrels produced over the period.

In terms of additions made in the next two decades since 1975, the advances made in India in particular, but also in Malaysia and China, are striking. From the 1975 estimate of just 0.9 billion barrels, India has added 8.8 billion barrels, leaving a 4.9 billion barrel rise in remaining reserve estimates net of two decades production. Within this overall improvement in the figures for remaining reserves, the poor performance of one country stands out. Starting from a 1975 estimate of 14 billion barrels, Indonesia has added only 2.4 billion barrels. Net of production, this resulted in the severe plunge in remaining reserves shown in Table 10.2, and left Indonesia at a lower reserves level than India as of 1995.[3]

The above might seem to imply that Eastern producers have little to learn from the North Sea. In 1975, excluding China, they had a lower estimated reserve base, yet produced more oil than the North Sea over the next two decades, and by 1995 had substantially more remaining reserves. While China produced less than the North Sea over the period, it had by 1995 also sharply improved its relative reserve base. However, Figure 10.2 makes it clear why any assessment of the performance of the upstream oil industry in the East, based on cumulative production and reserve addition arguments, needs to be qualified.

Figure 10.2 demonstrates that the relative cumulative production feature noted above is due to North Sea output having been just 0.2 mb/d in 1975, 2 mb/d lower than Asia and Australasia (excluding China). The UK and Norway thus fell behind in terms of cumulative production while output was still being ramped up. The North Sea caught up in 1983, and for the next nine years the two series remained close. The differential performance of the North Sea is then purely a feature of the 1990s. Between 1990 and 1995, output increased by 2.1 mb/d compared to just 0.36 mb/d in the East. Norway added 1.2 mb/d to reach 3.0 mb/d in 1995. Perhaps more significantly, the UK reversed its decline, adding 0.8 mb/d in reaching 2.75 mb/d in 1995 with virtually all of the increase occurring after 1992. The North Sea managed this surge in output primarily on the existing resource base. Even with the advantage of large reserve additions since 1985, the pace of output increases in the East has

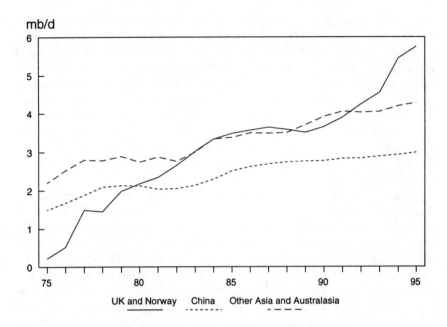

mb/d

Source: *BP Statistical Review of World Energy*, various years.

Figure 10.2: Crude Oil Production in the UK and Norway Combined, China, and Asia and Australasia Excluding China. 1975–95. Million b/d.

been comparatively lacklustre. This raises the question of why such a gulf in output performance should have opened up in the 1990s when it was absent before.

We have already dismissed geology as an explanation. The performance of the North Sea since 1985 has not been driven by new discoveries. Indeed, recent reserve additions have tended to come from existing fields rather than through discoveries, in the UK in the ratio of two to one.[4] Instead, we would advance three main reasons for the improvement, cost savings for any given state of technology, advances in technology, and (particularly in the UK) changes in the fiscal regime.

Cost reductions in the form of efficiency savings have represented a major factor. At the individual company level the degree of 'fat' in the system has been reduced. This was achieved through restructuring and reductions in corporate centre overheads, thus freeing up capital. However, cost savings arising from industry collaboration have perhaps been more important in boosting production. In the UK, this is typified by CRINE (Cost Reduction Initiative for the New Era), an initiative launched by government and industry intended to enhance the relative competitiveness of the North Sea in a low oil price environment. An aspect of CRINE is the attempt to get the UK industry to act in some respects as more of an organic system. In particular, it is the reflection of a belief that full blooded competition can lead to two undesirable externalities, which the initiative has intended to internalize. First, full competition promotes the hoarding of information which has positive externalities, for example the concealment of cost minimizing methods of operating discovered through learning by doing. Secondly, and particularly in a capital intensive industry, it does not tend towards any standardization of capital or method, creating extra costs for capital suppliers to pass on, as well as reducing the fungibility of the knowledge gained about production processes.

Reducing these two externalities by promoting, *inter alia,* information sharing and standardization, helps to produce tangible gains. It should be noted that it has also changed the nature of the competitive process. The UK industry had been one where companies were used to hiding information from competitors as a matter of commercial imperative. To move to the sharing of views about, for example, what constituted the best practice for a given procedure, represented a considerable change in the psychology of doing business in the upstream.[5] North Sea companies then utilized the economics of information externalities.

One further piece of theory that was embraced was that of agency economics, i.e. that optimal contractual arrangements involve a balance of risk and returns. For example, allowing contractors to share in the benefits of any undershooting of costs from the target for a given task, produces incentives on their part that were absent under the previous standard practices.

Beyond the forms of cost savings noted above, additional savings came from the application of new technology and the further development of older technologies. The elements of technological progress are myriad, but we would isolate the following as being the most important. Advances in drilling techniques enabled deviated wells to be sunk from existing platforms to exploit smaller pockets of oil in the vicinity. Horizontal drilling techniques helped to increase production per well. The development of 3-D seismic not only located oil more effectively, it also provided the information necessary for optimal use of the advanced drilling techniques. Latterly, the use of 4-D seismic (the fourth dimension being time) has enabled improvements in the science of the optimal management of producing fields. Improvements in floating production systems have reduced the need to build fixed platforms, as has the use of subsea production systems tied back to an existing platform. In total, if the development of the North Sea could have been made on the basis of the current technology and knowledge base, it would have been done at considerably lower cost, and resulted in output at an even higher sustainable level. As it is, the application of these techniques in the 1990s has resulted in a significant boost in output.

The trend in cost savings achieved is illustrated in Table 10.3, which gives the field lifetime unit cost (including exploration, development, pipeline and terminal equity shares, operating costs, final abandonment, and also a 10 per cent real return to capital). Cost reduction may have begun in the 1980s, but its real impact has been seen in the reduction in costs for fields developed during the 1990s. It should also be noted that the average size of field developed has been falling, and thus the unit costs for the earlier fields include economies of scale that were not realizable for later fields. In other words, the fall in the costs for any given undeveloped field has been greater than that indicated in Table 10.3.

Table 10.3: Unit Costs for UK Oilfields by Date of Production Start at 1995 Prices. £ Per Barrel.

Fields starting production 1980–5	14.0
Fields starting production 1986–90	13.0
Fields starting production 1990–5	9.0
Under development end 1995	8.5
Given development approval in 1995	8.0

Source: UK Department of Trade of Industry (1996), *The Energy Report, Volume 2: Oil and Gas Resources of the United Kingdom*, HMSO, London.

Cost savings and technology have provided a strong impetus to production growth in a mature oil province such as the UK, with spill-over effects into the still maturing Norwegian sector. However, their implementation has been facilitated by what has been the third driving factor behind production increases, and has been especially important in the UK, i.e. the fiscal regime. The key feature of the UK oil taxation system has been its endogeneity with respect to the development of the industry and its operating environment. Tax rates have been adjusted to maintain the international competitiveness of the sector, and to promote the full utilization of reserves. In the early 1980s, UK field revenues in excess of costs attracted PRT (Petroleum Revenue Tax), royalties, plus the standard rate of business taxation, resulting in a compounded marginal tax rate of 85 per cent. Since then the marginal rate for new developments has declined sharply. In 1982, royalties were abolished for new fields. In 1993, the rate of PRT was reduced from 75 to 50 per cent for existing fields, and to zero for new developments. The latter now have a marginal tax rate equal to the general business profits rate of 33 per cent.

Changes in the fiscal regime have produced powerful production incentives. As an example, consider field developments in 1981 and 1996, facing the respective lifetime unit costs (but net of the 10 per cent real return for capital) shown in Table 10.3. Assume that the 1981 development was (unwisely) predicated on the price of oil averaging $40 per barrel (in 1995 terms), with exchange rates taken to be constant. The 1981 fiscal regime, with a compound 85 per cent tax rate, applied to unit revenues net of cost, produces a net of tax profit of £2.50 per barrel. Using an $18 price assumption with the 1996 tax rate of 33 per cent and 1996 costs, produces a net of tax profit of about £3.20 per barrel. Shifts in the fiscal regime have thus more than compensated for the 1986 oil price crash. With the endogeneity of both government policy and industry behaviour in respect of prices, developments in the North Sea are thus more profitable in a low price oil environment than they were in the high price environment.

Part of the fall in costs noted above has also come from the fiscal regime, as the incentives to cut costs have been increased by lower tax rates. Under a regime of high marginal tax rates the conditions are right for lack of attention for costs and inflation of all input prices. For the North Sea experience to be repeated to its full potential in the East, fiscal conditions have to be conducive to gaining a strong position in the competition for international capital, and providing further incentives to capital. In general, fiscal conditions have not been conducive to these aims. In Table 10.4 we have shown the rankings given to fiscal conditions in Asia and Australasia by Barrows (1994),[6] in a survey of a total of 226 regimes. The ranking is the relative attractiveness of the conditions for exploration, based on rates of return adjusted for geological prospectivity.

Table 10.4 shows that there is not a single fiscal regime in Asia that ranks in the top 15 per cent worldwide. Most Asian countries are characterized by relative unattractive conditions, with a clustering in the middle and lower third of the rankings. The various Indonesian terms are all ranked very low, with the conventional terms ranked 222nd out of the 226 regimes. With current fiscal regimes, Asia is unlikely to attract a substantial share of international capital given the options available elsewhere. The changes in the UK detailed above, are such that any company deciding where to place development capital or capital for the introduction of new technology, will put Asia some way down on its list of priorities.

We would argue that the two most important determinants of the future path of Asian production are fiscal conditions and the industrial organization of the upstream sector. With these factors at work, as shown above, the path is not a function of the level of oil prices. Asian production is also not primarily limited by the reserve base (see Table 10.2). We have seen that the relatively poor performance of Indonesia in this context is very much a function of its extremely hostile fiscal regime. For there to

Table 10.4: World Rankings of Asian Fiscal Terms.

Top Third	Middle Third	Bottom Third
35 Pakistan (zone 1 Incentives)	81 Philippines (deep water)	160 Laos
72 Thailand (deep water)	82 Philippines	163 Brunei (offshore)
	85 Pakistan	172 Indonesia (EOR, frontier)
	109 Thailand (offshore)	175 India
	113 China (onshore)	177 Brunei (onshore)
	118 Thailand/Malaysia (JDA)	179 Indonesia (pre-tertiary, frontier)
	119 Malaysia (deep water)	195 Indonesia (pre-tertiary)
	122 China (onshore)	199 Indonesia (marginal, frontier)
	123 Thailand (onshore)	201 Indonesia (94 incentive terms)
	138 Vietnam	213 Malaysia (conventional terms)
	149 Timor Gap	216 Indonesia (marginal fields)
		217 Burma
		222 Indonesia (conventional terms)

Source: Barrows (1994), *World Fiscal Systems for Oil.*

be significant increases in Asian production, fiscal regimes need to be reactive to conditions. Without such reaction, given the relative positions shown in Table 10.4, there is a danger of capital outflow and the failure to realize the full potentialities of production from the reserves. One example of this is the case of Vietnam. Like offshore China, Vietnamese potential has so far failed to live up to many initial expectations. By 1996, the first signs of an exodus were seen with the withdrawal of Shell, and the onset of a decided ambivalence about their Vietnamese operations on the part of a number of other companies.

The Vietnamese taxation system is set on a field-by-field basis, which can lead to the need for renegotiation if reserve estimates prove overly optimistic, due either to geological surprises or a tendency to inflate reserve estimates when bidding for operating licences. An instance of this is the Dai Hung field, the second largest producing field as of 1996.[7] A downgrading of its reserves opened its commerciality to question, and, under the original tax regime, resulted in a complete lack of incentives for the operator (BHP) and its partner companies. Very early in the field's life, output began to fall. In cases such as this, it is not simply that an accommodatory fiscal regime directly helps production levels. A fiscal regime that is both punitive and rigid tends to send potential capital inflows elsewhere.

Government policy in terms of fiscal regimes and their flexibility is then crucial, but so is the structure of the upstream. As we saw in the UK, much has been achieved through competition with the additional utilization of information externalities, particularly in the development of best practice production and development methods. The optimal application of technical progress has had a strong impact, as has the optimization of commercial relationships and the incentives given to reduce costs and increase production. We believe that there would be the potential of very significant gains in production in Asia if these conditions were reproduced, i.e. if there were to be a liberalization of the Asian upstream. This is not a general principle. In Saudi Arabia, for example, considerations of the impact of the organization of the oil sector on production are a relatively minor concern. Saudi output is limited by government policy, and not in any way by the capability of the industry to produce significantly more. Reserves are low cost, and generally lie in exceptionally easy reservoirs, where the impact of new technology would be less dramatic than in other countries. These conditions do not hold in Asia. In particular, state companies with either monopolies or dominant positions in the upstream, without full commercialization and with constraints on access to capital, may not be the optimal way of organizing the industry.

In many cases the role given to foreign capital has been to develop marginal areas which the state company is unwilling, or unable, to develop themselves. For example, in terms of production increases and additions to remaining reserves, in Chapter 3 we suggested that foreign capital would be better used in China in joint ventures in existing fields and production areas in the east of the country, rather than being confined to (relatively unattractive) tracts of the Tarim basin. The combination of high risk and low return is not one that promotes any significant capital inflow.

3. Production Forecasts

Current forecasts of production all to some extent, but with varying degrees, exhibit supply pessimism. Table 10.5 shows four representative forecasts, three published in 1995 and one in 1996, with the sources indicated in the table.[8] At the regional level, the paths are different but mainly predict fairly minor changes up to 2005 and 2010. Forecast (i) sees a slight increase to the year 2000, and then a fall of 0.1 mb/d by 2005, while (ii) in its base case sees static production to 2000, and then a fall of about 0.3 mb/d by 2010. The high case for this forecast has a rise of about 1 mb/d by 2000 followed by decline, the low case a rather suspiciously symmetric fall of about 1 mb/d by 2000 followed by further decline.

The other forecasts are more optimistic, with (iii) putting the path as a gentle increase resulting in the addition of 0.9 mb/d in the fifteen years up to 2010. Forecast (iv) (which excludes Indonesia), sees a stronger increase of 1.05 mb/d in the ten years to 2005. A summary of the conventional wisdom is that average annual changes are projected to be very small, at least in projected base cases. At the level of individual countries, some wide divergences of view emerge, while for China and Vietnam there is a remarkable degree of consistency. The clearest differences lie in the forecasts for Indonesia. Forecast (i) implies a slight fall by 2000 continued to 2005. Forecast (ii) has a fall of nearly 0.5 mb/d by 2000, and then a further 0.3 mb/d by 2010. In contrast, forecast (iv) has no change by 2000, and then a gradual rise. The latter two forecasts treat condensate in a different manner. Adjusting for this, the difference in their figures for 2010 is of the order of 1.3 mb/d, a rather dramatic difference of opinion.

Source (ii), the most supply pessimistic, places Indonesian oil demand in 2000 at 1.25 mb/d, i.e. making Indonesia a net oil importer by this date. Interpolation of the forecasts from this source imply that the net import status begins in 1998. The low case for forecast (ii), while not presented disaggregated by the source, presumably implies an even more accelerated move to net importer status.

Table 10.5: Comparison of Forecasts of Oil Production in Asia and Australasia. 2000, 2005 and 2010. Million b/d.

	Forecast	1993	1995	2000	2005	2010	Notes
Total	(i)	6.76	-	7.0–7.2	6.9–7.1	-	
	(ii) base	6.72	-	6.77	-	6.50	
	(ii) high	6.72	-	7.70	-	7.50	
	(ii) low	6.72	-	5.70	-	4.50	
	(iii)	-	6.66	6.89	7.18	7.56	
	(iv)	-	5.68	6.42	6.73	-	Excludes Indonesia
China	(i)	2.89	-	3.1–3.3	3.2–3.4	-	
	(ii)	2.88	-	3.20	-	3.45	
	(iii)	-	3.00	3.29	3.61	3.95	
	(iv)	-	2.99	3.32	3.52	-	
Indonesia	(i)	1.39	-	1.2–1.4	1.1–1.3	-	
	(ii)	1.53	-	1.05	-	0.75	
	(iii)	-	1.34	1.34	1.55	1.83	
Vietnam	(i)	0.12	-	0.3–0.4	0.35–0.45	-	
	(ii)	0.13	-	0.36	-	0.45	
	(iv)	-	0.18	0.30	0.35	-	
India	(i)	0.55	-	0.6–0.7	0.6–0.7	-	
	(ii)	0.56	-	0.65	-	0.75	
	(iii)	-	0.66	0.63	0.53	0.44	Includes Pakistan
	(iv)	-	0.70	0.75	0.80	-	
Australia	(i)	0.52	-	0.4–0.45	0.2–0.25	-	
	(ii)	0.50	-	0.35	-	0.15	
	(iii)	-	0.54	0.46	0.39	0.32	Includes New Zealand
	(iv)	-	0.58	0.71	0.61	-	

KEY:
(i)   Koyama (1995)[9]
(ii)   Fesharaki, Clark and Intarapravich (1995)[10]
(iii) Considine and Reinsch (1995)[11]
(iv)   Varzi (1996)[12]

We believe that the forecasts presented above tend to be too supply pessimistic, and that, with changes in fiscal conditions and in some cases upstream liberalization, the scope for production increases is large. In fact, to suggest that regional production can quite conceivably rise from 7.3 mb/d in 1995 to 9 mb/d or more in 2010, seems considerably less outrageous than any suggestion made in 1980 that North Sea output could rise from about 2 mb/d to 6 mb/d by 1995 (even in the face of a substantial fall in real oil prices). The upside to production is potentially large, but only fully realizable if government policy becomes more accommodatory than it has been in the past.

In the specific case of Indonesia, we are sceptical of forecasts that see the country in the near future becoming the first member of OPEC to move to net oil importer status. The scope for government action is wide, on the demand side in fostering further fuel substitution and demand suppression, and on the supply side in providing a better commercial environment. Net importer status would represent a governmental policy failure on a massive scale. Table 10.4 is again crucial. While Indonesian terms remain so unattractive, additional capital flows will be limited, and existing capital will have no incentives to push forward what is possible on the basis of the existing reserve base.

It should be noted that territorial disputes are affecting exploration drilling in some offshore areas, particularly in the South China Sea.[13] For example, the area of and around the Spratly Islands is claimed, in full or in part, by China, Taiwan, Vietnam, the Philippines, Brunei and Malaysia.[14] In some cases, such as in the disputed area between Thailand and Malaysia, or in the Timor gap claimed by Australia and Indonesia, agreement has been reached for joint development of resources. In many other cases the allocation of exploration rights and sovereignty remains either undefined or contested.[15] We see these unresolved border disputes as primarily having implications for international relations, rather than at this point creating any major variable that impacts on the future production potential of the region.[16]

4.   Official Pricing Policies

The Asian crude oil market is characterized by the need to use prices, combined with a dearth of markets that generate prices. The bulk of exports of Asian crude oils are term contracts, requiring some method of price calculation. Taxation reference prices are required in countries where there is equity production by companies. Further, some sort of internal transfer and accounting price has to be used in even an imperfectly marketized national oil company.

The extremes of the solutions to this problem are to declare prices by governmental fiat, or to seek to create proxy or quasi market prices where none exist through reference to those that do. In addition, these prices or references can be created *ex ante* or *ex post*. We discuss these alternatives at greater length in the next chapter, in the context of Saudi Arabian pricing policy. Here, we note that official pricing systems in Asia are monthly and retrospective, and have tended to fall into an uncomfortable grey area between fiat and market responsiveness.

To illustrate this tendency, we consider Indonesian pricing. The matrix of prices to be derived is considerable, as the country produces a large number of crude streams. Table 10.6 shows the composition of exports by grade. By 1994, only Minas and Duri (for both of which the equity producer is Caltex) had an export level above 100 thousand b/d. Only four other streams had export levels above 50 thousand b/d, namely Arun (Mobil), Widuri (Maxus), Belida (Conoco) and Cinta (Maxus). Beyond these six, there are just over twenty other export streams.

The characteristics of Indonesian crude oils in respect of gravity and sulphur content are shown in Appendix Table A4.1. With the exception of Walio, they are very low in sulphur, including the unusual combination of a heavy but sweet crude represented by Duri. What sets many Indonesian crudes apart (including Minas), is a very high wax content. These waxy crudes are to a large extent non-fungible with other oil, and

Table 10.6: Indonesian Crude Oil Exports by Grade. 1985–94. Thousand b/d.

	1985	1986	1987	1988	1989	1990	1991	1992	1993	1994
Minas	230	315	330	368	356	325	271	243	211	243
Duri	52	64	60	50	76	94	130	117	145	152
Arun Condensate	103	103	99	109	120	110	96	94	95	92
Widuri	-	-	-	-	-	4	128	101	74	79
Belida	-	-	-	-	-	-	-	2	40	57
Cinta	81	80	66	61	69	68	38	37	45	50
Attaka	22	19	10	14	15	17	22	23	25	42
Handil	61	58	46	34	37	31	49	33	27	26
Bontang	-	-	-	-	-	-	-	-	-	23
Lalang	33	26	40	22	24	19	25	24	23	21
Badak	22	19	12	9	17	18	19	23	26	17
Ardjuna	41	18	13	9	6	8	22	18	15	13
Anoa	-	-	-	-	-	-	15	11	9	11
Kakap	-	9	15	10	6	21	21	11	6	9
Jatibarang	9	10	8	7	7	5	5	6	7	6
Sepinggan	8	6	5	3	4	7	8	8	3	5
Walio	31	30	28	23	22	21	17	17	4	5
Arimbi	3	2	5	6	6	4	5	5	5	4
Bunyu	8	7	6	6	6	6	6	4	4	4
Bekapai	9	7	6	3	5	4	4	6	4	4
Camar	-	-	-	3	-	-	2	8	3	2
Udang	13	9	11	8	6	6	3	3	2	1
Bima	-	-	19	13	12	10	7	5	-	-
Other	10	40	18	2	2	7	6	2	1	4
SMC	74	71	-	-	-	-	-	-	-	-
TOTAL	806	895	798	756	799	790	906	801	776	888

Source: Own calculations from *Petroleum Report Indonesia*, various years, Embassy of the United States of America, Jakarta.

in refining present the special demands caused by production of LSWR. Significantly, as we noted in Chapter 4, these crudes can be used for direct burning in power stations, meaning that the market conditions they respond to can be very different from other regional and extra-regional crudes.

The problem of price setting is compounded by the rather special nature of the waxy output streams. It is clear from Table 10.7 that Japan has consistently been the major market for Indonesian exports, with the amounts being reflected by those reported in Table 4.1 for Japanese power station crude burn. In their primary use in the major market, these exports are in competition with other fuel sources, and in particular act as a swing fuel in response to variations in both electricity demand and supply (in particular due to variability in hydroelectric availability). In short, this is not crude oil in the normal sense of the term.

Note also in Table 10.7 the growth in exports to China. As was detailed in Chapter 3, Indonesia has been the largest exporter to China. This is due to the close similarity (including wax properties) between Minas and Daqing, given the very limited ability of most Chinese refineries to switch to crudes other than the domestic crudes on which their operations have been predicated. If Minas can be pulled by the burning crude market, we might expect that the behaviour of Minas prices will be different from other crude oils. This is confirmed by Figures 10.3 and 10.4. In Figure 10.3, we have shown the monthly average spot price differentials between Tapis and Minas, and also between Tapis and the Australian grade Gippsland. Compared to the latter, note the sharp movements in the former, which often has very large monthly changes, and the downward

Table 10.7: Destination of Crude Oil Exports from Indonesia. 1985–94. Thousand b/d.

	1985	1986	1987	1988	1989	1990	1991	1992	1993	1994
Japan	372.4	382.5	405.6	426.6	444.4	476.8	505.0	421.5	366.2	405.8
Korea	55.5	49.0	44.1	39.6	30.1	52.5	93.9	101.0	93.6	103.5
USA	258.7	251.8	216.2	217.0	197.1	118.5	103.3	65.3	80.8	97.4
China	0.0	8.8	0.0	4.5	32.6	36.0	78.9	91.6	97.0	93.1
Singapore	23.0	75.3	45.9	8.0	26.8	27.0	12.0	12.5	16.3	80.0
Taiwan	23.0	19.8	26.0	26.2	31.1	45.1	42.8	39.2	43.2	56.1
Australia	9.8	21.5	32.5	27.1	24.2	22.4	58.6	61.5	56.8	36.9
Philippines	13.4	11.5	4.8	0.0	0.0	1.9	3.9	4.9	7.0	7.5
Caribbean	38.7	58.9	20.3	0.0	0.0	0.0	0.0	0.0	0.0	0.0
Others	11.7	15.2	2.2	6.8	12.2	12.5	7.1	3.4	15.3	7.4
TOTAL	806.2	894.1	797.5	755.9	798.5	789.9	905.5	800.7	776.1	887.6

Source: Own calculations from *Petroleum Report Indonesia*, various years, Embassy of the United States of America, Jakarta.

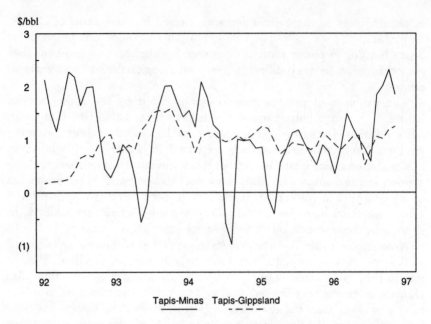

Figure 10.3:   Spot Price Differentials Between Tapis and Minas, and Tapis and Gippsland. Monthly Averages, January 1992 to November 1996. $ Per Barrel.

spikes which occasionally lead to Tapis trading at a discount to Minas. The behaviour of Tapis and Minas prices are clearly very different, certainly when compared to the gentler evolution of the Tapis to Gippsland differential.

Figure 10.4 compares Minas with two Middle East crude oils, namely Dubai and Oman. Compared to the less extreme movements in the Dubai to Oman differential, Minas swings wildly against Dubai, with the differential commonly moving over a $3 per barrel or more range within a twelve-month period. As we will see below, the choice of crudes used in these two figures is not random, and the swings in differentials between them and Minas are of some importance.

If one were to design a pricing system from scratch, it is rather unlikely that anything similar to the actual Indonesian system would come to mind. The rather convoluted method for computing the retrospective official prices was described in Horsnell and Mabro (1993)[17] as being 'clumsy'. In the period since that description was conferred, the system has been changed, and it is now even more complicated. The Indonesian Crude Price (ICP) for a given grade, call it X, in a given month is calculated as follows. The first element is the value of a basket of five equally weighted crude oils, namely Dubai, Gippsland, Minas, Oman and

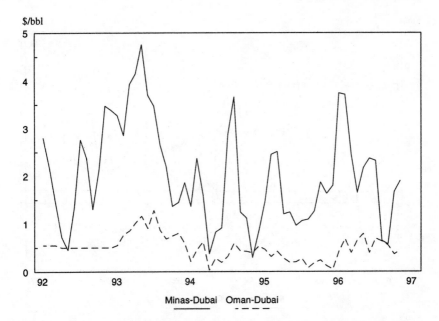

Figure 10.4: Spot Price Differentials Between Minas and Dubai, and Oman and Dubai. Monthly Averages, January 1992 to November 1996. $ Per Barrel.

Tapis (and hence the choice of crudes in the two preceding figures). The individual crude values used in the basket are derived from the APPI assessments for each. The 52-week average differential between the APPI assessment of X and the basket is then added to the monthly average of the basket, and the resulting price makes up 50 per cent of the ICP. If X does not have an APPI assessment, then a fixed differential is used.[18] The other 50 per cent is, if they exist, the average of other spot market quotes for X, with, as of 1996, Platt's and RIM being the agencies used.

If the above description seems complicated, that is simply a reflection of the reality. The ICP procedure is a true Frankenstein's monster, albeit consisting of slightly more body parts than the original. The formulation lacks any internal logic. First, consider the impact of the use of a 52-week average in the first part of the formula. We have seen above that the spot value of Minas swings dramatically against the other four components of the basket, and indeed it also swings sharply against the ICP basket as a whole. Figure 10.5 shows the differential between Minas and the ICP basket, expressed both as a monthly average and a yearly average.

If the differential against the ICP basket were to be stable, then the yearly average term need not introduce a major bias. However, as Figure 10.5 shows, the bias can often be considerable. The difference between

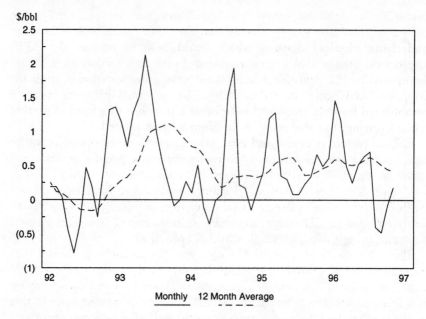

Figure 10.5:   Spot Price Differentials Between Minas and ICP Basket. Monthly and
Twelve Monthly Averages, January 1992 to November 1996. $ Per Barrel.

the two lines is a proxy for the amount that the first part of the formula
will differ from spot values. That amount is substantial, frequently more
than a dollar per barrel. The introduction of the second part of the
current formula was primarily a response to complaints that the yearly
average component was causing large biases. As such, bringing in spot
quotes, while reducing the weighting of the basket plus yearly average
differential, simply produces something that is only half as biased as it was
before the change.

There is very little logic in the choice of crudes in the ICP basket. For
instance, couching the discussion in terms of the Minas ICP, possible
reasons for the inclusion of the other crudes do not seem to bear
inspection. First, they could be included as price proxies because they are
deemed to be competitive with Minas. However, we have noted above
that Minas responds to different market forces, and Figures 10.3 and 10.4
clearly demonstrate the lack of price relationship. In addition, any effects
arising from competition should be contained already within the market
price of Minas.

Secondly, they could be included because of a fear that reliance on a
Minas assessment alone could leave it open to downwards manipulation.
However, any such consistent effect would be completely fed through into

the ICP through the yearly average adjustment term. Thirdly, the motivation could be simply a belief that the other crudes have a broader underlying physical market, which could help to anchor the APPI assessments. Again, this is not convincing. In the next section we find that (compared to 112 deals for Minas) there were no reported spot deals for Gippsland in 1995 (or indeed in 1994). The current ICP system is thus a complicated formula employed to indicate a pretence of a form of market related pricing, but the result is far from market related.

Official (where relevant) and term sales prices in other countries in the region are in general related by formulae involving premia or discounts either to the ICP or APPI quotes for Indonesian oil, or to APPI quotes for Tapis. The former category includes Vietnamese term sales (Minas APPI quotes), and Chinese term sales (average of Minas and Cinta ICPs and APPI quotes). The latter category includes Papua New Guinea and Australia, where producers link to the Tapis APPI.

The ICPs are at least derived using a transparent formula, even if that formula does not produce meaningful numbers. By contrast, the official selling price for Malaysian oil was, until 1995, set by Petronas according to a formula which was not made public. The lack of transparency in this 'black box' method, led to a suspicion on the part of buyers that there was more than an element of pure fiat in the resulting prices, and hence created pressure for a more demonstrably market related basis to the system. The change that arose involved deriving the Tapis OSP from the weighted average of a set of published Tapis assessments, the original basket comprising APPI and Platt's (both 30 per cent weight), and Reuters and Telerate (both 20 per cent weight). In addition, and by a less transparent method, a premium is added to the weighted average of the assessments to arrive at the OSP. Other Malaysian crudes have OSPs set as differentials to the Tapis OSP.

5.   Spot Trade in Asian and Australasian Crude Oils

Spot markets for Asian and Australasian crudes are extremely thin, with the bulk of crude oil production either moving through integrated channels, between state corporations in the same country, or through term contracts with regional refiners. Not only are spot deals few in number, the price level information they carry is all but negligible. Table 10.8 shows the number of reported deals from 1986 to 1995, divided by the country of origin of the crude.

In 1995 there was on average only one deal done per calendar day, or less than 1.4 per trading day. This translates to about 7 per cent of total production in the region. The low proportion of recorded spot deals out

Table 10.8: Spot Deals for Asian and Australasian Crude Oil by Country. 1986–95. Reported Deals.

	1986	1987	1988	1989	1990	1991	1992	1993	1994	1995
Indonesia	196	203	125	132	120	176	163	184	263	223
Malaysia	73	60	102	127	59	85	86	88	81	79
Australia	17	25	28	38	19	27	34	23	25	27
Papua New Guinea	-	-	-	-	-	-	22	47	30	19
Vietnam	-	1	-	-	-	1	-	2	6	14
China	5	1	6	2	-	9	10	6	3	-
Others	6	1	-	2	-	5	8	3	1	-
TOTAL	297	291	261	301	198	303	323	353	409	362

Source: Own calculations from Petroleum Argus database.

of total production is largely a function of the limited spot trade in Chinese crude, and the absence of spot trade in Indian crude. Chapter 5 detailed the fall in Chinese crude exports, with most of these moving under term contracts into Japan. As shown in Table 10.8, the bulk of spot trade is in Indonesian crude oil, with the figures implying that reported spot deals in 1995 were some 16 per cent of total production, and about 30 per cent of exports.

Spot trade by grade of crude oil is shown in Table 10.9 for the four main countries shown in the previous table. This demonstrates the extent of the deficit in Eastern markets in terms of centres of price formation. Only trade in Minas surpasses two deals per week, with Tapis following at one deal per week. No Australian grade has substantial spot liquidity, and even trade in Gippsland (as noted in the last section, quotations for which are used in the construction of the Indonesian ICP) fell away sharply from 1988 to very low, and sometimes zero levels, in the 1990s.

As we describe below, very few spot deals in regional crudes carry price level information. Most are priced through formulae, for example the Minas spot deals recorded above are virtually all transacted as differentials from the (unknown at the time of the deal) ICP for Minas. Spot trade in Minas has grown considerably, despite the fall in exports since the 1980s that was shown in Table 10.6. A major reason for this growth is the appearance of a major new buyer in the 1990s, namely China, whose use of Minas as a swing refinery feed meant sourcing from the spot market.

Table 10.10 shows the main buyers and sellers of spot Indonesian crude. Sinochem has emerged as the largest spot buyer, and note also the presence of Nicor as a buyer, a company that we identified in Chapter 6 as a user of refinery processing deals in China. Compared to the first

Table 10.9: Reported Spot Deals for Asian and Australasian Crude Oils By Grade. 1986–95. Number of Deals.

	1986	1987	1988	1989	1990	1991	1992	1993	1994	1995
Indonesia										
Minas	22	45	43	43	47	66	67	70	107	112
Belida	-	-	-	-	-	-	1	11	16	28
Duri	15	21	14	28	23	22	20	44	56	23
Widuri	-	-	-	-	-	38	29	37	42	23
Cinta	31	29	12	8	14	1	13	3	11	10
Attaka	27	18	12	9	3	15	-	1	7	9
Bontang Mix	-	-	-	-	-	-	-	-	1	5
Arjuna	5	3	3	6	5	5	3	2	8	4
Kerapu	-	-	-	-	-	-	-	-	-	3
Lalang	10	11	11	11	5	1	6	2	2	2
Bekapai	6	6	1	4	2	6	5	1	2	2
Arun Condensate	43	25	15	5	2	1	-	-	2	1
Bima	-	13	5	5	1	2	3	-	-	-
Walio	19	6	3	1	-	-	-	-	2	-
Kakap	3	12	4	4	15	13	2	1	1	-
Handil	7	1	-	1	-	4	4	5	-	-
Badak	-	1	-	-	-	-	3	6	5	-
Udang	4	6	1	3	-	-	-	-	-	-
Others	4	6	1	4	3	2	7	1	1	1
Malaysia										
Tapis	51	32	68	75	33	58	72	78	58	52
Labuan	18	24	23	41	18	13	11	10	16	14
Dulang	-	-	-	-	-	11	3	-	3	9
Miri	3	2	8	10	7	2	-	-	4	4
Bintulu	1	2	3	1	1	1	-	-	-	-
Australia										
Thevenard (Saladin)	0	0	0	0	4	9	10	5	7	11
Cossack	0	0	0	0	0	0	0	0	0	7
NW Shelf Condensate	0	0	0	1	0	2	0	0	1	6
Griffin	0	0	0	0	0	0	0	1	10	2
Gippsland	14	20	17	5	1	5	5	7	0	0
Jabiru	2	5	9	14	6	4	7	2	0	0
Skua	0	0	0	0	0	0	7	3	3	0
Talisman	0	0	0	9	2	0	1	0	0	0
Challis	0	0	0	1	4	5	1	1	0	0
Others	1	0	2	8	2	2	3	4	4	1
Papua New Guinea										
Kutubu	-	-	-	-	-	-	22	47	30	19

Source: Own calculations from Petroleum Argus database.

Table 10.10: Main Buyers and Sellers in Indonesian Spot Trade. Rank and Market Share. 1986–1995. Per Cent.

	1986–9		1990–2		1993–5	
Buyers						
1	Nippon Oil	9.9	Sinochem	11.6	Sinochem	12.6
2	Mitsubishi	6.9	BP	8.4	Exxon	8.8
3	Caltex	6.3	Exxon	5.8	Vitol	6.9
4	PRI	6.0	Nicor	5.3	Chevron	6.7
5	Mobil	5.8	Mobil	5.0	Mobil	5.2
6	BP	4.2	PRI	4.5	BP	4.8
7	Shell Australia	4.2	Yukong	4.5	Nippon Oil	4.5
8	Exxon	4.0	Caltex	4.5	Nicor	4.1
9	Kerr McGee	4.0	Unocal	4.2	CPC Taiwan	3.4
10	Lucky Goldstar	3.8	Mitsubishi	3.7	Caltex	3.3
11	Yukong	3.8	Nippon Oil	3.4	Mitsubishi	3.1
12	Cosmo	2.8	Taiyo Oil	3.2	Cosmo	2.8
13	Toyo Menka	2.8	Itochu	2.6	BHP	2.4
14	Union	2.8	Phibro	2.4	Taiyo Oil	2.4
15	Idemitsu	2.2	Marc Rich	2.4	Shell Australia	2.1
Sellers						
1	Pertamina	15.0	Itochu	13.7	Texaco	11.8
2	Natomas	11.1	Caltex	10.8	Itochu	11.3
3	Arco	10.5	Pertamina	9.2	Pertamina	10.0
4	PRI	7.1	Mitsubishi	7.2	Maxus	7.0
5	Itochu	6.1	CFP Total	6.3	Chevron	7.0
6	Union	5.7	Arco	5.8	Mitsubishi	5.4
7	Mobil	4.6	Tomen	5.3	Vitol	3.9
8	Toyo Menka	4.6	Chevron	4.8	Conoco	3.6
9	CFP Total	4.2	Marathon	4.3	Arco	3.3
10	Marathon	3.8	Maxus	4.3	Tomen	2.8
11	Mitsubishi	3.8	Kanematsu	3.6	Marubeni	2.5
12	Caltex	3.2	Texaco	3.4	Caltex	2.3
13	LASMO	3.2	BP	3.1	CFP Total	2.1
14	Exxon	2.6	LASMO	2.9	J Aron	2.1
15	Maxus	2.6	Marubeni	2.9	Nissho Iwai	2.1

Source: Own calculations from Petroleum Argus database.

period shown, the importance of Japanese and Korean companies in Indonesian spot trade has been greatly reduced. The most important sellers are Caltex (together with its two parent companies), Pertamina and its affiliates, and Itochu.

The information generated by the spot market in regional crudes carries very little price setting information. Table 10.11 shows the proportion of reported spot deals that were transacted on outright price terms, rather than agreeing a differential in a formula In 1986, virtually all trades were carried out in absolute prices. In April 1989, Indonesia introduced the ICP system (then given by the first of the two parts of the current formula detailed in the last section). Outright trades all but vanished immediately, and have remained at very low levels throughout the 1990s. Indonesian spot trade is now almost exclusively conducted in terms of differentials from ICPs.

Table 10.11: Outright Price Deals as Percentage of Total Spot Trades by Country of Crude Oil Origin. 1986–95. Per Cent.

	Indonesia	Malaysia	Other	Total
1986	99.5	97.3	96.2	98.6
1987	51.7	96.7	90.0	64.8
1988	86.4	99.0	74.3	89.7
1989	12.1	88.2	20.9	45.4
1990	1.7	71.2	5.3	22.7
1991	8.0	62.4	7.5	23.3
1992	4.3	20.9	1.4	8.0
1993	4.4	40.9	3.7	13.3
1994	2.3	35.8	1.5	8.8
1995	1.4	21.5	0.0	5.5

Source: Own calculations from Petroleum Argus database.

Malaysian trade has continued to be discussed in price levels, but as shown in Table 10.11, the proportion not done through a formula (primarily related to APPI quotes for Tapis) has fallen sharply over the 1990s. Spot trade in other regional crudes is now entirely done in formula terms, with APPI Tapis and ICPs being the dominant markers (following the division by country listed in the last section).

In 1986, the total number of outright price deals in regional crudes amounted to more than one per trading day. In 1995, the 5.5 per cent figure reported in Table 10.11, translates to just twenty in total over the year. A region, producing over 7 mb/d, then generates one piece of absolute price level information from actual trades every two and a half weeks. Asian crude prices may be market related, in that they are related to either assessment by market participants on pricing panels, or to the results of market talk gleaned by price reporters. However, they are certainly not *marketplace* related, in that the anchor of physical price level setting (rather than differential setting) actual deals is almost entirely absent from the process.

6.   Conclusions

We began this chapter by raising the issue of three deficits in the Asian oil market. The existing physical deficit is large, some 10.5 mb/d of crude oil and oil products in 1995. There is also a deficit at the margin, with regional crude oil supply falling ever further behind oil demand. We have however suggested circumstances in which the contribution of regional production in meeting incremental demand can be considerably greater than the dominant current view.

We have stressed that oil production is primarily driven by fiscal conditions, industry structure and geology, but not by prices. Asia's primary problem is not a lack of reserves (although clearly higher reserve addition rates would be helpful). The core problems are hostile and inflexible fiscal systems, a lack of commerciality, and deficiencies in the industrial structure of the upstream that lead to a lack of incentive provision. We have suggested that the region is capable of producing at a much higher level than at present. The future path of production is highly endogenous to governmental policy stances. We would add that a failure to achieve significant production increases would represent governmental rather than geological failure. As shown above, the superior performance in the 1990s of the North Sea compared to Asia has been driven by costs, incentives and fiscal conditions, and not by a superior reserve base or exploration success.

Oil refining is an industrial process, which tends normally to place it outside of most resource nationalism arguments. These factors do still have a strong presence in policy towards the upstream in several Asian countries. While the view taken of the role of the upstream is one of pure rent collection, i.e. extracting an amount of oil (at rates determined purely by nature) from easily accessible reservoirs, then the chances of major change are limited. The alternative view is to see the upstream as an industrial process itself. It has a production function, and output (and to a large extent the size of the recoverable resource base) is determined by the mix of factors of production and the type of technology employed, and the incentives given to those factors. The most important factor is capital, and Asia is currently faring extremely badly in the competition for global upstream oil capital, and failing to provide incentives for what capital it does have.

The deficit in crude oil price formation is total. We found in Part III that to some extent there is a similar void in oil product markets. However, while logistics were a factor for oil products, the primary source was a regional lack of marketization, and a reluctance (bar one notable exception) by traders to trade in absolute price levels. In crude oil markets, logistical factors alone are sufficient to preclude price setting physical markets.

Crude oil streams are either too small, too monopolistic or monopsonistic in trade, or affected by special market considerations. On top of those problems comes the smothering effect of retrospective tax and official price mechanisms, whose structures are obtuse and often unrelated to market levels.

In all, Asia has no crude oil stream capable of supporting even an informal forward market, let alone a futures market. By the mid-1990s, total production across the region of over 2.5 billion barrels annually, generated only twenty pieces of absolute price information from deals for physical oil. One swaps market, of rather limited liquidity, has arisen. The paper Tapis market does serve a useful purpose in providing some definition to the time curve of prices. However, it generates little useful information in terms of the absolute levels it trades at, being based on the (forward) assessments of a very thin underlying physical market. The price discovery deficit leads us in the next two chapters to search in other regions for the market that fills the Asian deficit.

Notes

1. The gravities and sulphur contents of major crude streams are given in Appendix 4.

2. For convenience, in this section we use the term North Sea to refer to the total of UK and Norwegian output. Two provisos should be noted. First, there are other producers in the North Sea, particularly Denmark whose output stood at 190 thousand b/d in 1995. Secondly, a small portion of UK output is in fact produced onshore.

3. Other than the four countries shown in Table 10.2, at the end of 1995 proved reserve estimates for the other Asian and Australian countries with estimates greater than 0.01 billion barrels (bb) were as follows. Australia 1.5 bb, Brunei 1.4 bb, Burma 0.05 bb, Japan 0.05 bb, New Zealand 0.1 bb, Pakistan 0.2 bb, PNG 0.4 bb, Philippines 0.2 bb, Thailand 0.2 bb and Vietnam 0.5 bb.

4. The UK Offshore Operators Association found this ratio in comparing their 1996 survey with that of 1989. See *Oil and Gas Journal*, 3 June 1996.

5. Such changes of course are not immediate. For example, when one major company opened up a new development to the full inspection of and dialogue with other companies, it was noticeable that many in the industry were still looking for a catch.

6. Barrows (1994), *World Fiscal Systems for Oil*, Barrows, New York.

7. As of 1996, Vietnamese output of about 150 thousand b/d is dominated by the Bach Ho field at 135 thousand b/d, with Dai Hung at 15 thousand b/d and the Rong field at below 3 thousand b/d. A series of other fields is in the development or appraisal stage, of which the Rang Dong field is currently the most significant. See *Weekly Petroleum Argus*, 20 May 1996 for a list of future prospects.

8. The various base year figures differ from each other and from those shown in Table 10.1 due to a combination of different primary sources and the treatment

of condensate production. Where country coverage varies in a significant fashion, this has been noted in the table.

9. Ken Koyama (1995), 'Outlook for Oil Supply and Demand in Asia-Pacific Region and Role to be Played by Japan's Oil Industry', *Energy in Japan*, March 1995, no. 132.

10. Fereidun Fesharaki, Allen L. Clark and Duangjai Intarapravich (1995), *Pacific Energy Outlook: Strategies and Policy Imperatives to 2010*, East–West Centre, Honolulu, Hawaii

11. Jennifer I. Considine and Anthony E. Reinsch (1995), *Battle for Market Share: World Oil Projections, 1995–2010*, Canadian Energy Research Institute, Calgary, Alberta.

12. Mehdi Varzi (1996), *OPEC and non-OPEC, The Battle for Market Share*, Kleinwort Benson Research, London, reported in *Middle East Economic Survey*, 6 May 1996.

13. See Mark J. Valencia (1985), *South-East Asian Seas: Oil Under Troubled Waters: Hydrocarbon Potential, Jurisdictional Issues and International Relations*, Oxford University Press, Oxford.

14. For a full discussion of the Spratlys issue see Mark J. Valencia (1995), *China and the South China Sea Disputes: Conflicting Claims and Potential Solutions in the South China Sea*, Institute for International Strategic Studies Adelphi Paper 298, Oxford University Press, Oxford.

15. Some disputes may also affect the pace of general cooperation and capital flows between countries. A notable example of this is the Kuril Islands, and the associated source of friction between Japan and Russia.

16. In particular, the potential of the Spratly Islands area should not be overstated. Given the lack of any hard evidence to date, a climate has been created where what would appear to be very wild claims can be made about potential reserves. For example, the highest figure the author has seen is 250 billion barrels, roughly equivalent to the reserves of Saudi Arabia. There is something similar in some of the claims made about Spratlys reserves to those made about offshore China in the late 1970s.

17. Paul Horsnell and Robert Mabro (1993), *Oil Markets and Prices*, op.cit.

18. See Appendix 1 for the list of Indonesian crude oils assessed by APPI, and for the details of APPI price methodology.

CHAPTER 11

THE MIDDLE EAST EXPORTERS AND ASIA

1. Introduction

In the last chapter we found three salient properties of the Asian crude oil market. First, indigenous production falls far short of demand, leaving the bulk of crude oil to be supplied from outside the region. Secondly, Asia has no market that serves as a centre of crude oil price formation, leaving price determination to the markets of other regions. Finally, we suggested that without a combination of changes in fiscal structures, the introduction of new technology after a redefinition of the stance of many governments on the role of foreign exploration capital, and also some geological serendipity, Asian supply was unlikely to make a major contribution to satisfying incremental crude oil demand. These three related gaps, i.e. crude oil flows in bulk and at the margin, and in centres of price determination, are further addressed in this and in the following chapter.

The first of these issues leads us naturally on to a consideration of the role of the Middle East in the Asian crude oil market. The Middle East OPEC members, together with Oman and Yemen, form the bulk of contemporary flows into Asia, and the next section provides an overview from the perspective of the OPEC members. The question of price determination brings us to two related discussions. Because of the absence, and indeed the impossibility, of an Asian price setting market, the focus falls on the market for forward Dubai, which is considered in Section 3. Dubai may be price setting in that it reflects market conditions, but the greatest impact on the market comes from the policies of the major exporters, and in particular Saudi Arabia. Section 4 considers the structure of Saudi Arabian pricing policy to the East, and its impact on the relative price of oil across world regions, with Section 5 considering other Gulf producers and the limited spot markets of the Middle East. Section 6 offers some conclusions.

The question of where the margin of crude oil supplies for the Asian market lies is perhaps the most important. Ultimately, the true centre of price determination lies with the marginal market regardless of its absolute size relative to intra-marginal flows. The hypothesis we address in the next chapter is that the margin no longer lies with Middle East crude oil, but with crude oils from other regions whose prices are Brent related. This represents a fundamental dilemma for Middle East OPEC producers. While production policy is primarily determined by the residual from world demand of non-OPEC supply, then strong growth of production outside OPEC acts as a constraint on OPEC growth. On the other hand, a bid to regain the margin by increasing production regardless of non-OPEC developments, carries strong implications for prices. The point is made here to note that our discussion of what constitutes the bulk of Asian imports carries no presumption that there is any one to one relationship between incremental Asian demand and incremental Middle East supply. Indeed, our discussion in Chapter 12 will show that the relationship may be weak in the absence of a fundamental shift in Middle East production and pricing policy.

2.   Oil Movements into Asia

In this section we concentrate on sales into Asia by the six Middle East members of OPEC, namely Iran, Iraq, Kuwait, Qatar, Saudi Arabia and the UAE. In addition to these countries, significant flows east come from non-OPEC Oman and Yemen. The economic boom in Asia has led to a strong increase in Middle East OPEC sales into the region; however the expansion of the market has been far from a linear process. To an extent, much of this growth was a reclamation of markets lost in the 1980s during the era of OPEC administered prices. Table 11.1 shows the total eastbound exports (i.e. exports to Asia and Australasia combined) from the six countries over the period from 1970 to 1995.

Table 11.1 shows that the eastbound flow was above 5 mb/d in 1975, in the context of which the 1995 figure of some 6.8 mb/d represents somewhat less than spectacular growth. The early 1980s saw a sharp retrenchment in exports east, which were squeezed by three main factors. First, overall output levels in the Middle East were cut sharply to attempt to defend high official prices. Secondly, those high prices dampened demand, particularly in Japan, where in addition explicit attempts were made to diversify away from oil in general and Middle East oil in particular. The third factor was a growth in indigenous Asian production as quantified in Chapter 10.

The total result of these effects was that exports east fell by 46 per cent between 1980 and 1985. It took six years from 1985 to recoup this loss; it was not until 1991 that these exports matched their 1980 level. Table 11.1 also shows that, in terms of volume of exports, for some countries the Eastern market has still not returned to its former importance. Most notably, in 1995, Iranian and Kuwaiti exports east were below their 1970

Table 11.1: Middle East OPEC Eastbound Crude Oil Exports. 1970–95. Thousand b/d.

From	1970	1975	1980	1985	1986	1987	1988	1989	1990	1991	1992	1993	1994	1995
Iran	1827	1482	332	419	465	470	480	578	616	782	940	987	1155	1014
Iraq	34	206	469	123	225	220	275	372	230	0	0	0	0	0
Kuwait	831	927	695	323	354	294	322	397	326	61	495	555	582	713
Qatar	61	49	197	220	238	208	290	283	294	295	338	334	318	330
Saudi Arabia	1025	1992	2882	999	750	804	984	1112	1675	2418	2668	2769	2671	2808
UAE	219	471	682	745	855	923	1123	1354	1516	1865	1735	1745	1830	1887
OPEC Middle East	3998	5128	5258	2829	2886	2919	3475	4097	4657	5420	6177	6390	6556	6752
To														
Japan	2787	3390	2923	1454	1622	1725	2335	2241	2505	2650	2878	3022	3206	3221
Other Eastbound	1211	1738	2335	1375	1264	1194	1140	1856	2152	2770	3299	3368	3350	3531

Source: *OPEC Annual Statistical Bulletin*, various years.

level, and Saudi Arabian just below their 1980 level. In contrast to the growth of exports of the 1970s and the early 1990s, the 1980s was the lost decade.

After 1987 exports east grew rapidly, rising between 0.55 and 0.6 mb/d in each of the next three years, and then by 0.8 mb/d in 1991 and by 0.75 mb/d in 1992. In just five years the flow had more than doubled, increasing by a total 3.25 mb/d. The geographical balance of exports had also changed, and in 1991 for the first time Japan was the market for less than half of the total flow. This fast growth in just five years, heavily biased towards the developing Asian economies, might in isolation in 1992 have looked like a trend. An interpolation would have suggested that by 1995 the total pull from the East would have increased by a further 1.8 to 2.2 mb/d, driven primarily by the growth of the tiger economies rather than the then faltering Japanese economy. As we have argued throughout this study, such trends (especially when predicated almost solely on Asian economic growth and no other factor), tend to be highly misleading.

As shown in Table 11.1, despite strong overall crude oil demand growth in Asia, exports of this oil from Middle East OPEC to Asia-Pacific outside Japan grew by just 0.23 mb/d between 1992 and 1995. The idea that Asian oil demand growth mapped neatly on to equivalent changes in the call on Middle East OPEC oil was already being shown to be weak. In addition, far from being the least promising area in the region, Japan provided 60 per cent of the growth in exports, increasing its Middle East OPEC purchases by 0.34 mb/d as its crude runs expanded and its economy remained resilient, at least compared to many prior expectations.

In earlier chapters we saw the high rates of supply dependence on Middle East supplies in three key Asian refining areas. Middle East supplies made up 84 per cent of refinery input in Singapore, and 78 per cent in both Japan and Korea. In addition, the share of Middle East crude oil in incremental crude runs had been close to, and in Singapore's case well above, 100 per cent. The reverse side to this supply dependence is the increasing demand dependence on Asia-Pacific of the Middle East exporters. This is demonstrated in Table 11.2, which shows by country the proportion of total crude oil exports moving east.

In 1992, for the first time, more than half of all Middle East OPEC crude oil exports went east. However, a key facet of Table 11.2 is that there is little uniformity of experience and policy within the six countries. At one extreme, Qatar and the UAE have steadily biased exports more towards the east, to the point of near exclusivity. At the other extreme, such sales were always a minor facet of Iraqi sales policy before the exclusion of Iraq from the market in August 1990, primarily due to the location of its potential export outlets. Iraq had three main export routes

Table 11.2: Eastbound Flows as Share of Middle East OPEC Crude Oil Exports. 1970–95. Per Cent.

	1970	1975	1980	1985	1986	1987	1988	1989	1990	1991	1992	1993	1994	1995
Iran	62.6	33.2	41.7	26.7	32.0	27.5	28.3	27.3	27.7	32.3	37.2	37.9	43.6	38.7
Iraq	2.3	10.6	18.9	11.4	16.1	12.8	13.1	16.5	14.4	0.0	0.0	0.0	0.0	0.0
Kuwait	32.2	51.4	53.6	69.0	46.8	48.5	46.2	46.7	50.5	71.4	71.2	38.5	46.1	60.6
Qatar	16.8	11.4	42.3	78.6	77.4	81.9	95.1	88.3	84.5	87.5	93.4	98.2	98.5	98.9
Saudi Arabia	31.9	30.2	31.2	46.5	22.9	33.3	32.5	33.4	37.2	37.0	40.5	44.0	42.9	44.6
UAE	31.6	34.4	40.2	76.2	75.6	73.8	83.5	82.1	80.0	85.0	84.2	88.6	93.6	98.0
OPEC Middle East	*35.5*	*30.9*	*32.9*	*43.3*	*34.7*	*36.7*	*37.9*	*38.9*	*41.6*	*46.7*	*50.3*	*50.3*	*52.5*	*54.4*

Source: Own calculations from *OPEC Annual Statistical Bulletin*, various years.

before the Gulf War. Crude oil was pipelined through Turkey for lifting out of the port of Ceyhan, and through Saudi Arabia for lifting at Yanbu on the Red Sea. On both of these routes on tanker freight grounds, vis-à-vis other Gulf producers, the oil was relatively disadvantaged for movement east, and tended to be sold in the Western markets. The third route was by pipeline south east through Iraq for loading at the port of Mina al Bakr, from which point eastern liftings were taken. The lack of spare throughput capacity at Mina al Bakr was a major factor behind the relatively low rate of expansion of sales into the East.

Iran, Kuwait and Saudi Arabia lie between the strong Eastern focus of Qatar and the UAE, and the, historically, Western focus of Iraq. Iranian sales were once very heavily biased towards the Asian market. Indeed, as shown in Tables 11.1 and 11.2, in 1970 nearly two-thirds of Iranian sales went east, with the 1.8 mb/d flow accounting for 45 per cent of all Middle East OPEC flows into the region. However, throughout the 1970s and the 1980s the emphasis was put on Western sales, particularly to Europe given the poor relationship with the USA followed by the explicit US embargo on Iranian oil made in 1988.[1] In the 1990s the emphasis has again shifted towards the East, with a redirection of crude from the European market into Asia.

Given the magnitudes involved, Saudi Arabian policy is a special case. The Kingdom's choice of how to allocate exports between regions can have a considerable bearing on relative regional price levels, as well as creating large accommodatory changes in the export patterns of other crude oil flows. We consider Saudi policy further in Section 4, where we note a tendency of higher pricing into the Asian market since 1992. We would note here that the premium paid in Asia is not merely the result of a relative price effect consequent on Saudi supplies being taken out of the East and into other regions. As Table 11.2 shows, over the first half of the 1990s Saudi Arabia has allocated a greater proportion of its exports into the higher priced region.

The overall increase in the proportion of Middle East OPEC exports moving east could simply be a function of higher oil demand growth rates in Asia compared to other regions. However, in addition there could be a process of the diversion or displacement of exports out of the Western markets. The evidence is that there has indeed been a displacement or diversion effect, as shown in Table 11.3. From 1990 to 1995 the total exports from the six countries have increased by 1.22 mb/d, while their exports to Asia increased by 2.1 mb/d. Further, the table shows that there have been long-term effects at work, most notably the loss of the market share of the Middle East in Europe. With the expansion of non-OPEC supplies (primarily from the North Sea, West Africa and also in the 1970s Russia), Middle East OPEC exports to Europe fell from 7.7 mb/d in 1975

Table 11.3: Destination of Exports from Middle East OPEC. 1970–95. Thousand b/d.

	1970	1975	1980	1985	1990	1993	1994	1995
North America	273	1084	2040	306	2013	1813	1801	1582
of which								
United States	186	718	1752	294	1921	1776	1715	1514
Latin America	277	1516	928	400	447	320	320	208
Eastern Europe	9	262	283	160	246	100	150	176
Western Europe	5857	7731	6822	2361	3332	3603	3148	3049
of which								
France	874	1673	1556	399	553	740	617	639
Germany	247	632	755	123	208	236	206	171
Italy	1176	1304	925	382	526	489	375	450
Netherlands	735	1212	760	284	559	716	631	467
United Kingdom	1082	1258	656	122	231	302	172	136
Middle East	333	312	370	369	301	299	313	304
Africa	521	565	260	105	208	178	196	338
Asia and Australasia	3998	5128	5258	2829	4657	6390	6556	6752
of which								
Japan	2785	3387	2925	1935	2505	3021	3204	3222
Unspecified	394	335	0	9	0	0	0	11
World Total	11,660	16,932	15,961	6538	11,204	12,703	12,485	12,419

Source: Own calculations from *OPEC Annual Statistical Bulletin*, various years.

to 3 mb/d in 1994, with particularly strong crowding out in the UK, French and Italian markets.

As Table 11.3 shows, in 1995, while the volume had fallen sharply from the level of about 10.5 mb/d in 1975, just over 5 mb/d was still moving into Europe and the Americas from Middle East OPEC members, i.e. into areas of relatively low crude oil demand growth, but high supply growth. There is then still some scope for further crowding out of Middle East oil from those markets, and the factors behind the future extent of this are considered further in the conclusions to the next chapter.

3.   The Dubai Forward Market

The question of how these large and growing crude oil exports from the Middle East to Asia are priced brings us to the consideration of the Dubai market. Dubai crude oil is produced by a consortium of companies, with the operator being Conoco.[2] Production peaked around 1990 at about 420 thousand b/d. Since then there has been a consistent decline in the output of the four mature fields (they were discovered in 1966, 1970, 1972 and 1973) that make up Dubai blend. By 1994 production had fallen to 300 thousand b/d, and in 1996 it is around 260 thousand b/d. Attempts are being made to halt the decline through horizontal drilling and enhanced oil recovery,[3] although the expectation must remain that the longer-term decline of Dubai can only be at best postponed. The significance of the decline during the 1990s is that the number of physical cargoes available to the market has been diminishing. At the production peak, loading programmes in each month involved about 25 cargoes of 500 thousand barrels each, with around twenty of these being sold through the forward market. By 1996, around fifteen cargoes per month in the loading programme was becoming the norm. The level of production could already be considered too low for a meaningful and liquid market, and the details below tend to confirm that suspicion.

Dubai emerged as the only forward market for a crude oil outside the USA and the UK, and therefore as a natural candidate for a marker crude oil, for a variety of reasons.[4] We consider the desirable characteristics for a marker crude oil in the next chapter, including a critique of Dubai's suitability judged against these criteria. However, in the context of the market's development in the mid 1980s, Dubai's major advantages compared to other Middle East crude oils were the following. First, it was a relatively large stream whose marketing was not dominated by term contracts. Secondly, rather than being produced by a state monopoly, it had a number of equity producers. Further, these producers were US and European companies amicable to the development of a free market, and who put no resale restrictions on their oil. It should also be noted that, while Dubai is a member of OPEC through the UAE's membership, there has been no perception that operations are influenced by the pursuit of OPEC policy or political control.[5]

The major area of potential stress for any informal forward market lies in its clearing mechanism, i.e. the process whereby deals for forward cargoes are resolved into the physical transfer of oil from producer to lifter. The creation of what may often appear to be clumsy procedures is a necessary result of the structure of informal markets. In a futures market, while deals are initiated as bilateral transactions between buyer and seller, once concluded the direct link is broken. The deal *de facto* becomes two, one between the buyer and the clearing house, and one between the seller and the clearing house. The futures market then exhibits the feature of automatic offset, the net buyer who balances their position with an offsetting sale of futures has cleared their position, and their deals play no role in the final resolution of net buys and net sells. In an informal forward market there is, by definition, no central clearing house. As a result there is no automatic offset, and the link between buyer and seller is preserved. There is then no equivalence (as there is in a futures market) between a balanced position and a cleared position.

As a result of the above, all deals made for a given delivery month are potentially still in play during the resolution of deals through the clearing mechanism. Deals may however be booked out, i.e. closed circles of transactions identified by participants may be removed with the consent of all parties in the process, with cash settlement being made between participants of the gains and losses in the circle. The bookout can be termed as non-automatic offset, non-automatic in that it first requires proaction from participants in the identification of potential closed circles, and secondly that it does require consent. Even with the possibility of bookout, the number of deals left in an active forward market when a given delivery month is resolved, will still be a multiple of the number of physical cargoes available to the market. As we saw in Chapter 7 in the context of the open specification naphtha market, clearance will then be accomplished by the creation, through a nominations procedure, of chains of transactions starting with the physical seller and ending with the eventual lifter.

As Dubai is an informal market, there is theoretically no need for contracts to be standardized. In practice they are highly standardized, and will consist of two parts. The first part will be the simple details of the deal, i.e. the parties involved, the delivery month and the pricing arrangements, with deals being made for a standard forward cargo parcel size of 500 thousand barrels. The second part is normally the detailed arrangements specified in the General Terms and Conditions (GTCs) used by Conoco.[6] Standardization has three main benefits. First, as was also noted in Chapter 8 in the context of the Singapore swaps markets, it facilitates the management of a portfolio of positions. Secondly, the homogeneity of deals makes the formation of delivery chains a more tractable and efficient process. Finally, the reliance on a single set of GTCs removes the need for negotiation of details and contingencies every time a deal is made.

The GTCs are sometimes revised by Conoco, after consultations with market participants on possible improvements. The last major changes as of 1996 were made with effect from the May 1994 loading programme, and the delivery mechanism in the current version can be summarized as follows. The provisional loading programme is published by the terminal coordinator around the 13th of the month. Equity producers can then start passing three-day loading ranges down the chains. Upon receipt of the dates, sellers have one hour to pass it on to their buyer in the chain. While further changes are possible later in the month, the schedule is then released around the 18th day of the month prior to loading.

The revisions made to the GTCs in 1994, arose out of problems with delivery chains the previous year. In particular, in June 1993 a company refused to lift a cargo, as it considered that the loading dates had not been given to it in time for it to arrange a tanker fixture. The delivery chains had proved rather cumbersome, and in particular the rules on passing nominated dates along a chain – i.e. simply that sellers had until the 18th to pass on dates – were rather ill defined. The GTC revisions added more definition to the seller's responsibilities, and greatly enhanced both the speed and efficiency of the clearing mechanism. As we noted above, the need for complicated mechanisms is an inescapable feature of the mesh of bilateral contracts that constitute an informal forward market, given the impossibility of automatic mutual offset of positions. It is then difficult to devise a mechanism that can be guaranteed to work completely efficiently in absolutely all circumstances, particularly given the possibility of human error.[7]

The evolution of the liquidity of trade in the market is shown in Table 11.4, which shows the forward Dubai cargoes traded and reported to Petroleum Argus by quarter from 1986 to 1995.[8] Reported deals are just a subset of the total liquidity. However, we have no reason to believe that the proportion of deals reported to Petroleum Argus has changed significantly over time, and thus believe that the patterns shown in Table 11.4 are a true picture of the market in total. The table shows the division of liquidity between outright deals, i.e. where the forward cargo was traded at an absolute price, and spread deals. Spread deals are further divided into those between different Dubai delivery months, and between Dubai and other crudes, namely Brent and WTI. An intra-Dubai spread deal involves the trade of two forward cargoes, and hence the number of deals made, as opposed to the number of forward cargoes traded, is half the number shown in the relevant column.

As shown in Table 11.4, the zenith of the Dubai market was reached in 1988, after which it went into a sustained decline. During 1993, when serious problems with the delivery chains arose, liquidity became extremely thin. The revisions in the Conoco GTCs, which as noted above improved the efficiency of the working of the chains, have been associated with the return of some liquidity. The improvement is noticeable when viewed from the lows of 1993, so that in 1995 as a whole the volume of trade reached its highest level since 1992. However it should also be noted that the 1995 level was still far below 1992, and by the last quarter reported liquidity was back down to just 62 cargoes, less than one per trading day.

Within the general decline in liquidity, there has also been a marked change in its composition. In particular, Dubai is no longer an absolute price market. Outright deals are now extremely rare, from their peak of three or four reported per trading day in early 1988 they have all but disappeared, with only one being reported to Petroleum Argus in the whole of 1995. As shown in Table 11.4, the demise of outright deals was hastened by the Gulf War, with a sharp decline in their number evident

Table 11.4: Composition of Forward Dubai Deals. Outright and Spreads. 1987–95. Number of Forward Cargoes.

Year	Quarter	Total Cargoes	Outright Deals	Total Spreads	Dubai Spreads	Brent-Dubai Spreads	WTI-Dubai Spreads
1986	1	39	39	0	0	0	0
	2	102	87	15	0	15	0
	3	292	273	19	4	15	0
	4	162	128	34	16	18	0
1987	1	255	184	71	28	38	5
	2	189	149	40	38	1	1
	3	251	204	47	26	18	3
	4	251	193	58	46	7	5
1988	1	444	276	168	120	38	10
	2	415	216	199	122	73	4
	3	469	197	272	140	124	8
	4	419	118	301	140	155	6
1989	1	422	85	337	194	137	6
	2	383	93	290	202	82	6
	3	418	103	315	232	79	4
	4	400	86	314	212	96	6
1990	1	391	91	300	194	102	4
	2	385	110	275	192	81	2
	3	304	62	242	156	83	3
	4	178	11	167	104	63	0
1991	1	269	12	257	166	89	2
	2	272	20	252	150	101	1
	3	312	11	301	214	86	1
	4	272	22	250	164	86	0
1992	1	279	15	264	198	66	0
	2	240	5	235	178	57	0
	3	210	9	201	154	47	0
	4	144	2	142	84	58	0
1993	1	113	4	109	62	46	1
	2	62	1	61	36	25	0
	3	59	0	59	36	23	0
	4	65	3	62	52	10	0
1994	1	112	3	109	80	29	0
	2	83	0	83	68	15	0
	3	83	0	83	60	23	0
	4	60	3	57	40	17	0
1995	1	112	0	112	90	22	0
	2	127	0	127	102	25	0
	3	142	0	142	98	44	0
	4	62	1	61	34	27	0

Source: Own calculations from Petroleum Argus database.

in the third and fourth quarters of 1990. However, even before 1990 the number of deals had been declining from the peaks of 1988, an effect we ascribe in large part to the scaling down of speculative position taking activity by the Japanese trading houses, the *sogo shosha*.

A further clue as to the reasons behind the decline in the Dubai market is given by consideration of the participants. Table 11.5 shows the main fifteen participants in Dubai forward trading and their market shares, for each year between 1987 and 1995.[9] Note the large number of Japanese companies and subsidiaries involved at the start of the period – seven in 1987 (including five of the top six), seven also in 1988, and six in 1989. From 1991 only Kanematsu has been a significant presence in the market. In 1987 *sogo shosha* provided about 40 per cent of the liquidity of the market, since 1991 they have provided less than 10 per cent. The trading activity of the Japanese companies in the late 1980s in all forward oil markets was highly speculative. Large open positions were taken, often being rolled in other months for long periods. Following a series of large losses, and undoubtedly after warnings from MITI, the level of activity was greatly reduced. To a large extent the exit of the *shosha* explains the decline in outright deals after 1988 noted above in the context of Table 11.4.

The *shosha* may have come and gone, but the bedrock of liquidity in the market has been the trades of US finance houses and their subsidiaries, the so-called Wall Street refiners. Three of these companies in particular have been evident. Table 11.5 ranks Phibro among the three largest Dubai traders in every year from 1987 to 1995. J Aron was predominant until 1993, and, while having reduced their involvement, still represent a significant force in the market. The last of the triumvirate is Morgan Stanley, increasing their relative share of activity in every year and becoming the largest Dubai trader in 1995. Since the exit of the bulk of the *shosha*, Table 11.5 shows that there has been a remarkable consistency in the companies that make up the major Dubai traders. If we add Shell, BP, Kanematsu and the equity holder CFP Total to the three Wall Street firms, we have a set of seven core Dubai market makers. Since 1989 the major five traders, and since 1990 the major six traders, have in each year all been drawn from these seven.

A basic weakness of the Dubai market is the difficulty of using it as a hedging mechanism. With pricing formulae based on monthly averages, having a large parcel size forward market and no small parcel size futures market meant that devising an efficient hedge for the monthly average was to say the least problematic. Attempts to launch a Dubai futures contract failed in both London and Singapore, as was discussed in Chapter 9. However, to a large extent the hedging problem was a non-issue. As we argued throughout Part II of this study, the bulk of Asian refiners have

Table 11.5: Main Participants in the Dubai Forward Market. Rank and Market Shares. 1987–95. Per Cent.

Rank	1987		1988		1989		1990		1991	
1	Phibro	8.7	J Aron	17.1	J Aron	16.3	Phibro	16.9	J Aron	12.1
2	Kaines	8.0	Phibro	12.8	Phibro	13.7	J Aron	13.2	CFP Total	12.0
3	Itochu	7.2	Kanematsu	9.3	CFP Total	6.3	CFP Total	7.9	Phibro	10.4
4	Kanematsu	5.7	Mitsui	6.8	Kanematsu	5.9	Kanematsu	6.4	BP	9.2
5	Mitsui	4.8	CFP Total	4.3	Morgan Stanl.	4.2	Morgan Stanl.	5.9	Morgan Stanl.	7.9
6	Nissho Iwai	4.3	Mitsubishi	3.4	Arcadia	3.9	BP	4.6	Kanematsu	5.2
7	CFP Total	4.1	Nissho Iwai	3.2	Mitsui	3.9	Arcadia	4.4	Cargill	5.0
8	Mitsubishi	3.7	Drexel	2.7	Cargill	3.9	Mobil	4.2	Mobil	4.4
9	J Aron	3.6	Shell Int	2.6	BP	3.7	Mitsui	3.4	Shell Int	4.4
10	Marubeni	3.5	Conoco	2.5	Marubeni	3.6	Neste	3.4	Marc Rich	3.6
11	BP	3.5	Marubeni	2.5	Conoco	3.3	Shell Int	3.1	Hess	3.1
12	Marc Rich	3.2	Kaines	2.4	Mitsubishi	3.1	Marc Rich	2.9	Arcadia	2.7
13	TWO	3.1	BP	2.3	Marc Rich	3.1	Nissho Iwai	2.6	Bear Stearns	2.3
14	Tradax	3.1	Morgan Stanl.	2.2	Elf	2.8	Hess	2.4	Neste	2.3
15	Elf	2.3	Elf	2.2	Itochu	2.5	Conoco	2.1	Conoco	1.5

	1992		1993		1994		1995	
1	J Aron	11.6	BP	15.1	Shell Int	15.6	Morgan Stanley	15.5
2	Phibro	11.3	Shell Int	13.9	BP	15.0	Phibro	15.2
3	BP	10.9	Phibro	13.2	Phibro	13.8	BP	14.3
4	Morgan Stanl.	8.4	Morgan Stanley	12.1	Morgan Stanley	12.9	Shell Int	8.2
5	Shell Int	7.4	J Aron	7.9	CFP Total	8.3	J Aron	6.3
6	CFP Total	7.1	CFP Total	7.3	Kanematsu	7.5	Kanematsu	5.4
7	Kanematsu	6.1	Kanematsu	7.0	Repsol	5.5	Repsol	4.7
8	Conoco	3.7	Neste	3.0	Mobil	3.6	CFP Total	4.1
9	Vitol	3.4	Mobil	2.6	Conoco	3.1	Glencore	4.0
10	Mobil	3.1	Dreyfus	2.1	URBK	2.8	JP Morgan	3.9
11	AIG	2.8	Koch	1.9	Neste	2.6	Arcadia	3.4
12	Neste	2.6	Repsol	1.9	J Aron	2.3	Neste	2.8
13	Cargill	2.5	Vitol	1.5	Caltex	1.0	Vitol	2.8
14	Repsol	2.4	Yukong	1.3	Sinochem	1.0	Conoco	2.1
15	Arcadia	2.3	Cosmo	1.1	Vitol	0.8	Koch	1.9

Source: Own calculations from Petroleum Argus Database.

been in highly regulated markets, and have had a very limited interest in risk management. Given this feature, the futures contracts were offering mechanisms that in the main were unwanted in the context of the physical operations of most buyers of Dubai related crude oil. On the other side of the market, the major state-owned oil producing companies also showed little interest in risk management.

Once the willingness to deal in outright terms had been so greatly reduced, the decline was self perpetuating and there was little chance of any resurgence in absolute price deals. Put simply, once liquidity in such deals had disappeared, there was no incentive to deal in outright Dubai given the existence of superior instruments. For instance, a trader who simply wished to take a position on oil prices can open a large position fairly discreetly using forward Brent. If a trader wished to take a position on or hedge the relative price of Dubai, then this can be accomplished through trading a Brent–Dubai spread. Alternatively, should the trader wish to lock in a particular Dubai price, this is more efficiently achieved by using an outright Brent position combined with a Brent–Dubai spread than by attempting to find a counterparty prepared to trade an outright Dubai. Dubai is then no longer a centre of absolute price discovery.

As shown in Table 11.4, Dubai is a relative price market, with the liquidity being in Dubai spreads, i.e. setting the time structure of Dubai prices, and Brent–Dubai spreads, i.e. setting the price of Dubai relative to Brent. In assessing the level of Dubai prices, price reporters can only fix it relative to Brent. From the second quarter of 1993 to the end of 1995, the number of these Brent–Dubai spread deals reported, as shown in Table 11.4, averaged 23 per quarter, i.e. one every three trading days. Given the low number of actual Brent–Dubai spreads reported, in reality the assessment has to be based on market talk, i.e. traders' perceptions of the differential as conveyed to the reporters. We have no objections to the use of market talk, and as noted in Appendix I, an extremely thin market is in fact easier to assess well than a market with a greater, but still limited, number of deals. However, the evidence of Table 11.4 is that the use of Dubai related pricing, if predicated on the basis that Dubai prices come from an active market for Middle East oil, can not now be justified. There is no longer any such market. Rather like the now defunct but once highly influential market for Alaskan oil into the US Gulf, the Dubai market throws up the façade of a market mechanism with no structure behind it.

In Table 11.1 we showed that in the mid 1990s over 6.5 mb/d of crude oil per day was exported into Asia by Middle East OPEC producers. As we see in the next two sections, the value of all of this flow is either directly or indirectly determined by the assessed prices of the Dubai market. At start of 1997 prices, the flow is worth nearly $50 billion per year. Added to this is non-OPEC Middle East and Mexican crude oil exports to Asia, all, with the exception of Yemeni crude, again directly or indirectly determined by assessed Dubai prices. Further, as we saw in the last chapter, Dubai plays a role in the determination of Indonesian and other Asian prices. Ignoring the secondary effect of changes in Dubai prices on other Asian crude oil prices, this results in a total of some 8.5 mb/d dependent on assessed Dubai prices for price determination.

Put in stark terms, the ratio of the number of barrels of physical crude oil priced from Dubai to the production level of Dubai is over 30. We would estimate the equivalent figures for the other two major marker crudes as about 20 to 25 for Brent and 10 to 15 for WTI. In terms of this ratio, the influence of Dubai does appear to be highly geared in comparison. However, consider now the amount of price information generated by each market. Brent has an array of forward, futures and derivatives markets, while WTI has futures, derivatives and a small forward market. Taking all components together, liquidity of 100 mb/d (50 per cent more than total world production) is common in *both* markets. On the basis of Table 11.4, allowing for reporting rates and also Dubai related derivatives, produces a level of trade in Dubai of less than 5 mb/d, of which as noted above, very little generates price level rather than differential information. Dubai, the most highly geared in terms of influence relative to production, then strikes a contrast in terms of the paucity of the size of the information set being generated.

4.   Saudi Arabian Pricing Policy

The evolution of the pricing mechanisms used for Middle East OPEC exports reflects the history of OPEC in terms of its market placement. A transition has been made from administered prices through to crude oil market related pricing, passing through a brief (and it should be said calamitous) phase of oil product market linkages.[10] From direct involvement, the producers have then placed themselves at one step's remove from price setting functions. Here we are concerned primarily with the varieties of pricing mechanisms that arose out of the wreckage of the administered price system and the oil price collapse of 1986. However, as we believe that some facets of the experience of administered prices still have a bearing on perceptions, particularly in concerns of pricing policy mechanism design, the *ancien regime* merits some consideration.

The reasons why OPEC, and in particular Saudi Arabian, sentiment switched against a proactive role in markets were made clear in Table 11.3. In attempting to defend an administered price, above that justified by market fundamentals, the exports of the six Middle East OPEC producers fell sharply over the early 1980s. As shown in Table 11.3, the total crude oil exports of the six fell from 16.0 mb/d in 1980 to just 6.5 mb/d in 1986. Within this sharp decrease, a disproportionate share of the burden was carried by Saudi Arabia. Figure 11.1 shows the level of Saudi Arabian crude exports from 1960 to 1994. From the high of 9.22 mb/d in 1980, exports fell to just 2.15 mb/d in 1985. High prices had brought a strong depressive effect on demand, and set the conditions for continuing growth of non-OPEC oil production. As Arabian Light was the marker

mb/d

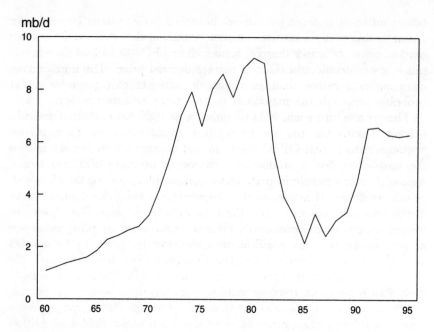

Source: *OPEC Statistical Bulletin*, various years.

Figure 11.1:   Saudi Arabian Crude Oil Exports. 1960–95. Million b/d.

crude oil, competitive discounting by other OPEC producers against the grade also helped to force Saudi Arabia even more firmly onto the margin of the market.

A policy attempting to defend the administered price left Saudi Arabia to cut back further with every lost demand barrel and every new non-OPEC barrel. As a result, the 1985 export level was the lowest since 1965. The value of oil exports fell from $112 billion in 1981 to $17 billion in 1986. While exports have increased again since 1985, by the mid 1990s they have still not regained the 7 mb/d level of 1973. In nominal dollars, export revenues have yet to regain even half of their 1981 level.

The course of events represented in Figure 11.1 has proved to be of major importance to the underlying psychology of Saudi Arabian policy. In particular, a stress has been put on not assuming an overt swing producer's role, and on avoiding the adoption of policies that might possibly facilitate either the perception or the reality of such a position. Given this imperative, proactive pricing policies have been ruled out, and more reactive policies favoured where price levels are set in the open market. Before considering these reactive forms of pricing, we would note that there are other forms of a more proactive policy beyond the rather brutal direct price setting of the administered price system. However, for Saudi Arabia they all involve a higher profile, in that they create a direct market price (or proxy thereof) for Saudi crude, rather than the current much lower profile and indirect market derived price. The implications, on a political rather than an economic basis, of this greater visibility probably represent the major bar to any more proactive policy.

The criteria under which Saudi policy post 1986 has evaluated potential pricing systems can perhaps be typified as the following. First, market responsiveness to non-OPEC prices, in such a way that no two-tier market emerges where Saudi production is undercut by non-OPEC producers. Secondly, a low profile in price formation avoiding any signal of explicit price leadership. Thirdly, and the simple reflection off the demand curve of the previous objective, no explicit quantity leadership. The above can be seen as a marginal producer with market power, attempting to behave as intra-marginally as possible, and desiring its price to be market determined rather than market determining. Given these priorities, the system of market related formula prices was adopted largely by default. Only this system can combine market responsiveness with a low profile. For example, creating a price for all Saudi crude through a, say, monthly tender or other auction procedures for a portion of the total flow, creates a visible absolute price that has to be declared. Further, setting an absolute price for a time period puts the onus of determining sales volumes onto lifters, i.e. carries with it the suggestion of a preparedness to swing, in sales if not necessarily in production, month by month.

Alternatively, the removal of resale restrictions on crude exports creates a base of transactions from which direct market values can be assessed, opening the possibility of linking the price of term contracts to the spot price of Arabian Light. An advantage would be that the price would be market responsive and also not determined by any direct Aramco participation or action in the market. In particular, the price would be called out by price reporters rather than Aramco personnel, and thus need not carry the visibility and hence pejorative of an officially declared price. Among the disadvantages is that, again, a highly visible absolute price arises. Further, lifting resale restrictions represents a considerable reduction in control. Beyond questions of resource nationalism, it removes the possibility of allocating volumes by region to optimize revenues.

In terms of degrees of market reactiveness, the pre-1986 system stands as the least reactive, with the increasing scale passing through auctions on to allowing spot markets to develop independently. However, we have suggested that, given the policy objectives, all of these would, at least to this point, have been considered undesirable. Instead, let us consider the other extreme to the scale, i.e. the most reactive form of pricing policy available. This would be a retrospective price based on the *ex post* observation of the market (as used by several other Middle East exporters as we note in the next section), and to be truly reactive it would need to be credibly based on actual market values with the minimum of discretion involved.

Such a system would certainly be highly reactive, but it still involves some potentially severe drawbacks. First, setting different prices according to regions becomes very difficult to operate when a *de facto* absolute price is being announced, i.e. giving rise to charges of discrimination. Secondly, there is a trade-off between operational parsimony of the system and the degree of market responsiveness involved. The use of a monthly retrospective price is problematic when selling into a market where the use of a monthly average is not the norm in pricing competing sources of supply. For example, five-day (and sometimes ten-day) averages are the *modus operandi* in Europe. If a potential buyer takes the view that prices will, say, rise over the month, they would prefer to lift crude in the latter part of the month (i.e. when the spot price is expected to be above the retrospective price), and will be less happy should the loading schedule require them to lift it earlier in the month. The problem could be reduced by announcing retrospective prices for shorter time intervals, but then the system becomes administratively far more clumsy. In total, balancing parsimony with market responsiveness, retrospective pricing is really only possible when selling into a region where monthly average pricing is commonly accepted. The only such region has been Asia, and the three main Middle East countries that employ retrospective prices (Oman, Qatar and Abu Dhabi), are all very heavily biased to Eastern buyers in their export patterns. Thus, in the context of Saudi Arabian sales, retrospective pricing would not appear to be a workable option.

The one perceived problem that is not solved by any system is that of competitive discounting. Preventing the emergence of any spot market assessment for Arabian Light and avoiding an absolute official price, does at least prevent other producers explicitly discounting from the Saudi price. However, while a formula price may be more discreet, there is nothing to prevent other countries from using the same markers in a formula and then discounting from the Saudi adjustment factor.

The structure of the formulae used by Saudi Arabia for sales to the USA, Europe and Asia is superficially the same, each consisting of a monthly adjustment term declared in advance (during the first week of the previous month) from the Platt's published assessments of the marker crude oil, each formula representing the value of fob Ras Tanura sales. For US sales, West Texas Intermediate (WTI) is used as the marker,[11] for European sales dated Brent,[12] and for eastbound liftings the average of Dubai and Oman.[13] However, the details of implementation produce significant structural differences between Asian formulae and other regions.

The major differences lie in the timing for the evaluation of the formulae, the time period over which the evaluation is made, and the treatment of freight costs.

Westbound liftings are priced after the loading date, forty days to Europe and fifty to the USA, consistent with sailing times for tankers out of Ras Tanura. The evaluation is then made over ten days around this date. Thus, for example, a cargo bound for the USA loaded on 15 August would be priced according to the average of WTI quotes for the ten trading days around 4 October with the application of the adjustment factor for July liftings to the USA. Similarly, if the cargo were bound for Europe, the ten dated Brent quotes around 24 September would be computed, and the adjustment factor for July liftings to Europe used. In addition, westbound sales have had an automatic freight adjustment factor which compensates the lifter if the freight rate rises above WS40. For eastbound sales the computation is made over the whole month of loading, and hence if our imaginary cargo in the example above was heading into Asia, it would have been priced at the average over August of the average of Dubai and Oman, plus the application of the relevant adjustment factor.

The difference between the application of eastbound and westbound formulae reflects differences in the regional markets. To a large extent the features are demand driven. In the USA and Europe, Saudi crude is moving into markets where spot crude oil competes actively with term sales, and where purchasers are highly conscious of risk. The application of the formulae turn what is formally a fob price into one that, *de facto*, exhibits the features of a delivered price. At the refinery gate, the buyer has term contract crude oil that is competitively priced at the time of arrival rather than the time of loading. Having a stream of crude oil through a term contract in some ways provides risk insurance in itself. Sometimes long haul crude priced at the loading date will appear overpriced at arrival at the refinery compared with short haul crude, sometimes it will be underpriced.

Over the whole life of the term contract there will be a degree of balancing out. However, given the widespread use of benchmarking in the internal evaluation of crude acquisition, and the judgement of actions against a weekly or monthly compilation of refinery value tables for various crudes, the smoothing out property of the flow may be less useful. In particular, there could be asymmetries. Refiners may be particularly averse to situations when the term contract cargo appears expensive on arrival, i.e. when crude oil prices have fallen between the loading date of the cargo and that of a short haul cargo. Under such circumstances, the method of implementation of the formula acts as a form of risk management reducing the need for any direct participation by the refiner in forward or futures markets. In addition, the formula has also contributed towards the management of the risk of freight charge movements over the life of the term contract.

Given the number of cargoes it sells as a flow over the year, Saudi Arabia tends to behave in a risk neutral manner to price changes between loading and arrival at refinery gate for any given cargo or set of cargoes. The westbound formulae then represent an economically efficient transfer of risk towards Saudi Arabia. There is however a further aspect to this. In a backwardated market (i.e. when prices are higher for closer compared to more distant delivery dates) oil near the market geographically is more valuable than oil further away. Short haul crude has a premium over long haul arising from the time cost of transit held in the backwardation. When the market is in contango (i.e. prices are higher the more distant in time is delivery), that premium *ceteris paribus* becomes a discount.[14] The timing delays used in the European and US formulae then effectively convert, in terms of pricing behaviour, long haul fob prices into c+f delivered prices, with the associated positive or negative adjustment in value according to the time structure of prices. We return to this point below in the context of the comparison between westbound and eastbound prices for Saudi Arabian crude oil.

In comparison with European and US markets, the Asian market has had far less spot traded oil competing with term contract oil. Further, in the past regulation has meant that risk management has been a minor concern, and supply security considerations have dominated. The implicit pressure from the preferences of customers, which helped produce the delayed timing aspect of the European and US formulae, has been absent. As a result, the eastbound formulae do not display the c+f proxy characteristic of sales to other regions.

Saudi Arabia currently produces five export grades of crude oil,[15] shown in Table 11.6 with their associated gravities and sulphur content.[16]

Table 11.6: Characteristics of Saudi Arabian Export Crude Oils and Planned Production. 1994, 1996 and 1998. Thousand b/d.

	Gravity Degrees API	Sulphur % by Weight	Production 1994	1996	1998
Arabian Super Light	48.2	0.04	0	150	150
Berri (Arabian Extra Light)	39.0	1.2	950	950	950
Arabian Light	34.0	1.8	5000	5150	5400
Arabian Medium	31.0	2.4	900	800	1150
Arabian Heavy	27.0	2.9	950	850	750

Source: *Middle East Economic Survey*, 14 February 1994.

Saudi Arabian export volumes increased due to the 1990–1 Gulf War and its aftermath, as was shown in Figure 11.1. However, after 1991, for reasons considered further in the next chapter, additional growth in export volumes without impacting severely on prices proved difficult, leaving Saudi Arabia effectively hemmed in. The emphasis was then put on revenue maximization on the basis of existing volumes, given that revenue growth through export growth had been precluded. Revenue maximization took two main forms. The first was to bias development expenditure and the pattern of exports more towards the higher value grades, and to bias the unutilized production capacity towards the lower value grades, i.e. to maximize light production at the expense of heavy. This aspect is shown by the internal Aramco projections of volumes by grade shown in Table 11.6, i.e. the bringing on stream of Arabian Super Light, expansion of Arabian Light and Medium, and the reduction of Arabian Heavy volumes.[17] The second facet to revenue maximization, discussed further below, was to attempt to charge what the market would bear in different regions.

The adjustment factors from the average of Dubai and Oman prices for eastbound sales are shown in Table 11.7 for the five grades. As noted above, the price paid for liftings is calculated as the addition of the adjustment factor for the relevant loading month to the average of Dubai and Oman prices over the course of the whole month. The adjustment factors for a given month are announced to customers by fax transmission in the first week of the previous month. Hence, in setting the adjustment factors for, say, April, the information set open to Aramco is market behaviour to the end of February plus the first few trading days in March.

We consider two sets of differential prices, namely between grades in the Asian market and between regions for the same grade. From the data of Table 11.7, plotting the differential from the Arabian Light formula price of both Arabian Medium and Arabian Heavy, results in the patterns shown in Figure 11.2. Differentials were held stable through 1989, before being widened over the first half of 1990. At the time of the Iraqi invasion of Kuwait, Arabian Heavy was set $1.80 below Arabian Light, and (with the surge in the absolute level of prices), the differential was moved out to a maximum level of $4. As shown in Figure 11.2, after the start of 1991, the general trend in the formula prices has been a narrowing of differentials. Note also the increased frequency of adjustments in the two differentials from 1994 onwards compared to previous years. The differentials are frequently fine-tuned with slight adjustments, compared to the less frequent and on average larger changes that characterize the earlier period.

The general collapse of light to heavy crude oil differentials and the underlying dynamic of this process is described in Chapter 13. Here we

Table 11.7: Saudi Arabian Eastward Sales Formula Adjustment Factors. 1989–96. Dollars Per Barrel.

		Arabian Light	Arabian Medium	Arabian Heavy	Berri			Arabian Light	Arabian Medium	Arabian Heavy	Berri	Arabian Super Light
1989	Jan	0.00	-0.70	-1.25	0.80	1993	Jan	0.40	-1.15	-2.30	1.35	
	Feb	0.00	-0.70	-1.25	0.90		Feb	0.40	-1.30	-2.65	1.50	
	Mar	0.15	-0.55	-1.10	1.15		Mar	0.30	-1.30	-2.65	1.30	
	Apr	0.25	-0.45	-1.00	1.25		Apr	0.30	-1.30	-2.50	1.25	
	May	0.35	-0.35	-0.90	1.35		May	0.30	-1.30	-2.50	1.15	
	Jun	0.35	-0.35	-0.90	1.15		Jun	0.30	-1.30	-2.50	1.15	
	Jul	0.35	-0.35	-0.90	1.15		Jul	0.30	-1.30	-2.50	1.15	
	Aug	0.35	-0.35	-0.90	1.15		Aug	0.20	-1.40	-2.60	1.10	
	Sep	0.35	-0.35	-0.90	1.15		Sep	0.20	-1.30	-2.40	1.20	
	Oct	0.35	-0.35	-0.90	1.15		Oct	0.20	-1.20	-2.25	1.30	
	Nov	0.25	-0.45	-0.90	1.05		Nov	0.30	-1.10	-2.15	1.45	
	Dec	0.25	-0.45	-1.00	1.05		Dec	0.30	-1.15	-2.25	1.50	
1990	Jan	0.25	-0.45	-1.00	1.05	1994	Jan	0.35	-1.30	-2.50	1.55	
	Feb	0.35	-0.35	-0.90	1.25		Feb	0.35	-1.05	-2.10	1.45	
	Mar	0.45	-0.45	-1.00	1.35		Mar	0.45	-0.70	-1.50	1.45	
	Apr	0.25	-0.65	-1.20	1.15		Apr	0.40	-0.80	-1.60	1.30	
	May	0.25	-0.65	-1.20	1.15		May	0.40	-0.80	-1.60	1.20	
	Jun	0.25	-0.65	-1.20	1.15		Jun	0.65	-0.40	-1.10	1.35	
	Jul	0.00	-1.35	-1.80	0.90		Jul	0.50	-0.30	-0.90	1.10	
	Aug	0.00	-1.35	-1.80	0.90		Aug	0.55	-0.25	-0.85	1.10	
	Sep	0.35	-1.00	-1.45	1.25		Sep	0.40	-0.10	-0.50	0.80	
	Oct	0.65	-0.60	-1.85	2.05		Oct	0.50	-0.55	-1.20	1.20	
	Nov	1.00	-1.00	-3.00	2.50		Nov	0.50	-0.55	-1.20	1.20	
	Dec	1.00	-1.00	-3.00	2.50		Dec	0.50	-0.45	-1.10	1.20	
1991	Jan	1.00	-1.00	-3.00	2.50	1995	Jan	0.50	-0.40	-1.00	1.20	
	Feb	0.85	-0.85	-2.15	2.35		Feb	0.50	-0.45	-1.10	1.10	
	Mar	0.85	-0.85	-2.15	2.35		Mar	0.50	-0.15	-0.60	0.95	
	Apr	0.85	-0.85	-2.15	2.35		Apr	0.50	-0.05	-0.45	0.85	1.80
	May	0.25	-1.45	-2.75	1.25		May	0.40	-0.05	-0.45	0.75	1.65
	Jun	0.35	-1.35	-2.40	1.35		Jun	0.50	-0.05	-0.55	0.90	1.65
	Jul	0.35	-1.35	-2.40	1.35		Jul	0.50	-0.05	-0.55	0.90	1.65
	Aug	0.35	-1.35	-2.40	1.60		Aug	0.50	-0.15	-0.85	0.90	1.65
	Sep	0.35	-1.35	-2.40	1.60		Sep	0.50	-0.05	-0.65	0.80	1.65
	Oct	0.50	-1.35	-2.50	1.80		Oct	0.50	-0.20	-0.70	0.90	1.70
	Nov	0.50	-1.35	-2.80	2.00		Nov	0.50	-0.20	-0.70	0.90	1.70
	Dec	0.50	-1.35	-2.80	2.00		Dec	0.60	-0.10	-0.40	1.05	1.80
1992	Jan	0.50	-1.25	-2.70	1.80	1996	Jan	0.60	-0.20	-0.55	1.20	1.90
	Feb	0.50	-1.25	-2.70	1.80		Feb	0.60	-0.20	-0.55	1.40	2.10
	Mar	0.50	-1.25	-2.70	1.50		Mar	0.55	-0.25	-0.60	1.40	2.10
	Apr	0.50	-1.10	-2.40	1.50		Apr	0.55	-0.35	-0.75	1.40	2.10
	May	0.50	-1.05	-2.20	1.30		May	0.55	-0.35	-0.75	1.40	2.40
	Jun	0.50	-1.05	-2.20	1.30		Jun	0.55	-0.35	-0.75	1.40	2.40
	Jul	0.50	-1.05	-2.20	1.30		Jul	0.55	-0.60	-1.25	1.40	2.40
	Aug	0.50	-1.15	-2.20	1.50		Aug	0.55	-0.50	-1.00	1.40	2.40
	Sep	0.50	-1.15	-2.20	1.50		Sep	0.55	-0.50	-1.20	1.40	2.40
	Oct	0.50	-0.95	-1.95	1.35		Oct	0.55	-0.40	-1.10	1.40	2.40
	Nov	0.35	-0.95	-1.95	1.15		Nov	0.60	-0.30	-1.05	1.55	2.40
	Dec	0.35	-1.05	-2.05	1.25		Dec	0.75	-0.30	-1.15	1.85	2.75

Sources: *Middle East Economic Survey, Weekly Petroleum Argus,* various issues.

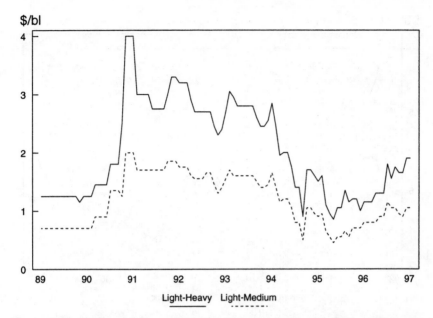

Figure 11.2: Differentials Between Arabian Light and Arabian Medium, and Between Arabian Light and Arabian Heavy in Saudi Arabian Formula Prices for Eastbound Sales. January 1989 to January 1997. $ Per Barrel.

wish to consider to what extent the pattern shown in Figure 11.2 is market responsive, and to what extent discretion has been used in setting the differentials. In order to do this we have calculated a proxy for differential between the refining values of Arabian Light and Arabian Heavy, based on Singapore oil product prices, which is shown, together with the formula differential, for the period 1992 onwards in Figure 11.3.[18]

The refining value difference shown in Figure 11.3 is lagged by two months. As noted above, formula prices are set at the start of the month prior to their application, when the only market information available is that up to and including the month two months prior to application. If the differentials were set reactively by looking at the available information set, changes in the formula differential should then be correlated to the two-month lagged changes in the refining value differential as shown.

There are several important features shown in Figure 11.3. First, the tendency of more frequent changes in the formula differential from 1994 onwards, noted in the context of Figure 11.2 and evident also from Table 11.7, is shown to be a result of a much closer correlation between the refining value and formula differential in that period than in earlier years. The refining value differential moved in a far greater range between 1992

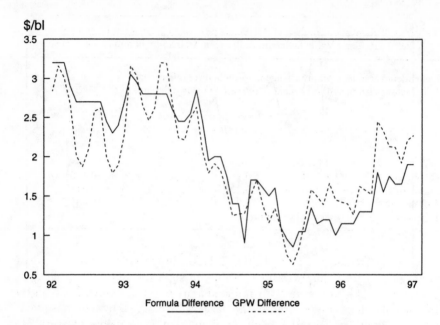

Figure 11.3: Formula Price and Refining Value (lagged by two months) Differences Between Arabian Light and Arabian Heavy for Eastbound Sales. January 1992 to January 1997. $ Per Barrel.

and 1994 than the more stable formula differential. This might be considered suggestive of a pricing policy that has grown more market responsive, and that is primarily driven by oil product market conditions.

From inspection of Figure 11.3, it is clear that the formula price structure from 1994 onwards does bear a relationship with the refinery value differential information available at the time of announcement, and in particular their turning points generally coincide. This is confirmed by a simple regression analysis. Table 11.8 shows the results of regressing four variables on the monthly changes in the formula price differential for the period from the start of 1994 onwards. We show as regressors in column (a) the monthly change in the refinery differential lagged twice (i.e. the information set available at the time of formula price announcement), in column (b) the same change in differential lagged once (i.e. the change in the month at the start of which the formula prices were announced), in column (c) the unlagged value (i.e. the change that actually occurred during the month of application of the formula prices), and finally in column (d) the value lagged one month forward (i.e. in the month after application).

The results shown in column (a) of Table 11.8 demonstrate that there

Table 11.8: Regression Analysis of Relationship Between Changes in Formula Price and Refining Value Differentials Between Arabian Light and Arabian Heavy.

Dependent variable: monthly change in formula price differential.
Explanatory variable: monthly change in refining value differential.

| | Explanatory Variable | | | |
	Lagged Twice	*Lagged Once*	*Unlagged*	*Lagged Once Forward*
	(a)	(b)	(c)	(d)
Coefficient	0.749	0.112	-0.041	-0.195
Standard Error	0.166	0.222	0.225	0.211
R-Squared	0.439	0.009	0.001	0.031
Intercept terms (not shown) all insignificant				

is a strong, significant and positive relationship between formula price changes and the realized refinery value changes (i.e. the change lagged twice) observable at the time when formula prices are announced. Changes in formula differentials are to the largest extent reactive to observed market conditions, with the results implying that 75 cents out of every $1 monthly change in refinery values differentials are reflected in adjustment terms. The other columns of Table 11.8 have been calculated as tests of other hypotheses. Column (c) considers whether there is a relationship between formula differentials and realized refinery value differentials in the month of application, and shows that there is no significant relationship. In other words, formula price changes do not forecast refinery value changes. Columns (b) and (d) test whether there is any causality that runs from formula changes to refinery value changes, using two possible specifications of how that causality might operate.

In the first specification of this causality, formula price announcements have no signalling role, create no changes in the expectations of traders, and only cause changes in the physical market according to the following mechanism. A narrowing of the light to heavy differential in the formula leads refiners to substitute away from the heavier grades. As a result less fuel oil is produced, relative fuel oil prices rise and the refinery value margin also narrows. Most cargoes would only arrive in the East in the month following that of formula application, and be run through refineries in that and later months. If, say, the formula adjustments had been changed in an announcement in early March for April loadings, the physical impact would not be felt in product markets until May at the earliest. We would then expect a significant relationship to be shown in column (d) of Table 11.8. However, there is no such relationship, and hence this specification of causality from formula prices differentials onto product markets can be rejected. The above assumed no changes in market expectations. If the physical mechanism described was valid, then in an informationally efficient market prices would change at the time of announcement of formula price adjustments. The relevant test would be that shown in column (b) of Table 11.8, which again shows no significant relationship. This specification of the causality from formula prices can then also be rejected.

Formula price adjustments have been highly market reactive, with no causality running from them towards relative refining values. However, they are not totally reactive. The final major feature of Figure 11.3 is that it does indicate a degree of discretion having been used in the narrowing of the light to heavy differential. In particular, the narrowing through formula is of a greater extent than that implied by refining values. Comparing average 1992 with average 1995 figures, our proxy for the refining value differential fell by $1.17 per barrel, while the formula price differential fell by $1.60 per barrel. The light to heavy differential per formula seems then to have been narrowed by more than 40 cents more than refining values would suggest. This would be consistent with the intent of weaning customers away from the heavy oil in the context of the crude production composition aspect of the revenue maximization strategy noted above. The additional 40 cents can be taken as an extra payment for Arabian Heavy given the tightening of its relative scarcity among Saudi exports, combined with a lack of perfect substitutability with other grades.

We noted that the second aspect of the revenue maximization strategy was to allow prices to vary by region. The conditions for this to be successful are twofold. First, there can be no possibility of resale between markets, or oil originally destined for the lower cost market will simply be diverted and resold into the higher price market. With resale and destination restrictions being utilized, this condition was met for Saudi Arabian oil. In addition, even with a price gap of $1 per barrel, freight costs provided a buffer against non-Saudi oil leaking out of Western markets into a higher priced Asian market. Further, other Middle East producers preferred to maintain a fairly constant position relative to Saudi prices, and thus did not discount any further into Asia.

The second condition is that the elasticities of demand should differ between markets, i.e. simply that the willingness to pay should vary. The perception of Saudi Arabia on this point at the start of 1992, when a significant price gap first opened up, could have run as follows. US and European customers want comparability with their own short haul crude oils, and in particular have been unwilling to pay any premium simply to have a term contract. By contrast, Asian customers are concerned with matters of supply security and also securing a favourable position for incremental supplies in later years, and have apparently been prepared to pay a premium for obtaining a term contract. Their willingness to pay more for crude oil is reinforced by higher oil product prices in Asia than in other regions, which has produced higher refinery profit margins than in the West. In addition, non-OPEC oil was not as of 1992 crowding out Middle East production in Asia, while in Europe and the USA its strong growth was forcing discounting to remain competitive. Justified in this way, the second condition was apparently met. Quantification of the degree of the price gap is made in Figure 11.4, which shows, on a monthly basis and as a twelve-month rolling average, the differential between the realized fob price of Arabian Light for eastbound and European sales.

The twelve-month average shown in Figure 11.4, began 1992 slightly negative, i.e eastbound loadings had realized slightly lower prices than European loadings in 1991. The eastbound realization was then higher in each of the 50 months from January 1992 to February 1996, before falling lower than the European in March and April 1996 for reasons that are discussed below. The average size of the differential (represented by the twelve-month average) increased to a maximum of about $1.20 per barrel, and maintained a steady level above $1 per barrel through 1994 and 1995.

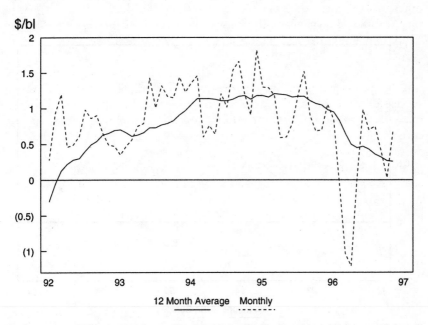

Figure 11.4: Differential Between Realized Price for Eastbound and European Sales of Arabian Light. Monthly and Twelve-Month Rolling Average. January 1992 to November 1996. $ Per Barrel.

In Figure 11.5 we have replicated the twelve-month average shown in Figure 11.4, together with the equivalent difference between US and European bound loadings. While, compared to eastbound loadings, it was of a smaller degree in that the twelve-month average only reached a maximum of 50 cents, there was a premium paid by US customers relative to European over the course of 1993. However, the conditions for successful revenue maximization listed above do not hold in the context of the US market. Since the start of 1994, while the $1 plus premium has been paid for eastbound loadings, the prices of US and European loadings have effectively been equalized.

We would stress that the existence of a higher realized price for eastbound liftings does not in itself imply that Asian customers were overpaying for crude oil. Refinery profit margins will be considered in Chapter 13; however, we would note in advance that the profit margin in Singapore from refining Arabian Light remained higher than that in Rotterdam until 1995, when it was roughly equalized.[19] To some extent, higher crude prices to Asia effectively captured a rent. In other terms, complaints from Asian refiners that they were paying more for their crude oil, could be met by European complaints that they were making less money from refining it.

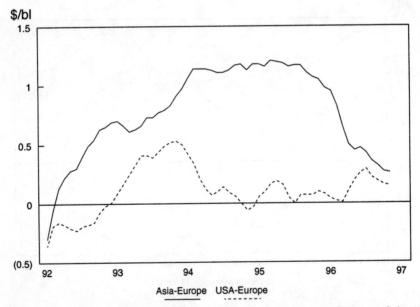

Figure 11.5: Differentials Between Realized Price for Eastbound and European Sales, and Between US and European Sales of Arabian Light. Twelve-Month Rolling Averages. January 1992 to November 1996. $ Per Barrel.

The maintenance of higher relative prices for eastbound loadings since 1992 is clear – however what is unclear on the basis of Table 11.7 is precisely how it was engineered. Considering Arabian Light, the formula price adjustment from the Dubai/Oman average has stayed within the range of plus 20 cents and plus 65 cents since the start of 1992, and has shown no significant overall increase over the period. The $1 plus per barrel differential was not achieved by increasing the price of Arabian Light into Asia relative to that of Dubai. In fact the effect was achieved by discounting exports in other regions against the Dubai price, as is demonstrated in Figure 11.7.

In Figure 11.6 we have shown the discount of Arabian Light to dated Brent, as represented in the announced formula price adjustment factors into Europe, together with the market price discount of the Dubai/Oman average from the dated Brent price. For clarity, both series are shown as twelve-month averages. The key feature is that the adjustment factor in relation to dated Brent has been narrowed far less than has the open market discount of the Dubai/Oman average to dated Brent. The twelve-month average of the latter narrowed by about $1.90 per barrel from the start of 1992 to the end of 1995. The corresponding average for the adjustment factor from dated Brent for Arabian Light sales into Europe

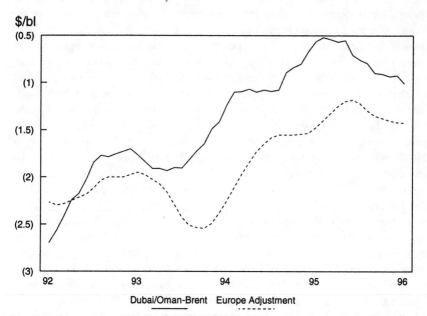

Figure 11.6: Adjustment Factor for European Sales of Arabian Light and Differential Between the Dubai/Oman Average and Dated Brent Prices. Twelve-Month Rolling Averages. January 1992 to December 1995. $ Per Barrel.

narrowed by just 85 cents per barrel over the same period. Overall, Arabian Light was discounted by $1.05 per barrel against the Dubai/ Oman average into the European market (and, by interpolation from Figure 11.5, into the US market as well). This $1.05 per barrel accounts for virtually all the difference between the realized prices of eastbound and westbound liftings.

Thus, the rise in the relative price to Asia was more a result of discounting into the West than overtly raising Asian prices. The growth in non-OPEC production in the first half of the 1990s was heavily biased towards light low sulphur supplies from the Atlantic basin, resulting in an increase in the relative price of longer haul and higher sulphur crude oil such as Dubai. The market signal being sent by the narrowing of the Brent–Dubai differential was one that facilitated the displacement of Middle East oil out of the West by the incremental North Sea and Latin American production. The decision not to pass on the full increase in the relative price of Dubai into the West, was aimed at limiting the displacement of Saudi crude oil from the West. The higher relative price in Asia was the indirect effect of not wishing to cede volumes in the West, rather than the effect of a decision to raise the Asian price directly.

The apparent reduction in the relative price to Asia in 1996 was also indirect rather than an explicit policy implementation. The European formula has delayed pricing from the time loading, which captures the value of a backwardation and proxies a c+f price. By contrast the Asian formula is closer to a pure fob price, being the monthly average in the month of loading. Figure 11.4, while an accurate reflection of relative realized prices, does not compare like with like. In particular, during market backwardations the price of Dubai tends to weaken relative to Brent as it is longer haul crude and thus its price bears the time cost of transit. To remove this effect, forty days of the time structure of Brent prices needs to be stripped from the European price. The first quarter of 1996 exhibited extremely large backwardations. For March 1996, we calculate that the average value of forty days Brent backwardation was of the order of $1.90.[20] When this is accounted for, Asia's apparent advantage disappears.

Figure 11.4 showed market outcomes which, particularly on a monthly level, could have varied considerably from the Aramco intention regarding relative price levels. For example, the adjustment factors for March 1996 were set at the beginning of February on an information set that covered the period to the end of January. If we were to assume that markets were efficient, then the rational expectation at the start of February of the March dated Brent to Dubai differential would be that which prevailed in January. In Figure 11.6 we have recalculated Figure 11.4 using this rational expectation rather than the actual market outcome, i.e. the data for, say, March 1996, involves the use of the March adjustment factors in conjunction with the January market outcomes. The result is a rational expectation of the degree of price discrimination given that no market view is taken. In addition, we have adjusted the series for the time structure of prices to produce the proxy for a fob price comparison.

As shown in Figure 11.7, the result of this process is that there is no downwards trend in the differential in late 1995 and early 1996, and that the twelve-monthly average remains at around the $1 per barrel level. Unless Aramco had taken a market view that foresaw a sharp increase in the dated Brent to Dubai spread, we must conclude that the fall in the differential shown in Figure 11.4 is purely a function of *ex post* relative market movements rather than any *ex ante* intention to narrow the degree of that differential.

We have therefore found no evidence of an intentional reduction in the price to Asia relative to that in the West over the period covered. However, as noted above, the pressures arising from the change in Asian markets are still coming into play. The general liberalization, and continuing deregulation in Japan in particular, mean that the difference in the nature of Eastern and Western markets is becoming less pronounced. The

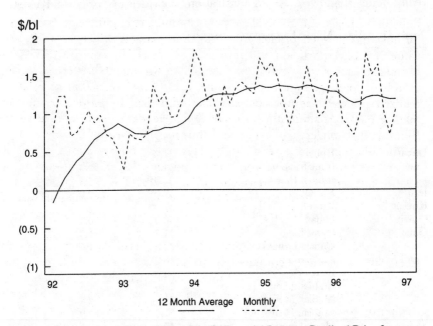

Figure 11.7: Rational Expectation of the Differential Between Realized Price for Eastbound and European Sales of Arabian Light Adjusted for the Time Structure of Prices. Monthly and Twelve-Month Rolling Average. January 1992 to November 1996. $ Per Barrel.

distinction made is also becoming less severe with a volume of crude oil, primarily from West Africa but also from Alaska, operating at the margin of the market. The implications of these changes for Saudi policy are considered in more detail in the next chapter, but we would note here that a differential of the order of over $1 per barrel may not be sustainable.

5. Pricing Policies of Other Gulf Producers

The methods used to price term contracts for crude oil in the Middle East fall into two categories. The first is marker crude related formulae pricing with adjustment factors announced in advance, as we have seen employed by Saudi Arabia, but also used by Kuwait, Iran, Iraq and Yemen. The other method is retrospective pricing, used by Abu Dhabi, Oman and Qatar. We consider both forms in turn. The qualities of the Middle East crudes considered in this section are shown with their country of origin in Table 11.9.

Formula related pricing for other countries in the Gulf tends to involve a distinction between the formal marker and the crude oil whose value is really being shadowed. As an illustration, consider the price of Kuwait

Table 11.9: Major Middle East Export Crude Oils.

		API Gravity	Per Cent Sulphur
Abu Dhabi	Murban	40.4	0.8
	Lower Zakum	40.1	1.1
	Upper Zakum	33.9	1.8
	Umm Shaif	37.4	1.5
Dubai	Dubai	31.0	2.1
Iran	Iranian Light	34.0	1.4
	Iranian Heavy	31.0	1.7
Iraq	Kirkuk	37.0	2.0
Kuwait	Kuwait	31.0	2.5
Oman	Oman	36.3	0.9
Qatar	Qatar Land	41.7	1.3
	Qatar Marine	35.3	1.6
Saudi Arabia	Arabian Super Light	48.2	0.04
	Berri	39.0	1.2
	Arabian Light	34.0	1.8
	Arabian Medium	31.0	2.4
	Arabian Heavy	27.0	2.9
Yemen	Marib Light	40.3	0.1
	Masila	30.5	0.6

Sources: Various.

exports to the East. Kuwait produces a single export blend, with a quality (as shown in Table 11.9) that is very close to that of Arabian Medium. The formal marker is the monthly average of the Dubai/Oman average, and the adjustment factors from 1993 into 1996 are shown in Table 11.10. We also show the implied differentials from Arabian Medium, derived from the Saudi Arabian adjustment factors shown in Table 11.7. This makes the principle of Kuwaiti pricing clear, i.e. to shadow the Arabian Medium formula price.

In the previous section we noted that one of the problems, from the Saudi Arabian perspective, with the OPEC administered price system using Arabian Light as a marker, was what was known as the problem of the differentials. In particular, there was a possibility of competitive discounting by other OPEC members. As Table 11.10 shows, this facet did not disappear with the adoption of formula pricing, leaving Saudi Arabia with the role of price leadership. In other words, producers can wait until they have observed the Saudi adjustment factors before setting their own, choosing in each month to shadow, discount or mark up from changes in Saudi adjustment factors according to policy and market conditions.

Iran sets prices on a quarterly basis, with provision for changes to be made during the quarter should circumstances change. The latter qualification could be interpreted as meaning that changes in Saudi adjustment factors will engender a reexamination of the Iranian factors. Prices are based on Oman for Iranian Light and Dubai for Iranian Heavy, with a further component based on the differential between Oman and Dubai. As of 1996, that component is 77 cents minus the Oman/Dubai differential, and is added in the Iranian Light formula, but subtracted in that for Iranian Heavy.

Yemen currently represents something of a special case among Middle East exporters, in that its sales to the East use dated Brent as a marker. This puts it in a strong position when the dated Brent to Dubai spread is narrow, as it undercuts other Gulf producers of light sweet crude and retains competitiveness with Brent related West African crude oil attracted into Asia. Of course when the spread is wide, Yemeni exports become relatively expensive compared to neighbouring countries, and term contract holders become dissatisfied.[21] A Dubai or Oman linkage might then in the short term seem attractive in terms of contract relations, but, as the argument of the next chapter demonstrates, could be seen as a retrograde move in terms of longer-run trends.

No official producer price is set in a vacuum, in the sense that there is an interplay of views between producers and their customers. Oman has formalized this communication through what can be described as a pseudo panel price mechanism in the derivation of its monthly

Table 11.10: Adjustment Factors for Kuwaiti Eastbound Exports. January 1993 to December 1996. Dollars Per Barrel.

		From Dubai/Oman Average	From Arabian Medium
1993	Jan	-1.25	-0.10
	Feb	-1.40	-0.10
	Mar	-1.40	-0.10
	Apr	-1.40	-0.10
	May	-1.40	-0.10
	Jun	-1.40	-0.10
	Jul	-1.40	-0.10
	Aug	-1.50	-0.10
	Sep	-1.40	-0.10
	Oct	-1.30	-0.10
	Nov	-1.20	-0.10
	Dec	-1.25	-0.10
1994	Jan	-1.40	-0.10
	Feb	-1.15	-0.10
	Mar	-0.80	-0.10
	Apr	-0.90	-0.10
	May	-0.90	-0.10
	Jun	-0.50	-0.10
	Jul	-0.40	-0.10
	Aug	-0.35	-0.10
	Sep	-0.20	-0.10
	Oct	-0.65	-0.10
	Nov	-0.65	-0.10
	Dec	-0.55	-0.10
1995	Jan	-0.50	-0.10
	Feb	-0.55	-0.10
	Mar	-0.25	-0.10
	Apr	-0.15	-0.10
	May	-0.15	-0.10
	Jun	-0.15	-0.10
	Jul	-0.15	-0.10
	Aug	-0.25	-0.10
	Sep	-0.15	-0.10
	Oct	-0.25	-0.05
	Nov	-0.30	-0.10
	Dec	-0.20	-0.10
1996	Jan	-0.30	-0.10
	Feb	-0.30	-0.10
	Mar	-0.35	-0.10
	Apr	-0.45	-0.10
	May	-0.45	-0.10
	Jun	-0.45	-0.10
	Jul	-0.70	-0.10
	Aug	-0.60	-0.10
	Sep	-0.60	-0.10
	Oct	-0.50	-0.10
	Nov	-0.40	-0.10
	Dec	-0.40	-0.10

Source: *Weekly Petroleum Argus*, various issues.

retrospective price. Customers are requested to provide a valuation for Oman crude, from which an average price is derived. This average is itself an average of three others, one produced from the inputs, one from those of the five most important customers, and one produced using a rejection criterion for outliers. A further average of all the averages derived during the month then provides a guidance price, which serves as an element in the ministerial decision of the value to be announced. The announced value is known as the MPM price, after the Omani Ministry of Petroleum and Minerals.

Qatar sets its retroactive prices in terms of a premium to the MPM price. However, a broad range of discretionary pricing is employed rather than merely keeping within a narrow range to the MPM price. Table 11.11 shows the MPM price, together with the announced differentials made by Qatar for Qatar Land and Qatar Marine, expressed as premia to the MPM and in absolute terms. The retroactive prices for the four Abu Dhabi grades are also shown. While Abu Dhabi price changes are very broadly related to those in the MPM, attention is apparently paid to maintaining competitiveness with the prices that had been generated by Saudi Arabian formulae, and in particular Berri.

There is a spot market for Middle East oil, beyond the sale of wet Dubai cargoes, but it is limited in scale. Table 11.12 shows the number of deals reported to Petroleum Argus from 1986 to 1995 by country of origin and crude oil type. The spot market for Iranian crude oil is almost exclusively for delivered cargoes into the Mediterranean market at dated Brent related prices, and is not included in the table.

The level of trade is extremely slight, except in the cases of Oman (as of 1995 around 3.5 reported deals per week), Murban (1.5 reported deals per week), and Lower Zakum (one reported deal per week). Even these deals do not in any way represent a centre of price discovery, and in fact carry very little price information. The problem lies with the retroactive official price systems. Since, by definition, those prices are unknown at the time a deal is made, for risk insurance purposes deals are transacted as a differential to the official price, with actual settlement occurring after the prices are announced. Thus, deals for Oman are done against the MPM price, and those for Murban and Lower Zakum against the relevant ADNOC (Abu Dhabi National Oil Company) price. For absolute price discovery purposes, agreed differentials against an unknown (at the time of the deal) official price carry no information at all. While up to 1987 the majority of spot deals were transacted at absolute prices, of the 830 reported in Table 11.12 during 1994 and 1995, only one was done at a fixed price. In Section 3 we saw that absolute price information was no longer generated by the Dubai market; now we find that there are no further sources in the Middle East.

Table 11.11: Retroactive Prices for Omani, Qatari and Abu Dhabi Crudes. January 1993 to December 1996. Dollars Per Barrel.

		Oman MPM	Qatar Land		Qatar Marine		Murban	Lower Zakum	Upper Zakum	Umm Shaif
1993	Jan	15.84	16.34	0.50	16.19	0.35	16.90	16.80	15.40	16.55
	Feb	16.80	17.30	0.50	17.15	0.35	17.90	17.80	16.40	17.55
	Mar	16.95	17.45	0.50	17.30	0.35	17.95	17.85	16.40	17.55
	Apr	17.05	17.55	0.50	17.40	0.35	17.85	17.75	16.30	17.45
	May	16.70	17.20	0.50	17.05	0.35	17.55	17.45	15.95	17.15
	Jun	16.46	16.96	0.50	16.81	0.35	17.25	17.15	15.55	16.85
	Jul	15.21	15.71	0.50	15.56	0.35	16.00	15.90	14.15	15.60
	Aug	15.46	15.96	0.50	15.81	0.35	16.50	16.40	14.70	16.10
	Sep	14.60	15.25	0.65	14.95	0.35	15.90	15.80	14.10	15.50
	Oct	15.26	15.98	0.72	15.70	0.44	16.65	16.55	14.80	16.25
	Nov	14.12	14.95	0.83	14.70	0.58	15.40	15.30	13.55	15.00
	Dec	12.68	13.43	0.75	13.18	0.50	14.00	13.90	12.10	13.55
1994	Jan	13.78	14.58	0.80	14.35	0.57	15.10	15.00	13.30	14.65
	Feb	13.24	14.14	0.90	13.94	0.70	14.60	14.50	12.80	14.15
	Mar	12.35	13.33	0.98	13.13	0.78	13.75	13.65	12.05	13.30
	Apr	14.20	15.02	0.82	14.84	0.64	15.45	15.35	13.90	15.00
	May	15.05	15.80	0.75	15.62	0.57	16.10	16.00	14.65	15.60
	Jun	16.08	16.68	0.60	16.51	0.43	17.00	16.90	15.65	16.50
	Jul	16.90	17.38	0.48	17.18	0.28	17.60	17.50	16.40	17.10
	Aug	16.21	16.79	0.58	16.59	0.38	17.00	16.90	15.75	16.50
	Sep	15.71	16.40	0.69	16.20	0.49	16.70	16.60	15.30	16.20
	Oct	15.84	16.58	0.74	16.40	0.56	16.90	16.80	15.40	16.40
	Nov	16.64	17.22	0.58	17.03	0.39	17.55	17.45	16.05	17.05
	Dec	16.04	16.62	0.58	16.44	0.40	16.85	16.75	15.50	16.35
1995	Jan	16.55	17.16	0.61	16.98	0.43	17.35	17.25	16.05	16.90
	Feb	17.10	17.69	0.59	17.54	0.44	17.90	17.80	16.75	17.50
	Mar	16.60	17.15	0.55	17.01	0.41	17.40	17.30	16.35	17.00
	Apr	17.70	18.20	0.50	18.00	0.30	18.45	18.40	17.50	18.05
	May	17.49	17.86	0.37	17.72	0.23	18.20	18.15	17.35	17.80
	Jun	16.43	16.66	0.23	16.52	0.09	17.00	16.95	16.20	16.60
	Jul	15.18	15.44	0.26	15.30	0.12	15.80	15.75	15.00	15.40
	Aug	15.56	15.87	0.31	15.73	0.17	16.30	16.30	15.35	15.90
	Sep	15.60	15.98	0.38	15.84	0.24	16.50	16.50	15.45	16.10
	Oct	15.00	15.44	0.44	15.32	0.32	16.00	16.00	14.80	15.60
	Nov	15.96	16.38	0.42	16.26	0.30	16.95	16.95	15.65	16.55
	Dec	17.48	17.80	0.32	17.68	0.20	18.40	18.40	16.95	18.00
1996	Jan	17.26	17.63	0.37	17.53	0.27	18.20	18.20	16.65	17.80
	Feb	16.50	16.90	0.40	16.80	0.30	17.45	17.45	15.90	17.05
	Mar	17.77	17.93	0.16	17.83	0.06	18.45	18.45	16.95	18.05
	Apr	18.63	18.80	0.17	18.68	0.05	19.25	19.30	17.70	18.85
	May	17.89	18.15	0.26	18.04	0.15	18.60	18.65	16.95	18.25
	Jun	18.24	18.45	0.21	18.37	0.13	19.00	19.10	17.30	18.70
	Jul	18.77	19.05	0.28	18.96	0.19	19.55	19.65	17.85	19.25
	Aug	19.60	19.93	0.33	19.84	0.24	20.45	20.50	18.75	20.15
	Sep	21.20	21.75	0.55	21.66	0.46	22.30	22.35	20.55	22.00
	Oct	22.50	23.11	0.61	23.03	0.53	23.80	23.85	21.95	21.20
	Nov	21.80	22.49	0.69	22.42	0.62	23.10	23.15	21.20	22.85
	Dec	22.80	23.50	0.70	23.43	0.63	24.20	24.25	22.10	23.95

Sources: *Middle East Economic Survey* and *Weekly Petroleum Argus*, various issues.

Table 11.12: Spot Deals for Middle East Crude Oil by Grade. 1986–95. Reported Deals.

Country	Crude Oil	1986	1987	1988	1989	1990	1991	1992	1993	1994	1995
Iraq	Kirkuk	15	19	25	17	10	-	-	-	-	-
	Other	2	2	4	10	7	-	-	-	-	-
Kuwait	Kuwait	47	22	12	2	4	2	6	1	-	-
Neutral Zone	Hout + Khafji	1	4	2	-	-	-	-	-	-	-
Oman	Oman	188	138	118	219	241	237	213	200	225	189
Qatar	Qatar Land	43	17	21	20	13	7	12	24	20	27
	Qatar Marine	52	22	22	34	34	24	21	17	11	15
	Other	-	-	-	-	-	-	-	-	-	1
Saudi Arabia	Arabian Medium	1	19	21	36	11	9	8	12	23	16
	Arabian Light	7	21	13	18	4	7	4	1	-	-
	Arabian Heavy		23	6	1	3	3	1	-	-	-
	Other+Mixes	1	2	1	2	-	5	-	-	2	-
UAE	Murban	59	54	24	65	58	112	90	106	50	81
	Lower Zakum	19	11	14	20	22	43	44	40	33	58
	Umm Shaif	17	10	7	11	16	13	10	8	4	13
	Upper Zakum	22	6	9	2	2	-	-	1	1	1
	Other	15	5	12	1	-	-	-	2	7	1
Yemen	Marib Light	-	-	1	14	1	21	12	7	11	7
	Masila	-	-	-	-	-	-	-	8	14	20
	Other	-	-	9	1	-	-	-	-	-	-
TOTAL		489	375	321	473	426	483	421	427	401	429

Source: Own calculations from Petroleum Argus database.

The development of any meaningful alternative price marker in the Middle East is thus hampered by existing pricing mechanisms. Even if retroactive pricing was discontinued, further obstacles remain in the form of official stances towards spot markets. For instance, the attitude of the Omani Ministry of Petroleum and Minerals (MPM) to the spot market could be described as ambivalence combined with periods of outright hostility. Attempts have been made at points to prevent term contract holders from offering cargoes onto the spot market.[22] These points tend to coincide with general market weakness, and in particular MPM concern seems to grow when cargoes are offered at a discount to the MPM price. The implicit assumption is that the spot market during weakness creates downwards pressure on the official price, which could be removed by suppression of the market. The argument is somewhat less than convincing. First, the mechanism is a two-edged sword, strong markets provide a signal for official prices to rise. Secondly, removing the spot market does nothing to affect the pressure from the market as a whole. In short, trying to smother the messenger does not affect the message. We examine the suitability of Oman as a marker crude oil in the next chapter, judged against a set of criteria. However, we would note at this point that the swings in official policy towards the spot market do constitute a further strong negative factor in addition to the problems caused by retroactive pricing.

7.   Conclusions

We began this chapter by listing the deficits in the Asian crude oil market. Among these, the Middle East certainly fills the bulk of the deficit in crude oil requirement. However, in questions of price formation, we have found a Dubai market with limited trading volume (and very limited physical production underpinnings), and spot trade in other crude oils either highly suppressed or distorted by retroactive pricing mechanisms. The deficit in markets does not seem to have been adequately filled.

If there are doubts regarding the suitability of the Dubai market to price a volume more than thirty times larger than Dubai's own production, the question is raised as to why it has continued to be used as the marker for Middle East exports into Asia. We would suggest two main explanations. Given its *de facto* price leadership in the region, the key role is that of Saudi Arabia, with other countries preserving the Dubai link primarily because Saudi Arabia does. A possible reason for Saudi Arabia's use of Dubai is the attractiveness of being able to affect relative price levels across regions, and in particular in the post Gulf War mode of revenue maximization, to be able to price discriminate. Using the same marker for

Europe and Asia would make price discrimination extremely transparent *ex ante*, and probably very difficult to maintain for an extended period. Using separate markers makes price discrimination a less transparent and *ex post* phenomenon. In this sense any perceived failings in Dubai are compensated for many times over by the extra revenue extraction abilities given by using three different markers. The second possible reason is inertia: put simply, few doubts about Dubai related pricing have been expressed by Asian refiners, who have been very oil acquisition price insensitive in comparison to their Western counterparts.

If these are the relevant reasons, then factors are at play to diminish greatly their force. The maximum extent of price discrimination (assuming revenue maximization) is limited by the extent to which demand elasticities vary between markets and by the possibility of resale. Considering the first of these factors brings us again to the market deregulation processes described in Part II of this book. We suggested that liberalization was causing greater price sensitivity, in other words behaviour of the East was becoming more like that in the West. Liberalization has also been accompanied by an expansion of refining capacity, indeed an over-expansion in the short run, helping to reduce the buffer that higher product prices created. The second factor is changing through a general weakening of Western relative price levels following continuing increases in oil production, increasing the possibilities of arbitrage into the East. The latter point constitutes the major part of the argument in the next chapter.

Notes

1. In 1995 US sanctions on Iran were tightened, with the expansion of the embargo to the refining operations of US companies anywhere in the world.
2. As of 1996, Conoco has a 32.5 per cent share, CFP Total 27.5 per cent, Repsol 25 per cent, DEA 10 per cent, and Wintershall 5 per cent.
3. Various EOR schemes have been considered, see Ian Seymour (1994), 'Dubai: Making the Most of Remaining Reserves', *Middle East Economic Survey*, 17 October 1994. In 1995 the US government blocked a Conoco project in the Sirri field in Iran (close to the marine border with Dubai) that would have given Conoco the option to reinject Sirri gas into the Dubai fields to help arrest the decline.
4. As it is an informal market with deals conducted by telephone, it has of course no physical location. However, in terms of the location of traders, Dubai is a London market, with the vast majority of deals being transacted between London trading rooms.
5. Formally, it was Abu Dhabi who joined OPEC before the creation of the UAE. The UAE's policy within OPEC, and actions consequent to that policy, have historically been led by Abu Dhabi rather than the other oil-producing emirates within the union.

6. Formally, the document is entitled 'Dubai Crude Oil General Terms and Conditions for fob Oil Sales'.

7. For instance, during nominations for the September 1995 loading programme, a dry chain appeared, i.e. one where no cargo is actually lifted. Confusion over dates of two cargoes meant that the allocated dates being passed down the chain from the terminal were retained in the middle of the chain, and hence the end of the chain was left without a cargo.

8. In Table 11.1, and in the other tables in this section, we have included only trades of forward cargoes, and not those of wet cargoes. Our results then differ from those in Paul Horsnell and Robert Mabro (1993), *Oil Markets and Prices*, op.cit. which considered the whole market including wet cargoes to the end of 1991.

9. Market share is defined as a proportion of both buying and selling positions, i.e. a share of 10 per cent indicates an involvement in 20 per cent of all trades. As noted above, the table differs from that shown for the period up to 1991 in Paul Horsnell and Robert Mabro (1993), *Oil Markets and Prices*, op.cit. in that it includes only forward and not wet Dubai cargo trades.

10. A further discussion of OPEC pricing policy is to be found in Paul Horsnell and Robert Mabro (1993), *Oil Markets and Prices*, op.cit.

11. WTI replaced Alaskan North Slope as the marker for the US sales at the start of 1994.

12. Some companies buy on a first month rather than dated Brent related basis.

13. In addition, separate (WTI based) formulae are used for sales of crude oil out of Aramco leased storage facilities in the Caribbean.

14. This factor then creates a strong incentive, and indeed an observed tendency, for average tanker speeds to be higher in backwardations than in contangos.

15. A sixth stream, Shaybah, is due to be introduced in 1998 as a blend with Arabian Extra Light.

16. The details of Arabian Super Light are those contained in an interim blend assay published in *Middle East Economic Survey*, 24 October 1994. The gravity shown is the stabilized gravity, i.e. after debutanization.

17. The Aramco production figures shown in Table 11.6 are shown as a demonstration of the intent, as of 1994, of Aramco to bias production towards light grades, rather than as a projection or description of reality. In particular, Arabian Super Light production has run well ahead of the figures, reaching 200 thousand b/d as early as the start of 1995.

18. Our refining value difference is based on a simple three cut calculation of gross product worths based on the methodology employed by Platt's. We make no claim for it to represent the 'correct' value of the differential, for instance it makes no allowance for differential refining costs, and is extremely simple compared to the multivariate refinery models employed within Aramco. However, we do believe that it serves as an effective indicator for the range of the changes in relative refining values.

19. See Figure 13.5 and the related discussion.

20. This figure is arrived at as follows. Over the course of a month, the difference between the forwardness of a dated Brent Platt's assessment and a second month

Brent assessment is 38 days. Forty days backwardation can be first proxied by 1.05 times the monthly average differential between dated Brent and second month Brent. This produces a figure of $1.40. However, use of the tolerance clause in Brent contracts tends to produce a natural discount for dated Brent. While this is extremely variable, and dependent on market volatility and the structure of Brent traders' portfolios, 50 cents would appear to be an absolute minimum for the average size of the effect in March 1996. Removing this effect, and so having a measure of pure time structure, increases the forty day backwardation measure to $1.90.

21. See 'Yemen's Term Customers Wage War over Pricing', *Energy Compass*, 17 May 1996.
22. For the text of one such telex to this effect see *Middle East Economic Survey*, 23 May 1994.

CHAPTER 12

THE MARGIN OF THE ASIAN CRUDE OIL MARKET

1. Introduction

While nature may abhor a vacuum, economic systems abhor a disequilibrium. Distortions across markets can only be maintained by transactions costs (primarily, in the case of oil, freight costs) or by the maintenance of barriers to entry. The lesson of the OPEC experience of the early 1980s detailed in the last chapter, was that the only barriers to entry in oil production, given the right price, technology and fiscal system, are geological. This chapter is concerned with changes in the flows of oil into Asia facilitated by two main disequilibria.

The first disequilibrium has been the relatively higher acquisition cost of crude oil in Asia since the 1990–1 Gulf War, brought about by Middle East crude oil allocation as discussed in the last chapter. With relative prices remaining away from parity for a significant period, the likelihood of arbitrage by moving crude oil in from other regions increased. The start of deregulation processes which *inter alia* reduced the premium Asian refiners were prepared to pay for a long-term supply arrangement, further increased the potential for that arbitrage. The distortion certainly brought forward the time when marketers would begin to consider the possibility of trade flows (Norway to Taiwan is a good example in relation to a term contract signed in 1996) that had previously been considered as pure flights of fancy. However, even without that distortion, a second more powerful factor would have brought the time about in any event.

The second factor, both more important and sustainable, began as an asymmetry before creating a disequilibrium. Over the 1990s the geographical location of the most important elements of incremental crude oil supply (notably the North Sea and Latin America), has been distant from that of the most important elements of incremental demand (i.e. Asia). A depressive effect has been put on European and North American markets, while Asian demand has moved the fundamentals of the Eastern markets in the opposite direction. Under this asymmetry something has to relocate. Either Middle East OPEC oil must be diverted out of the West, or more non-OPEC oil must move out of the Atlantic basin markets and into Asia, or, as has been the reality, both adjustments occur together.

Our view is that this asymmetry will result in the growth of Brent related crude oil exports into Asia, and, further, that these flows will be at the margin of the market, and therefore price determining. The logic of this development runs as follows. The US Gulf Coast market has become the most competitive in the world. Extra cargoes arriving from Colombia, Venezuela and Argentina have a clear advantage over North Sea and West African oil. They load only a few days sailing time from the market, with an associated benefit for refiners, who can minimize their stocks, face less price risk, and achieve greater flexibility. In competition with other imports, Latin American output will always be considered as base load supplies for the Gulf coast. Thus, with the long-term increase in the USA's demand for imports unable to absorb all the extra Latin American material (particularly with the development of new US domestic supplies from deep water Gulf of Mexico fields), a volume of North Sea, West African and Middle East oil will be displaced. The absorptive capacity of the European market for incremental supplies is also limited. In addition to displacements from the US Gulf, there is also the burgeoning production of the North Sea. Again, there must be displacement, with West African supplies (similar in quality to North Sea oil) being prime candidates for further expulsion.

The consequences of the geographical asymmetry between incremental crude oil supply and demand are the subject of this chapter. In particular we take up the story of a desirable marker crude oil for Asia. In Chapter 10 we found no centre of price discovery in Asia. In Chapter 11 we found the shell of one in the market for Dubai crude oil, but noted that the market itself was in effect defunct, while its marker role continued unabated. In this chapter we introduce a further candidate for price discovery in Asia. Perhaps surprisingly at first glance, that candidate is the North Sea market.

In the next section we document how crude oil flows are changing, and in particular advance the hypothesis that the margin of the Asian market is now made up of Brent rather than Dubai related crude oils. Section 3 considers the potential of Alaskan oil in Asia, in the context of the history of US restrictions on exports. In Section 4 we consider what constitutes desirable characteristics for a marker crude oil, and use this categorization as a benchmark for possible markers in the Asian context. A final section provides some conclusions, and some implications for the pricing policy of the Middle East exporters.

2. The Changing Geography of World Oil Trade. New Western Flows to Asia

By the end of the Gulf War in March 1991, the position of OPEC, and in particular its Middle East members had been transformed compared to the situation just eight months previously, before the ill starred Iraqi attack on Kuwait. A very soft crude oil market had been transformed into a tight one, with almost all the usable OPEC excess capacity removed. That still left the question of 5 mb/d of the unusable capacity represented by Kuwait and Iraq, but at the time there were many both inside and outside OPEC who could rationalize any bearish instinct away. Two or three years of world demand growth would be enough to enable the reincorporation of both countries to be accomplished with little or no cutback by others, and without significant price weakness. Indeed, in the interim there was the prospect of supply tightness and higher prices. After all, by common consensus, non-OPEC production was not just peaking – due to the collapse of the oil industry in the former Soviet Union (FSU) it was actually falling away fast.

Although the optimistic scenario may have been justifiable in 1991, history has not been kind to it. By 1997, other than limited humanitarian sales, Iraq was still out of the market, yet no significant buffer had been built up to reabsorb it. Kuwait came back on stream very swiftly. Recessions in major oil-importing countries dampened demand growth. FSU demand followed FSU supply down, and then fell further, with the two effects resulting in an increase in the exportable surplus. Most importantly, non-OPEC supply outside Russia showed no sign of a peak being reached. OPEC was being hemmed in, faced with the dilemma that the call on OPEC crude oil could only be increased significantly by accepting lower prices, and all the time faced with the proposition that one day Iraq would rejoin the market for more than just UN sponsored humanitarian oil sales.

These processes are shown in Table 12.1, based on IEA data for 1992 to 1997.[1] The large declines in FSU supply and demand tend to make figures based on the global level rather misleading in their implications. We have accounted for world demand and non-OPEC supply exclusive of

Table 12.1: Global Oil Balances. 1992–7. Million b/d.

	1992	1993	1994	1995	1996	1997
World Demand (excl. FSU)	60.4	62.0	63.9	65.2	67.5	69.4
Of which Asia and Australasia	15.5	16.3	17.1	18.0	18.9	19.9
Non-OPEC Supply (excl FSU)	32.3	32.9	34.3	35.3	36.4	38.4
FSU Net Exports	1.8	2.2	2.4	2.4	2.7	2.8
Residual	26.3	26.9	27.2	27.5	28.4	28.2
Residual – OPEC NGLs	24.2	24.7	24.9	25.0	25.8	25.3

Source: Calculations from IEA *Monthly Oil Report*, February 1997.

the FSU, using the net FSU surplus as a separate item. Using this disaggregation, Table 12.1 shows demand increasing strongly over five years, rising by a total of 9 mb/d with Asia and Australasia providing one-half of the total increase. Had the bulk of this been supplied by OPEC, the above scenario could have easily been fulfilled. Indeed, if under this scenario Iraq had not returned to the market at full capacity by 1996, the upwards pressure on prices could have been extremely severe.

In reality, OPEC's share of the 9 mb/d increase in demand proved to be slight. As shown in Table 12.1, non-OPEC supply outside the FSU rose strongly, up by 6.1 mb/d over the five years. Adding in the significant contribution from an expansion in the FSU surplus, and deducting the result from the demand figures, gives the residual shown. Finally, subtracting OPEC natural gas liquids (NGLs) gives the last line of Table 12.1, i.e. the call on OPEC crude oil plus stock changes. Increases in this item account for just one-eighth of incremental non-FSU demand between 1992 and 1997.

The above demonstrates that the rise in non-OPEC supply played the major role in constraining OPEC, and preventing the possibility of a relatively painless swift full return of Iraq. Non-OPEC supply has always been a source of disagreement between geologists, reservoir engineers and economists. The former two categories tend to aggregate production profiles, and usually project imminent declines of non-OPEC production. Economists concentrate on fiscal regimes and technological change, resulting in later peaks and the projection of oil production from reservoirs that for the geologist do not yet exist. However, the one common feature between both approaches has been a repeated tendency over many years to underestimate non-OPEC production. As noted in Chapter 3, outside of the China Seas, the surprises have tended to be heavily biased towards the upside.

The pattern of non-OPEC production outside the FSU since 1973 is shown in Figure 12.1, both inclusive and exclusive of the USA. We have at various points throughout this book warned against the dangers of straight line interpolations. We would however admit that the case for the prosecution is weakened by Figure 12.1, with the tendency for non-OPEC production outside the FSU and USA to move linearly for often prolonged periods. Note that the rise has gone on in a remorseless near linear fashion, regardless of shocks and discontinuities in the oil market. The price shocks of the 1970s hardly register, the price collapse of 1986 appears to flatten the curve temporarily before it reassumes its early 1980s gradient in the early 1990s. Without having to make reference to geology, the salient features of the production shown in Figure 12.1 are that it has never fallen, and that it has averaged a yearly growth of about 850 thousand b/d over a period of nearly a quarter century.

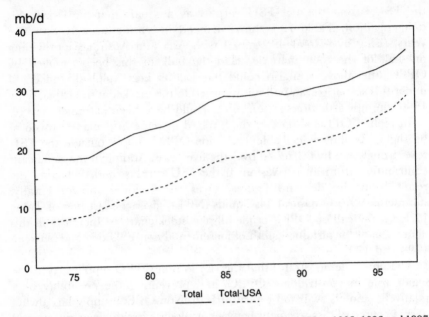

Sources: *BP Statistical Review of World Energy* (various years) for 1973 to 1995. 1996 and 1997
based on IEA data converted to BP consistent figures.

Figure 12.1: Non-OPEC Production Outside the Former Soviet Union, Inclusive and
Exclusive of the USA. 1973–97. Million b/d.

The factors behind the resilience of non-OPEC output were discussed
in Chapter 10 in the context of the North Sea. Increases are being driven
by a combination of accommodatory fiscal regimes, cost savings and
efficiency gains for given technology, plus the development of new
technologies. There are of course limits to the scope of the first two
factors. For example, the fiscal regime for new fields in the UK can not
now become much more liberal except by the provision of tax breaks and
effective subsidization relative to other industries. There is a limit,
represented by a cost curve, to the extent of cost saving and efficiency
gains achievable for given technology. However, there are a large number
of non-OPEC oil provinces (indeed, the vast majority) where the scope for
fiscal relaxation is large, and where cost saving along the lines of the
North Sea model has hardly begun. In addition, technological change has
become rapid and has produced considerable advances in production
possibilities. There seems then to be no necessary change in the dynamic
behind Figure 12.1, with the ultimate constraint being set by geology
alone, and the nature of that constraint constantly being redefined by
technological progress.

Onshore US production in the lower 48 states presents something of a special case. A relatively liberal tax regime has meant that there has been little scope for fiscal incentives to provide additional impetus to production. New seismic and drilling technology offer comparatively little when, during a long history of exploration, nearly all potential oil bearing structures have been so intensively drilled as to resemble a slice of Swiss cheese. Enhanced oil recovery offers some returns, but overall the factors that have contributed to production increases elsewhere have been of little import. Offshore US production and Alaska are however different. Technology and efficiency savings have made deep-water Gulf of Mexico production viable, and bring new production such as the substantial Mars field. In Alaska, there remains the prospect of the Arctic National Wildlife Refuge (ANWR), should exploration there ever gain political acceptance.

There remains a tendency in forecasting to predict short- to medium-term relative output paths primarily on the basis of current estimates of relative reserve bases. Part of the problem is the treatment of the estimated reserve base as a purely geological rather than an economic variable. There are various quantities involved. There is the total amount of oil in place, say X, which is purely determined by nature; and the total amount actually discovered, say Y, from which we attempt to estimate X. Estimating X, a completely fixed quantity, on the basis of Y which is a function of highly variable factors, most particularly technology, has always been a fraught and rather unsatisfactory process. There is also the recovery factor, say α, which is the proportion of Y that can be economically developed and then recovered, and is historically only about 30 per cent.

Particularly in the short to medium term, predicting relative output paths on the basis of relative αY is not a useful methodology. Both Y and α are predominantly economic quantities, in that they are affected by fiscal conditions, industrial structure, technology, capital market conditions and (to a more limited extent than is often recognized given the endogeneity of the other factors), by oil prices. Of the two, only the size of Y has a long-term correlation with (and in the very long term convergence onto) the size of X. The decline of non-OPEC production view seems to be confusing αY sometimes with Y, and sometimes with X, and forgetting the impact of the residual $(1-\alpha)Y$. Non-OPEC output is partly being driven by changes in Y, but it is also very strongly driven by α. We saw that in the case of the UK North Sea, the process is almost entirely dependent on the latter, with fiscal conditions and technology enabling the development of fields that were already part of Y, but were not previously economic, and in getting more out of existing reservoirs. A small increase in α results in considerable extra production. In the longer term, larger increases of α, perhaps up to 0.4 or beyond, are perfectly feasible with the consequent implications for the sustainability of non-OPEC production.

After twenty years of non-OPEC oil continuing to rise despite all claims of imminent falls, adjustments have been made in thinking to start to correct the constant undershooting of expectations.[2] An extremely good example of the new realism about non-OPEC production, and in particular the expansion of the time frame before it is expected to fall, is the forecasting of Kleinwort Benson, shown in Table 12.2.[3] The forecasts for 2005 are intended by their author to be cautious, but their major importance is that a watershed in psychology has been reached, whereafter forecasters feel confident to project non-OPEC production to rise for at least another ten years.

The geographical implications of the continuation of strong increases in non-OPEC supply arise when we consider the demand side. For purposes of illustration, we use the forecasts of the IEA, who use two sets of assumptions.[4] A fairly bearish view on demand is embodied in the 'Energy Savings' case. In this scenario the total increase in world demand between 1995 and 2000 amounts to 5.8 mb/d. Table 12.2 shows a total 6.1 mb/d increase in non-OPEC output. In this case, the totality of incremental Asian demand (some 2.5 mb/d under the scenario) would have to be met by non-OPEC oil, or by large-scale diversion of Middle East OPEC oil out of Western markets. There is a lack of perfect fungibility between incremental non-OPEC production and the generally heavier and higher sulphur OPEC flows. Further, the diversion is from high conversion refining areas into the lower conversion capabilities of the Asian market. Putting these two aspects together, there would appear to be scope for a significant movement from the Atlantic basin of non-OPEC

Table 12.2: Non-OPEC Production. 1995 and Forecast 2000 and 2005. Million b/d.

	1995	2000	2005
Total	40.6	46.7	49.2
of which			
Non-OPEC (excl. FSU and USA)	24.9	30.1	31.0
of which			
Norway	2.9	3.7	3.2
UK	2.8	3.6	3.2
Latin America	6.1	7.8	8.6
Africa	2.2	3.0	3.3
Asia-Pacific	5.7	6.4	6.7
USA	8.6	8.5	8.6
FSU	7.1	8.1	9.7

Source: From Mehdi Varzi (1996), *OPEC and non-OPEC*, op.cit.

oil into Asia, rather than the pure diversion of Middle East material. As suggested in the introduction to this chapter, the swing into Asia would then be made primarily by West African but also by North Sea crude oil.

The alternate IEA scenario, 'Capacity constraints', is more bullish about demand, and shows a 7.5 mb/d increase in world demand between 1995 and 2000. Even in this case, the call on OPEC only increases by 1.4 mb/d, again less than growth in Asian demand. Note also that a 1.4 mb/d increase does not cover the potential for Iraqi production following a full return to the market. The implication of Table 12.2 is that, in spite of the scenario's title, in fact no capacity constraints are likely to be reached, and the moderating effects on demand of rising prices in the IEA scenario would be absent. Demand would then increase faster. However, assuming that Iraq's potential as of 2000 would be at least 3 mb/d, demand would have to rise by 9 mb/d in five years before the other OPEC members could increase production without causing price weakness.

The combination of forecasts based on the new realism on non-OPEC supplies and the IEA demand figures, produces a scenario where OPEC is heavily constrained throughout the 1990s. One option for OPEC is perhaps to take the long view, after all the two IEA scenarios produce a further increase in world demand between 2000 and 2010 of the range of 16 to 19 mb/d. Could they just wait on the assumption that within that decade incremental demand would accumulate to them? This would perhaps be a triumph of hope over experience, since it is the same scenario that was commonly used in the 1980s in regard to the 1990s.

Given that there is no reason to see the underlying process behind non-OPEC increases as changing in character or effect, we would see the combination of demand optimism with the belief that non-OPEC production will fall soon after 2000 as being by no means realistic. In Table 11.3 we showed that the exports of the six Middle East OPEC producers had not regained their 1975 level of 16.9 mb/d by the mid 1990s, and the scenarios above suggest that they will still fall short of the 1975 level by 2000. Within this context, projections that place OPEC output in 2010 between 45 and 50 mb/d (and imply Middle East OPEC exports of well over 30 mb/d), seem to be predicated on some major discontinuity from the last quarter of the twentieth century.[5]

An alternative to the wait and hope strategy, is to increase field development and output in the expectation that a period of lower prices will rein in longer-term non-OPEC production by dampening exploration and development.[6] There are multiple problems with this strategy. The first is that at best the effect is long term. To accept a prolonged period of low prices to attempt to depress long-term non-OPEC output, would suggest that producer countries apply a very low discount rate to future revenue streams. However, the pressures of low current revenues, increasing population, increasing calls on government spending and, in some countries, political instabilities and uncertainties, all suggest that the discount rate is in reality fairly high. More serious, however, is the possibility that low prices may do little or nothing to impact on non-OPEC output.

Because any effect can only be long term, in most countries there is scope for endogenous changes in fiscal regimes to mute, or even reverse, the price impact. Low prices also increase the incentives for technological research and development effort, and for the application of new technology. If costs are not already being minimized, low prices also increase the incentives to make cost savings and cut into the slack within the system. In total, prolonged periods of low current prices may have little impact on the long-term level of non-OPEC supply. Particularly, rather than the level of current prices, the trinity of technology, fiscal and geological conditions tend to be the major determinants of non-OPEC production growth.[7]

The presence of flows of West African and North Sea crude oil at the margin of the Asian market, displaced by the pace of western hemisphere non-OPEC production growth, marks a radical change in the behaviour of the crude oil market. The behaviour of these flows is illustrated in Table 12.3. This covers movements of spot West and North African oil into Asia for loading months between September 1995 and March 1996. The degree of swing is large, with the flow reaching as high as 879 thousand b/d for February 1996 loadings. It should be noted that this table only shows reported spot trades, and not term contracts, and thus represents a minimum estimate of the size of the marginal stream. Only one of these cargoes went to Japan, and only four to Singapore, with the most active buyers being Korean refiners, China, Taiwan and (through the monthly IOC crude tender) India.

The degree of monthly swing in the volumes shown in Table 12.3 is to some extent a function of Middle East pricing policy. First, the $1 per barrel price differential for oil exports to Asia we found in the last chapter, has enhanced the viability of moving African crude oil east. Secondly, West African crude oil is priced on a five- or ten-day dated Brent related basis, with Middle East exports to Asia being monthly average Dubai related. The flows are then related to the dated Brent to Dubai differential, and also in part to the behaviour of prices over the month. For example, in the latter half of a month during which prices have been falling, five-day dated Brent averages will be more attractive relative to the expected level of the Dubai monthly average.

When the marginal crude supply is Brent related, and can swing in volume to the degree shown in Table 12.3, the case for monthly average Dubai related pricing is severely impacted on. We noted in the last chapter

Table 12.3: Eastward Movements of African Crude Oil. Source, Destination, Grade and Volume. September 1995 to March 1996. Thousand Barrels.

Loading Month	Destination	Source	Grade	Volume	Loading Month	Destination	Source	Grade	Volume
Sep-95	India	Egypt	Suez Blend	1800	Feb-96	Australia	Nigeria	Brass River	1000
	India	Nigeria	Quae Iboe	900		Australia	Nigeria	Brass River	910
	India	Nigeria	Quae Iboe	900		China	Angola	Cabinda	950
	Indonesia	Nigeria	Brass River	900		China	Angola	Nemba	650
	Korea	Gabon	Cam Lokele	900		China	Angola	Cabinda	300
	Korea	Nigeria	Forcados	1800		China	Nigeria	Palanca	985
	Philippines	Nigeria	Bonny Lt/Med	900		China	Nigeria	Palanca	985
Oct-95	India	Nigeria	Qua Iboe	900		China	Nigeria	Palanca	950
	India	Nigeria	Qua Iboe	900		India	Nigeria	Qua Iboe	1800
	Japan	Nigeria	Bonny Lt/Med	1800		India	Nigeria	Qua Iboe	1800
Nov-95	Korea	Nigeria	Bonny Lt	900		India	Nigeria	Qua Iboe	1800
	Korea	Nigeria	Bonny Lt/Med	900		Indonesia	Nigeria	Qua Iboe	900
	Korea	Nigeria	Bonny Med	900		Japan	Nigeria	Bonny Light	475
	Korea	Nigeria	Forcados	900		Korea	Gabon	Mandji	900
Dec-95	China	Angola	Cabinda	1000		Korea	Nigeria	Bonny Light	1400
	China	Angola	Cabinda	900		Korea	Nigeria	Bonny Light	1400
	China	Egypt	Es Sider	900		Korea	Nigeria	Forcados	900
	India	Nigeria	Qua Iboe	1800		Korea	Nigeria	Odudu	900
	India	Nigeria	Qua Iboe	1800		Korea	Nigeria	Bonny Medium	425
	Korea	Nigeria	Bonny Med	900		Korea	Nigeria	Bonny Medium	425
	Korea	Nigeria	Forcados	900		Singapore	Nigeria	Brass River	910
	Korea	Nigeria	Forcados	900		Singapore	Nigeria	Brass River	900
	Singapore	Egypt	Suez Blend	1500		Taiwan	Congo	Djeno	850
	Singapore	Egypt	Suez Blend	400		Taiwan	Nigeria	Forcados	2000
Jan-96	Australia	Gabon	Rabi	900		Taiwan	Nigeria	Palanca	980
	China	Angola	Cabinda	950	Mar-96	Australia	Gabon	Rabi Blend	900
	China	Angola	Cabinda	900		Australia	Nigeria	Brass River	600
	China	Angola	Cabinda	900		China	Angola	Cabinda	950
	India	Nigeria	Qua Iboe	1800		India	Nigeria	Quae Iboe	1800
	India	Nigeria	Qua Iboe	1800		India	Nigeria	Qua Iboe	1800
	Indonesia	Nigeria	Qua Iboe	1800		India	Nigeria	Qua Iboe	1800
	Korea	Nigeria	Bonny Light	900		Indonesia	Nigeria	Escravos	920
	Korea	Nigeria	Forcados	900		Korea	Nigeria	Forcados	1200
						Korea	Nigeria	Bonny Light	900
	Totals	(Thousand b/d)				Korea	Nigeria	Odudu	900
	Sep-95	270.0				Taiwan	Angola	Palanca	980
	Oct-95	116.1				Taiwan	Congo	Djeno	900
	Nov-95	120.0				Taiwan	Nigeria	Forcados	1000
	Dec-95	354.8							
	Jan-96	350.0							
	Feb-96	879.1							
	Mar-96	472.6							

Source: Bloomberg.

that the ability to achieve a price gap between regions was the only remaining virtue of using Dubai, and that liberalization in the Asian industry would make this harder to maintain. The effect shown in Table 12.3 is more powerful in reducing this scope, i.e. there is now scope for arbitrage from other regions.

An attempt at creating a price gap then simply becomes an increase in relative prices to which an alternative source of supply can respond by undercutting. In terms of the criteria used for the design of Saudi Arabian pricing formulae noted in Chapter 11, the principle of competitiveness with alternative supplies is foregone. When price discrimination began in 1992, the structure of relative prices was such that a large magnitude of discrimination could be maintained without inducing flows from outside the region. The growth in western hemisphere non-OPEC production since 1992 has meant that the same degree of compartmentalism of the world oil market can no longer be achieved.

Beyond the push factors of displacement, it should be noted that there are also strong pull factors that bring Brent related crude oil into Asia. In particular, in Part II we noted the tightening of the sulphur specifications for oil products throughout Asia, as well as a relative lack of both conversion and desulphurization capacity in several countries (most notably Korea). To some extent, low sulphur West African supplies are being bought so as to enable sulphur specifications to be met. At the margin, particularly for Korea and Taiwan, there is a volume of West African crude oil that can not be substituted for by higher sulphur supplies. We also noted in Chapter 5 that, while China has cracking facilities, only a limited number of refineries can run the bulk of Middle East crude oil due to the widespread use of untempered steel. Again there is a niche market for a quantity of low sulphur West African material.

The refinery economics of using Brent related crude oils, beyond West Africa even from the North Sea, can be highly favourable. Table 12.4 shows the results of comparing the value of two Norwegian crude oils, Oseberg and Gullfaks,[8] against swing crude oils used in China, Japan and Singapore. The Norwegian crude oil employs the economics of VLCC transport of cargoes of 2 million barrels against the smaller tankers assumed for the swing crude oils, with prices being representative for the period October 1995 to February 1996. Differences are shown in refining value, freight cost, price and overall, with positive numbers representing an advantage for the Norwegian crude oil against the swing crude oil.

Further considerations beyond the calculations shown in Table 12.4 arise from the cost of breaking the VLCC cargoes into smaller parcels, the consideration of the increased sailing time and its implications, and switching costs. However, the table does seem to demonstrate that North Sea supplies on VLCCs can be competitively placed in the Asian market.

Table 12.4: Economics of Processing Norwegian Crude Oil versus Swing Crude Oil in Selected Refining Centres. $ Per Barrel.

Country	Versus	Refining	Freight	Price	Overall
Oseberg					
China	Minas	+0.30	-0.54	+0.92	+0.68
Japan	Murban	+0.30	+0.50	-0.31	+0.49
Singapore	Tapis	+0.00	-0.46	+1.53	+1.07
Gullfaks					
China	Minas	+0.24	-0.54	+0.99	+0.69
Japan	Murban	+0.05	+0.50	-0.24	+0.31
Singapore	Tapis	+0.00	-0.46	+1.60	+1.14

Source: Statoil.

The market conditions that affect the competitiveness of Brent related crude oil are as follows. First, the time structure of prices plays a critical role. In particular, backwardation (i.e. a premium for prompt delivery) reduces the scope for moving oil into Asia. A backwardation tends to increase the differential between dated Brent and Dubai (producing the effect noted for March 1996 in Figure 11.4 in the context of Saudi formulae prices), as well as increasing the opportunity cost implicit in longer sailing times. The second factor is movements in freight rates. It should also be noted that crude movements from the Atlantic basin can take advantage of the favourable economics of backhaul tankers[9] as well as the scale economies of VLCCs and ULCCs.

As mentioned earlier in the context of Norwegian oil, the costs of breaking up large cargoes provide another determining factor. A fourth factor, as noted above, is intra-month volatility and trends in prices when arbitraging Brent related spot cargoes into a market where both Middle East and Asian crudes are priced from monthly averages. The fifth factor is relative product prices through their impact on refinery valuations. The final major factor is the underlying Brent–Dubai differential net of the time structure effects noted. This in part reflects the factors behind light-heavy crude differentials, which are discussed in the next chapter.

3.   New Oil Flows from Alaska

There is a natural trading link between East Asia and Alaska, one which for over twenty years had been ruled out for oil by US policy. In 1996 that link became possible with the relaxation of export restrictions on Alaskan North Slope (ANS) crude oil. For more than twenty years the free market solution has been blocked by US legislators at an economic cost to the USA, and now there is the possibility for Alaskan oil to enter what has always been its natural market. The export of Alaskan North Slope (ANS) crude oil had been effectively banned under US legislation. With a relaxation of the ban, but not as yet complete liberalization due to some residual protectionism, new trade possibilities have been opened. We consider the origins and implications of the ban, and some of the effects of the liberalization.

In the last week of 1967, oil was struck in Prudhoe Bay on the north coast of Alaska. What proved to be the largest US oilfield had been found in the most inaccessible area of the country. Surrounded by permanent ice, transporting the oil was an immediate problem. Eventually the option chosen was to build a pipeline, the Trans-Alaska Pipeline System (TAPS), to the south coast and the port of Valdez in Prince William Sound. The Trans-Alaska Pipeline Authorization Act finally came to a vote in both houses of Congress in November 1973, and was signed by President Nixon on 16 November. To put it into the context of contemporary opinion, less than a month earlier on 18 October 1973 the OAPEC oil embargo on the USA had been announced. The pipeline issue had become a part of economic warfare as a defence against the OAPEC oil weapon. Given the conditions of the time, it is perhaps not too surprising that political rather than economic criteria came to the fore. What is more surprising is that it took more than twenty-two years for the resulting policies to be reversed.

At a late stage of deliberations on the Alaskan pipeline during the summer of 1973, foreign policy had become explicitly involved. First, amendments were raised by a group (led by the future Vice-President Walter Mondale) wishing to reconsider an option to put the pipeline through Canada. This option would have allowed Alaskan oil to enter the network of Canadian and US pipelines, and to reach mid-West refineries. No terminal for loading oil onto tankers would be necessary with this option, and hence the question of exports would be an irrelevant one. The attempts to resuscitate the Canadian route failed, and a concern that Alaskan oil would be exported remained.[10] These concerns then manifested themselves in the pressure for an overt export ban.

The ban was without doubt aimed specifically at stopping Alaskan oil going to Japan. At issue was a fear that the US West Coast would not be able to absorb all the production, and that large-scale exports would occur. According to Hoyler (1984), US politicians did not want to be part of the building of a pipeline all the way across Alaska, only for the oil to go to the Japanese market. A charitable explanation would be that they believed public opinion would be unable to understand why the USA was exporting oil to Japan, at a time when the supply of oil to the USA was under threat. A less charitable explanation would be that they could not themselves understand. Whatever the reason, the effect was the same; the pipeline act signed by Nixon had an anti-export clause inserted. The result was a triumph of irrationality, in the words of Bradley, 'as much an economic perversity as a special interest coup'.[11]

The clause was not strictly a straightforward ban, but oil could only be exported if the president deemed it to be in the national interest, and even then Congress could reverse the decision. The restriction was reinforced by a ban on all US exports of crude oil under the Export Administration Act. ANS was then kept within US boundaries, with the exception of shipments to the Virgin Islands and Puerto Rico. In regard to Alaskan oil, US policy was to be based on political and nationalistic principles and not on the operation of free markets. This was made clear when soon after signing the pipeline act, Nixon launched 'Project Independence' through a television broadcast. In his words, 'in the last third of this century, our independence will depend on maintaining and achieving self-sufficiency in energy'.[12]

Economic irrationality has historically tended to be associated with the treatment of trade issues. To use a famous example, during the US Civil War the Confederacy decided to ban sales of its major export, i.e. cotton, to the Union, on the grounds that the North should be deprived of strategic commodities. Of course this simply meant that the Union found its cotton elsewhere, albeit at a higher price, while the Confederacy was bankrupted. There are certain elements of this reverse logic in the Alaskan export ban, at least in the idea that supplying a resource (available elsewhere) to another country would not be in your best interests. As Mead (1978)[13] points out, Japan had been involved in the pipeline project from its inception, and was the natural trading partner for Alaska in every other area.

The supply security argument that American oil should go to the US market to help prevent oil dependence was fallacious. If the economics of transport suggest that it would prove more efficient to export from Alaska and import from elsewhere, the policy carries an economic cost. But even in terms of dependence, having a large export stream is likely to be an advantage. Given any disruption in imports there is an immediate domestically available substitute that can be diverted to fill any import gap. For example, consider the threat is one of embargo, and the possible trade flows consist of an export of 1 mb/d with an offsetting import volume. Only if the additional imports were obtained entirely from countries joining the embargo would the overall shortfall be the same as under an export ban. We would argue that far from reducing the vulnerability of the USA to oil supply shocks, the Alaska export ban actually increased it.

In reality, as markets became more flexible the likelihood of any real shortfall was reduced, and in addition the US built up a Strategic Petroleum Reserve. Further, as any supply shock was most likely to impact mainly on the US Gulf, there would usually be far cheaper alternative sources than bringing in the Alaskan oil. In other words the Alaskan oil would probably not be needed even in the case of a supply disruption. In short, the supply security argument was never valid, and it became less and less defensible with the passage of time.

The Trans-Alaska pipeline opened in 1977, with Alaskan production reaching 1 mb/d in 1978 and peaking at 2 mb/d in 1988. With the suppression of exports, ANS was absorbed into the US market. Most of it moved into the US West Coast, and the rest was taken into the US Gulf Coast and the Caribbean. The latter supply route was long and expensive. From Valdez tankers went to Panama, where the oil crossed the isthmus by pipeline, and was then reloaded into tankers for the last leg of the journey. The tankers involved were all Jones Act tankers, that is to say US built, US flagged and US crewed, or in other words extremely expensive.[14] For producers who moved the oil on their own tankers, the element of economic profit assigned to their tanker operations also reduced the netback value of the oil in Alaska. To a limited extent there was a reduction in upstream taxation to compensate for some of the loss in wellhead price caused by the ban. Table 12.5 shows the distribution of ANS deliveries in 1990 by first port of call, and by eventual destination for shipments through Panama.

Table 12.5: Volume of Alaska North Slope by Destination, 1990. Thousand b/d.

Alaska		68
West Coast		1358
of which		
	Puget Sound	*478*
	San Francisco	*250*
	Los Angeles	*630*
Hawaii		48
Virgin Islands		19
Panama		252
and then to		
	Gulf Coast	*174*
	East Coast	*2*
	Virgin Islands	*72*
	Puerto Rico	*4*
TOTAL		1745

Source: Marks (1992).[15]

The fact that the US West Coast absorbs most, but not all ANS, and that flows at the margin continued into the US Gulf has meant that West Coast ANS prices are bounded in the following manner. Flows into the US Gulf go into an extremely competitive market, which, were relative prices to be favourable, could absorb substantially more ANS. Thus US West Coast prices are bounded on the lower side by the netback to California of flows to the US Gulf. Should the California price fall below this bound, i.e. US Gulf price net of freight exceeds the Californian price, ANS would be reallocated into the US Gulf. On the upper side, the price of California ANS is bounded by the marginal availability of other crude oil, whether domestic Californian or Latin American (and in particular, Mexican) supplies.

Focus has tended to be put on the lower rather than the upper bound, and in particular a belief in the prevalence of a 'West Coast discount', whereby ANS was undervalued due to the implicit netback from the US Gulf Coast. Hence, a major part of the argument put forward in favour of removing the ban ran as follows. To the extent that lifting the ban diverted supplies out of the US Gulf and into the Far East, together with the (assumed) higher realization of crude oil in Asia, the West Coast discount would be removed. Thus Californian crude oil prices would rise, resulting in gains for domestic oil producers. Note that the argument in these terms is one of supply security versus the profitability of domestic producers. General principles on open markets or the spirit of GATT and NAFTA did not figure very highly in the debate.

The argument is considerably less straightforward than the above implies. There are a series of variables that determine the scope for ANS exports into Asia, and also the impact on Californian prices. A major factor in the relative price of crude oil in Asia and in California is Middle East pricing policy. We noted in the last chapter that the general level of crude oil prices has been higher than in Europe and the USA since 1992. We have also noted that this has meant that arbitrage of Brent related crude oil into Asia has been facilitated by the price discrimination, weakening the ability to discriminate and the justifications for doing so. In addition, we have suggested that liberalization in Asia causes greater price sensitivity, again making the price discrimination policy less sustainable. The degree of that price discrimination will play a role in whether ANS is competitive, and the reverse relationship also means that potential ANS flows represent further downwards pressure of the relative crude oil acquisition price in Asia.

A further set of considerations arises from the degree to which Asian refiners, and those in Japan in particular, are prepared to buy ANS on grounds that are not directly economic, and the extent to which government encouragement has a trade-off with refinery economics. For example, from the viewpoint of MITI and the Japanese government as a whole, the purchase of ANS would have some useful side benefits. It would help reduce the trade surplus with the USA, and simultaneously reduce reliance on the Middle East and increase the diversity of supply sources. From a government point of view, it would be worth paying a premium for these effects. In the context of the analysis in Chapter 4, the direct extent to which refiners could be encouraged to pay this premium has certainly been reduced by liberalization. However, we do not believe that it has been reduced to zero. Further, there are enough swings and roundabouts in the relationship between Japanese refiners and the government for some accommodation to be reached, should the government place a value on the trade surplus and supply security effects.

At the macroeconomic level the trade surplus argument could be persuasive. Imagine that the difference between the refining value of ANS and that of Arabian Light is x cents, and the ANS price is \$17 per barrel. The payment of x cents then reduces the deficit by \$17. If x is positive, the ANS purchase option is attractive given that the overall cost of other measures (e.g. operations to encourage further yen appreciation) which could reduce the deficit is greater than x/17 cents for every \$1 of surplus reduction. Given that there are few other options which can so directly target solely the trade gap with the USA, and not as a side effect allow greater penetration into the Japanese market by other countries, the limiting value of x might not be negligible.

A series of studies has considered the competitiveness of ANS in the Asian market.[16] A general conclusion is the significantly higher netback value of ANS exports east compared to the US Gulf Coast and Caribbean, suggesting a minimum volume of these exports east equal to about 150 thousand b/d (the volume of the Panama and other long haul west flows in 1996). The economics of exports east compared to California are more marginal, although not impossible, leaving the possibility of some further displacement out of the USA. The effective end to the ANS export ban came on 28 April 1996, when President Clinton instructed that the Commerce Department should issue a general licence allowing exports.

A variety of conditions were attached, related to both protectionist and environmental concerns. Among several other inspection and operational requirements, tankers to be used for ANS exports need to be US flagged, equipped with satellite communications, and must remain outside a designated exclusion zone designed to protect shorelines.[17] The first contracts for ANS were signed between BP and CPC Taiwan, with pricing on a Dubai/Oman basis (i.e. an indirect Saudi Arabian price relation). CPC were followed by first Hyundai, and then other Korean refiners. However, the natural market for ANS now, as in the 1970s, remains Japan, with Idemitsu and then Cosmo being the first Japanese refiners to purchase ANS cargoes. Given a favourable coincidence of the factors noted above, we believe that the scope for the penetration of ANS into the Asian market as a whole is considerable, with an eventual range of up to 250 thousand b/d being perfectly feasible.

4. The Characteristics of a Marker Crude Oil

We have advanced the hypothesis that Brent related crude oil now sits at the margin of the Asian market. Further, trends in global supply-demand balances will tend to reinforce that role, and we have suggested that Brent therefore has primacy over Dubai as a marker crude oil for Asia. This begs the question of whether Brent would serve as an adequate marker, and in particular what are the sources of its advantages in relation to Dubai and Asian crude oils. To consider this we now provide a set of criteria for judging a crude stream in relation to a marker crude role.

We define twelve characteristics which it would be desirable for a marker crude oil to have. They are sufficient but not necessary conditions, and a crude oil could perform as a marker without meeting all of them. The conditions do however serve as a form of checklist in discriminating between possible candidates for a marker role for oil moving in international trade.

(a) It should be freely tradable.
Often constraints are put on the tradability of oil. For instance, a crude oil that can not be resold without the producer's permission (for example Saudi Arabian grades) will fail this condition.

(b) It should have an adequate tradable physical base.
The relevant definition of physical base here excludes internal transfers within integrated companies, and term contracts with oil that are not for resale, i.e. the definition includes only oil that is available for trading within the market.

(c) It should neither have a monopoly or dominant seller, nor a monopsony or dominant buyer.
Even without evidence of direct intervention, the presence of potential monopoly or monopsony power does create problems of market credibility.

(d) It should have adequate loading facilities, and known loading schedules.
Facilities unable to guarantee loading within narrow date windows, primarily because of weather difficulties, are likely to produce markets where prices will be distorted by those difficulties. This tends to weigh

against many offshore loaded crudes. Likewise the size of a given month's loading programme needs to be generally known in advance, otherwise traders will consider the loading programme a potential source of manipulation (i.e. cargoes being held back, or added, for market purposes).

(e) It should be capable of being loaded onto VLCCs.

This condition is aimed at ensuring that the crude oil can be moved between regions in arbitrage, and thus its behaviour can be linked to that of other markets. Without this, the crude may prove to be a poor marker for others that are capable of being used in arbitrage.

(f) The fiscal regime the market operates under should have tax certainty, and there should be no official constraints on trading prices.

This condition rules out markets in countries where the taxable price or official selling price is either not known at point of sale or is unrelated to the prices of realized arm's length sales. We have already noted the problems with this in terms of the price information generated in the Middle East spot market. In addition, some countries choose to put restrictions on the range of acceptable sales prices in relation to tax prices.

(g) The country of origin should be non-embargoed, have low risk of being embargoed, be politically stable, and have commercial practices that are not considered to be far from general norms.

These conditions rule out Libya and Iran (embargoed by the USA), and perhaps Russia and Angola among many others where political stability is perceived to be a potential problem, or countries in which transfers are not solely accomplished through the invoiced price.

(h) The crude should be deliverable in waterborne cargoes.

Virtually all world trade (excepting exports from Canada and some from Russia) is carried in tankers. Pipeline crudes, as well as sometimes presenting logistical problems, are considerably less fungible with oil delivered in tankers.

(i) The crude should not have any special characteristics that reduce the number of refineries able to run it, or easily substitute towards it, to low levels. It should also not have significant non-refining uses that could affect its price determination.

This rules out crude oils with extremely high sulphur or metals contents, or with uses away from mainstream crude oil. An example would be Indonesian burning crudes, such as Minas, that can be used in power stations as a substitute for fuel oil.

(j) There should be no perception that the market could be influenced by political control.
Trade in many OPEC crude oils in particular could be affected by the perception, even without the reality, that political involvement was possible.

(k) The market should be subject to a stable regulatory regime, and not subject to any regulatory risk.
Ideally, transnational legal jurisdictions should be well defined, and regulatory change should be gradual, consultative and not enforceable retrospectively.

(l) Price behaviour should serve as a proxy for marginal conditions.
There should be some logical connection between the marker and the marginal conditions in the market in which it operates. Otherwise, agents using the marker crude oil as a reference, may find themselves undercut and uncompetitive in relation to marginal flows.

This represents a fairly stringent set of criteria and matching potential markers against them is to a large part subjective. However, solely in the context of the Asian market, we would identify the following major failures from the set of potential markers consisting of Dubai, Oman, Tapis, Minas, WTI and Brent. Criterion (a) (tradability) is met in Asia by all bar WTI (US export laws) and by Oman during the occasional clamp-down on the spot market. Combinations of large term sale volumes and/or low physical production, rule out Dubai, Oman, Tapis and Minas in the context of criterion (b) (adequacy of the number of tradable cargoes). Oman, Tapis and Minas also fail (c) (large degree of monopoly or monopsony power).

While some cases are arguable, we would pass all six on the basis of (d) and (e) (adequate loading facilities and the use of VLCCs), given that (e) is not applicable in the case of WTI. Criterion (f) (transparency of official pricing and taxation) is failed by Oman and Minas, and to some extent by Tapis. Criterion (g) (stability and commerciality) may be arguable in some cases. WTI is ruled out of (h) by being a pipeline crude that is never loaded onto tankers. Minas fails (i) through its use as a direct burning crude, and arguably (j) (possibility of political control, particularly via OPEC). To some degree, all except WTI fail (k) (regulatory regime) due to the tendency towards supra-nationality in US law. For example, the question of the degree of jurisdiction claimed by US law over the Brent market has never been completely resolved. Likewise, the problem is potential to any oil market in which US companies are active. WTI passes this criterion since there is no potential clash of national jurisdictions. Oman fails this condition firmly, given the swings in official policy towards market mechanisms noted in the last chapter. Finally, we would argue that only Brent now meets (l), the marginality condition in the context of the Asian market, although, should Alaskan oil play a key swing role, this would add an element of a WTI relationship into the margin.

On the basis of this exercise, the absence of any meaningful indigenous Asian marker becomes clearer, with Minas in particular failing a gamut of key conditions. Criteria (b) and (c) alone are enough to rule it, together with Oman completely out of the frame. Dubai now performs badly because it fails (b) and (i), aside from empirical deficiencies in liquidity of trade and in the price discovery functions observed in the last chapter. The deficiencies in Dubai suggest to us that, despite failing several of our criteria, WTI would be superior as a marker in Asia. However, as is slightly given away by the title of this chapter, overall we believe that the most appropriate marker for trade in Asia is now Brent. In particular, Brent only fails one criterion (one which any non-US market must fail due to the distinct lack of the basic notion of foreign sovereignty in US competition and anti-trust law).

The basic institutions of the Brent market were described in Chapter 9, and other sources provide a more detailed description.[18] However, we make the further point that the participants in the Dubai market are just a subset of Brent market participants, not surprising given that both Dubai and Brent are predominantly traded out of London offices. In particular, a switch to Brent from Dubai does mean a radical change in the composition of the companies at the heart of price setting functions, and in fact means a significant expansion in the total number of companies involved. Within the Brent complex, we can illustrate in relation to the forward Brent market. The twenty largest traders of forward Brent are shown in Table 12.6, together with their market shares. The seven companies we identified in Chapter 11 as being the dominant traders of Dubai figure prominently.

Brent has become a market of increasing importance to Asian price determination. It is thus somewhat ironic that the involvement of Asian companies in the Brent market has declined over time. In Table 12.6, note the strong presence of *sogo shosha*, most notably Nissho Iwai, Mitsui, Marubeni and Kanematsu, in the late 1980s. The withdrawal of the *sogo shosha*, with the exception of Kanematsu, from what was primarily purely speculative trading in Brent, parallels their withdrawal from the Dubai market noted in the previous chapter. During the 1990s, Sinochem has also been active in Brent, again in largely speculative trading, but this involvement has been sharply scaled down. By the mid-1990s, price formation in forward Brent, and therefore as we have argued also at the

Table 12.6: Main Participants in the Brent Forward Market. Rank and Market Shares. 1986–95. Per Cent.

	1986		*1987*		*1988*		*1989*		*1990*	
1	Phibro	8.8	Phibro	10.0	J Aron	14.7	J Aron	15.7	J Aron	16.0
2	J Aron	7.4	J Aron	9.7	Phibro	12.5	Phibro	11.7	Phibro	12.7
3	Nissho Iwai	6.1	Shell Int	5.3	Drexel	4.7	Drexel	5.7	Shell Int	5.6
4	Shell UK	5.5	Nissho Iwai	4.7	Shell Int	4.7	Bear Stearns	4.0	CFP Total	4.6
5	Drexel	4.3	BP	4.0	BP	4.0	Cargill	3.9	Morg. Stanley	4.6
6	Marubeni	4.2	Marubeni	3.6	Nissho Iwai	3.8	Shell Int	3.9	Cargill	4.3
7	BP	4.2	Shell UK	3.6	Kanematsu	3.4	Nissho Iwai	3.5	Statoil	4.0
8	Internorth	3.5	Drexel	3.5	Shell UK	3.2	BP	3.3	Neste	3.7
9	Shell Int	3.3	Exxon	3.5	Morg. Stanley	2.8	TWO	2.9	Nissho Iwai	3.2
10	Kanematsu	2.9	Kanematsu	3.2	Exxon	2.4	Morg. Stanley	2.9	Shell UK	3.1
11	TWO	2.7	Marc Rich	2.8	Mitsui	2.2	Shell UK	2.6	BP	2.9
12	Conoco	2.1	TWO	2.7	Mobil	2.2	CFP Total	2.5	Marc Rich	2.8
13	Nichimen	2.1	Avant	2.5	Cargill	1.9	Hill Petroleum	2.3	Sinochem	2.5
14	Enron	2.0	Mobil	2.1	Avant	1.8	Sinochem	2.0	Nova	2.3
15	URBK	2.0	Morg. Stanley	2.0	Bear Stearns	1.8	Marc Rich	1.9	Hess	1.9
16	Gotco	1.9	Enron	1.9	Dreyfus	1.8	Elf	1.9	TWO	1.7
17	Sun	1.9	Mitsui	1.9	TWO	1.5	Mitsui	1.9	Arcadia	1.3
18	Exxon	1.9	Kaines	1.7	Nova	1.4	Avant	1.7	Mobil	1.3
19	Mobil	1.7	Dreyfus	1.6	Marc Rich	1.4	Statoil	1.6	Kanematsu	1.3
20	Nova	1.6	Elf	1.6	Scan	1.4	Scan	1.5	Sun	1.3

	1991		*1992*		*1993*		*1994*		*1995*	
1	J Aron	15.8	J Aron	14.3	J Aron	11.6	Phibro	10.0	JP Morgan	11.5
2	Phibro	10.6	Phibro	9.1	Koch	9.3	BP	9.6	BP	11.1
3	Cargill	5.9	BP	7.0	BP	8.8	Shell Int	8.5	Phibro	10.8
4	Shell Int	5.7	Cargill	6.8	Phibro	6.1	J Aron	8.4	J Aron	8.6
5	CFP Total	5.4	Shell Int	6.3	Shell Int	5.7	Koch	7.4	Shell Int	7.2
6	BP	5.2	Shell UK	4.7	Cargill	5.1	Cargill	5.4	Statoil	5.8
7	Statoil	4.7	Morg. Stanley	4.6	Elf	5.1	Statoil	4.8	Elf	5.1
8	Morg. Stanley	4.3	Statoil	4.5	Dreyfus	5.1	Morg. Stanley	4.5	Morg. Stanley	4.9
9	Hess	3.4	AIG	4.2	Shell UK	4.8	Chevron	4.1	Shell UK	3.8
10	Marc Rich	3.4	Dreyfus	3.9	Chevron	4.4	Elf	3.8	Koch	3.8
11	Elf	3.2	Elf	3.3	Morg. Stanley	3.9	CFP Total	3.1	Cargill	3.6
12	Shell UK	3.1	Exxon	2.9	Exxon	3.1	JP Morgan	3.1	Vitol	3.1
13	AIG	2.8	Chevron	2.8	AIG	3.0	Mobil	2.9	Arcadia	3.0
14	Neste	2.7	CFP Total	2.7	CFP Total	2.8	AIG	2.8	AIG	2.4
15	Dreyfus	1.9	Neste	2.7	Statoil	2.8	Shell UK	2.6	Exxon	2.0
16	Chevron	1.9	Koch	2.6	Mobil	2.5	Dreyfus	2.4	CFP Total	1.7
17	Nova	1.7	Bear Stearns	1.7	Kanematsu	2.2	Vitol	2.1	Chevron	1.3
18	Arcadia	1.6	Vitol	1.7	Neste	1.7	Arcadia	2.1	Sinochem	1.2
19	Bear Stearns	1.5	Nova	1.6	Texaco	1.6	Kanematsu	2.0	Mobil	1.2
20	TWO	1.4	Kanematsu	1.6	Nova	1.5	Sinochem	1.6	Texaco	1.1

Source: Own calculations from Petroleum Argus Crude Oil Deals database.

margin in Asia, was primarily resultant from the activity of a group of Western oil companies and financial trading houses. The names shown in Table 12.6 are in the main not only the same companies who constitute the bulk of Dubai trading, they are also those that play the leading role in the Singapore oil swaps market (discussed in Chapter 8). In total, a switch from Dubai to Brent related pricing does not mean that Asian prices would be determined by insular North Sea based traders, but by the same subset of traders who dominate activity in most of the other markets we have considered in previous chapters.

5.   Conclusions – The North Sea and Asia

We have noted a basic asymmetry in the world oil market, namely that oil demand has become biased towards the East, while incremental supply has been biased to the West. The import dependence at a regional level of Western markets is falling, while it is rising in the East. In this chapter we have focused on one set of accommodatory flows, i.e. the movement of Brent related crude oil out of West Africa and Europe into Asia, to sit at the margin of the Asian market. Two other possibilities would achieve the same balancing function. The first is a diversion of Middle East OPEC exports out of the West, following the trends noted in the last chapter in Table 11.3. The second would be a switch towards output maximization in OPEC, i.e. a refusal to be constrained by the growth of non-OPEC production, leaving incremental Asian demand to be supplied from the Middle East and leaving prices to fall. We consider the impact of these two alternatives in turn.

We do not believe that Middle East supplies can return to the margin of the Asian market simply by diversions out of other regions, without a change in pricing policy. The maintenance of monthly average Dubai related pricing still leaves the possibility of arbitrage against five or ten day Brent related prices. Even removing the oil export gap between Asia and the West, and thus making the arbitrage more difficult, still leaves the possibility of the right coincidence of intra-month volatility and backwardations that will facilitate the flow at certain times. Logically, the only method for effectively competing with the Brent related crudes at all times is also to adopt Brent related pricing. If, as we have suggested, price discrimination is not a sustainable policy, the only major motive for persevering with a Dubai relation is removed. Add to this the general decline in the Dubai market, and the case for Brent becomes stronger. Under this scenario, the only realistic alternative for Brent related West African crude oil to be at the margin is for that margin to be taken by Brent related Middle East term exports. The net effect is the same, the

Brent market would provide the centre of price discovery for Asia.

The second scenario is the OPEC breakout strategy, i.e. output maximization in the hope that lower prices will reduce the impact of non-OPEC supply growth. We would refer back to the discussion in Section 2, which argued that as a strategy it is high risk and would be extremely slow to operate. At its heart is the idea carried over from the 1980s that low cost OPEC oil can always displace high cost non-OPEC oil. The lesson of the 1986 price fall is that non-OPEC fiscal regimes are accommodatory and that the degree of cost reduction enabled by efficiency savings and by technological change is large.

The balance of non-OPEC versus OPEC production growth over the next decade is central to the evolution of trade flows and price relationships. The argument we have followed in this chapter is one where the geography of world oil trade is subject to major changes. As described, those changes lead the North Sea market to exert a growing influence at the margin of Asian price formation. In Chapter 9 we described the IPE–SIMEX link, and the establishment of Brent futures trading in Singapore. The processes described in this chapter, together with the implications of market liberalism in Asia, would suggest that the long-term prospects for Brent in Singapore are potentially bright. In particular, there is a strong logic to the use of Brent as a marker for Asia, a logic which extends to the Middle East producers.

Notes

1. The 1997 figures are derived from IEA projections as of February 1997. Ecuador and Gabon, both of whom left OPEC during the period shown, have been included under non-OPEC for the whole period to prevent discontinuities in the data.
2. See John V. Mitchell (1994), 'Oil Production outside OPEC and the Former Soviet Union', *The Energy Journal*, and Paul Stevens (1996), 'Oil Prices: the Start of an Era?', *Energy Policy*, vol. 24 no. 5.
3. Mehdi Varzi (1996), *OPEC and non-OPEC, The Battle for Market Share*, Kleinwort Benson Research, London, reported in *Middle East Economic Survey*, 6 May 1996.
4. International Energy Agency (1996), *World Energy Outlook*, International Energy Agency, Paris.
5. The IEA's two scenarios in the 1996 edition of the *World Energy Outlook* give OPEC production in 2010 of 48.6 mb/d (Capacity Constraints) and 48.6 mb/d (Energy Savings). The CERI forecast is 44.6 mb/d, see Jennifer I. Considine and Anthony E. Reinsch (1995), *Battle for Market Share: World Oil Projections, 1995–2010*, Canadian Energy Research Institute, Calgary, Alberta.
6. Once developed, non-OPEC output is extremely price insensitive. For example, for prices to be insufficient to cover variable operating costs for the major developed fields in the North Sea, they would have to fall well below $5 per barrel. Even in these very extreme circumstances, the production would only

be shut in, ready to come back on stream when prices rose.

7. See also Paul Stevens (1996), 'Oil Prices: The Start of an Era?', op.cit.

8. Oseberg is 36.3 degrees API with 0.3 per cent sulphur by weight, and Gullfaks 29.9 degrees API with 0.4 per cent sulphur by weight.

9. The principle of backhaul tanker economics is as follows. A crude tanker making the round trip to the West from the Middle East is empty for half its voyage. If, however, after discharging its cargo it loads in the North Sea or West Africa, continues to Asia for discharging, and then returns to the Middle East, the proportion of its journey spent empty has been minimized. An additional rent has been created, which through the bargaining process will be partially reflected in a lower chartering rate.

10. James P. Roscow (1977), *800 Miles to Valdez: The Building of the Alaska Pipeline*, Prentice-Hall, Englewood Cliffs, New Jersey.

11. Robert L. Bradley Jnr. (1996), *Oil, Gas and Government, The US Experience*, Rowman and Littlefield, Lanham, Maryland, USA.

12. Quoted in Peter A. Coates (1991), *The Trans-Alaska Pipeline Controversy: Technology, Conservation and the Frontier*, Lehigh University Press, Bethlehem.

13. Robert Douglas Mead (1978), *Journeys Down the Line: Building the Trans-Alaska Pipeline*, Doubleday, New York.

14. The only exception is for shipments to the Amerada Hess refinery at St. Croix in the Virgin Islands, which can be transported by foreign vessels in a 5000 nautical mile journey around Cape Horn. As shown in Table 12.5 the bulk of shipments to the Virgin Islands use the Panama route and Jones Act tankers.

15. Roger Marks (1992), 'Estimated Prospective Tanker Rates for Alaska North Slope Crude Oil', *OPEC Review*, vol. XVI, no. 4.

16. See US Department of Energy (1994), 'Exporting Alaskan North Slope Crude Oil : Benefits and Costs', US Department of Energy, Washington D.C.; Frank C. Tang and Ronald D. Ripple (1996), 'Potential Value of Alaskan North Slope Crude Oil to China : A Refinery Modelling Analysis', *OPEC Review*, June 1996, and Purvin and Gertz estimates reported in, *inter alia*, *Weekly Petroleum Argus*, 3 June 1996.

17. *Platt's Oilgram Price Report*, 30 April 1996.

18. See Paul Horsnell and Robert Mabro (1993), *Oil Markets and Prices*, op.cit.

PART V

Refinery Profitability

CHAPTER 13

REFINERY MARGIN DYNAMICS

1. Introduction

In previous chapters we have evidenced a number of structural shifts in the 1990s in both Asian and world oil markets. In Asia, liberalization in some countries together with fast demand growth has brought a capital inflow, refinery expansions and a more competitive and less regulated downstream. In many countries it has led to realignments in relative domestic product prices and a shift to greater reliance on price setting by the Singapore market, either directly or through import competition. In China, oil policy has become a macroeconomic control variable. Oil demand in Asia has increased sharply, its composition has changed, and government directives have brought about significant alterations in product specifications. The link between Asian and European markets has become closer, with the development of large-scale arbitrage trading. Relative crude oil prices across regions have been affected by the policy of a major producer. The balance between OPEC and non-OPEC supplies has also changed, bringing new trade patterns and impacting on crude oil price differentials.

These are diverse changes, but all have in varying degrees affected the operations of the Asian oil industry and its dynamics. This chapter seeks to draw together some of the threads listed above, in the context of their impact (both past and potential) on just one variable, namely the profitability of refining operations in Asia. The refinery profit margin serves as the bridge between the developments in downstream liberalization, product markets and crude oil markets which have been a focus of this book.

At the start of the 1990s, the dominant view of the near future of Asian refining was overwhelmingly bullish while prospects in other regions seemed distinctly limited. The view held by capital markets of the US oil industry was soured by the ever increasing financial and logistical demands of environmental legislation, and by poor rates of return. The European market was flat and demand prospects were poor. The European industry faced an increasing burden of taxation on its products, increasing environmental demands, and a retailing sector where margins had become wafer thin in many countries through competition, and especially from hypermarkets. In addition, Europe was beginning to see the scale of the excess refining capacity that was to dampen the market over the decade.

By contrast, the outlook for Asia seemed extremely positive. Demand was booming, and there was a widespread perception that refinery capacity shortages would be an increasing feature of the market. Liberalization offered new opportunities to tap into the expanding markets. Even China seemed to be opening up, with the eventual prospect of the industry regaining a large but long lost market. The gloom surrounding other regions was lightened by the prospect of greater profitability and opportunities in Asia. However, the reality has not lived up to the expectation. Refinery margins have weakened throughout the first half of the 1990s. In 1995, they were as weak in Singapore as in Europe, even at a time when European margins were poor given the weight of excess capacity.

Calculating the Gross Product Worth (GPW) of straight run refining of Dubai using Singapore product prices, minus the price of Dubai and the freight cost from the Middle East, produces the path of refining margins shown in Figure 13.1. The period of the Gulf crisis (during which simple margins exceeded $6 per barrel even allowing for the sharp escalation in freight rates) is excluded from the figure for the sake of clarity.

The margin shown in Figure 13.1 remained relatively strong after the

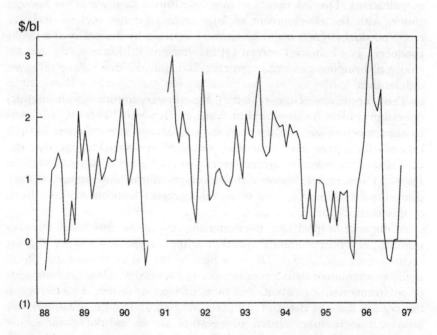

Figure 13.1:   Simple Refining Margin for Dubai Crude Oil Using Singapore Prices. Monthly Average. January 1988 to November 1996. $ Per Barrel.

Gulf War until March 1994. From this point it remained depressed for some eighteen months, before recovering sharply at the end of 1995 (for reasons discussed further below), and then collapsing again in the summer of 1996. The start of weak margins in 1994, equates to the start of tighter import restrictions in China, as was discussed in Chapter 3. China may have turned the sentiment in oil product markets towards the bearish, but a series of other factors has been depressing margins for the rest of the period shown. We would identify three main factors; globalization, Kuwait, and regional capacity. While these have pushed margins down, tightness in world markets causing deep backwardations (i.e. when prompt prices are above those for more forward delivery), has helped to support and sometimes inflate the measure shown in Figure 13.1, reinforced by occasional tightness and dislocations in the gasoil market. A steep decline in the margins for more complex refinery installations has been exacerbated by a number of other factors, which will be discussed later.

2.   The Factors that Depressed Refinery Margins

Globalization. We noted above that at the start of the 1990s there were very different expectations about the future of the downstream in Europe and in Asia, with the implicit assumption that the two regions could be ring-fenced, that each could be isolated from developments in the other. However, in oil market terms, to a considerable degree there are no longer two regions called Europe and Asia, just the one enlarged region of 'Eurasia'. While the two component parts can diverge within certain bounds, the extent of those bounds has become considerably narrower. Arbitrage trading (as discussed in Chapter 7), which has been facilitated by the growth of the Singapore swaps market, has strengthened the link between Singapore and Rotterdam gasoil price developments. Through this link, Europe's downstream weakness has been partly exported to Asia. In Chapter 7 we contended that this effect was more important than the failure of China's demand for oil product imports to live up to expectation. In particular, had Chinese import demand been stronger, the price impact would not have been particularly great. Additional volumes would have been obtained from Europe and limited the extent of price rises.

The extent to which Singapore prices can move ahead of European prices is governed by two main factors. The most obvious is freight rates, but note that the move from ordinary product tankers to VLCCs, which became a major factor from 1993 onwards, means that the bound is now at a lower level. The second major factor is the time structure of prices. In working the arbitrage, gasoil is effectively being bought spot in Europe and sold forward in Singapore, the degree of forwardness being the transit time of the cargo. The bound is therefore not the differential between spot prices but the differential between Singapore prices for delivery about a month forward and Rotterdam spot prices. The greater the degree of backwardation (i.e. when prompt prices are above forward prices), the greater the extent to which Singapore spot prices can still rise above European levels. Before active paper markets, arbitrage was accomplished purely by physical flows, and at that time by transfer of gasoil in ordinary product tankers. The arbitrage worked at higher differentials than currently, due to both the higher unit freight costs of non-VLCC movements, and the need to have a risk premium to cover the non-hedgeable risks involved. As it worked through physical transfer alone, the time lag over which an inter-regional dislocation of prices could last was potentially a matter of weeks.

With the development of paper markets in Singapore, and the presence of a gasoil futures market in London (giving the opportunity of locking in differentials between Europe and Singapore), together with the now common practice of VLCC transfer, the arbitrage works at lower differentials. More importantly, if the arbitrage opens up at any point along the time curve of prices, the trading response is extremely fast. Indeed, we suggested in Chapter 8 that the speed with which dislocations are corrected is now so fast that it has been a factor behind the fall in swap market liquidity. When the arbitrage is possible, the price response can be measured in hours if not minutes. Price behaviour for the most important refinery product in Asia has become more closely linked to the European market. Our first explanation of weaker than expected refining margins in Asia during the 1990s, is then that Asia has been weak because Europe has been weak.

Kuwait. The second explanation we advanced above was the impact of Kuwait. As we saw in Chapter 7, Kuwaiti oil product exports stood at 690 thousand b/d in 1989. The invasion of Kuwait in August 1990, and the subsequent damage to the country's infrastructure, removed the output of three highly upgraded refineries from product markets. The fall in crude oil output from Iraq and Kuwait was easily made up by the utilization of spare production capacity elsewhere. However, there was no equivalent spare complex capacity in the downstream to replace the output of three hydrocracking units. The temporary loss of Kuwaiti output was then a factor that helped to bolster refinery margins during and immediately after the Gulf crisis, an effect that, as we see below, was particularly pronounced for complex margins. Gradually, the Kuwaiti refineries were brought back on stream, and the hydrocrackers were repaired. To some extent, the weakening of margins in the early 1990s reflects the resurgence of Kuwaiti exports, as shown in Table 7.12. The temporary vent for surplus that Asian refiners enjoyed in the Indian market was removed as Kuwait reestablished itself, first on the west coast of India and then increasingly on the east coast.

Regional Capacity. The third factor we advanced above as a reason for weakening margins was regional capacity. After 1994, there was a series of tranches of large new units coming on stream. As we saw in Chapter 6, the major market for Singapore gasoline exports in the early 1990s had been Malaysia. With the Melaka refinery coming on stream in 1994, regional demand for gasoline imports had been considerably reduced. This was followed by the spate of new units coming on stream in Korea over 1995 and 1996, adding some 0.8 mb/d to regional capacity. Shell's new Tabangao refinery in the Philippines also started up in 1995. In early 1996, the Star and Shell refineries in Thailand came on stream, producing another major and discrete jump in capacity. Added to these large increments was a series of minor increments elsewhere, together with the continued process of capacity creep through debottlenecking in the entire region.

3.   The Complex Refining Margin

While simple refining margins became weaker over the first half of the 1990s, the returns to more upgraded refineries have been even more severely affected. The monthly average of the margin for a complex refinery (assumed for illustration to have vacuum gasoil fed catalytic cracking) is shown in Figure 13.2, again for clarity excluding the period of the Gulf crisis.

Figure 13.2 shows a very strong downward trend in the differential (i.e. the upgrading margin) after the Gulf War through to the start of 1996, when after a brief recovery, the margin slumped back to a record low. Those who had complex units up and running in the first few years of the 1990s at least achieved high returns for a while, but those that came on stream after 1993 found that the early returns were far lower than they had projected. Figure 13.2 represents a source of considerable under-performance in the *ex post* analysis of the viability of a series of large refinery investments. In terms of the explanations given above for the decline in simple margins, the upgrading margin has been more severely hit by the return of Kuwaiti refineries to full production.

Not all complex units have fared as badly as that shown in Figure 13.2. In particular, margins have been higher for those (such as hydrocrackers and RFCCs) that can minimize their gasoline output and maximize their gasoil output. The severe decline in the upgrading margin as shown in Figure 13.2 since the Gulf War is a function of the relative strength of fuel

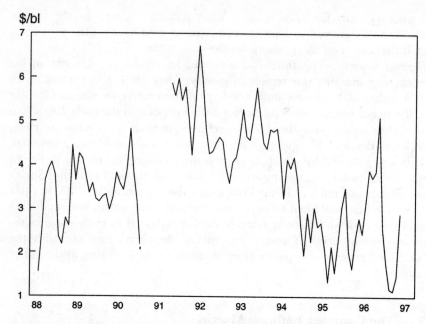

Figure 13.2:   Differential Between Complex and Simple Refining Margins for Dubai Crude Oil Using Singapore Prices. Monthly Average. January 1988 to November 1996. $ Per Barrel.

oil prices against lighter products. However, since 1994 relative gasoil prices have recovered, while the relative price of gasoline has continued to fall. The differentials between gasoline and fuel oil, and between gasoil and fuel oil, are shown in Figures 13.3 and 13.4 respectively.

The premium for gasoline over fuel oil, as shown in Figure 13.3, stood at some $16 per barrel immediately after the Gulf War. Over the next nearly seven years, this premium has followed a strong downwards trend, reaching a low of just $2 per barrel at the end of 1996. Any conversion process that reduces fuel oil yields and increases gasoline yields has experienced a dramatic collapse in its profitability. While the relative price of gasoline has fallen consistently since the Gulf War, as shown in Figure 13.4, the relative price of gasoil remained fairly robust until the end of 1993, after which it plunged from about $14 per barrel over fuel oil to a low of $4 per barrel, and then remained below $8 per barrel until the end of 1995.

The two years from the start of 1994 were extremely poor in terms of the profit margin for any fuel oil conversion process. However, while gasoline continued to weaken relative to fuel oil in 1996, gasoil rebounded. The trigger was an unexpectedly large tender for gasoil from India at the turn of the year, which was followed by months of consistently tight

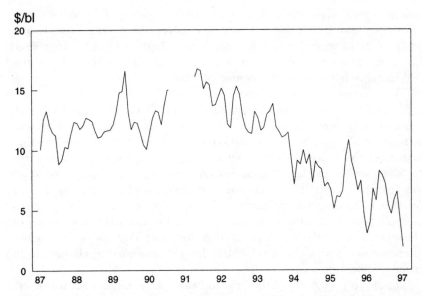

Figure 13.3: Differential Between Gasoline and 180 Centistokes Fuel Oil Prices in Singapore Market. Monthly Average. January 1987 to November 1996. $ Per Barrel.

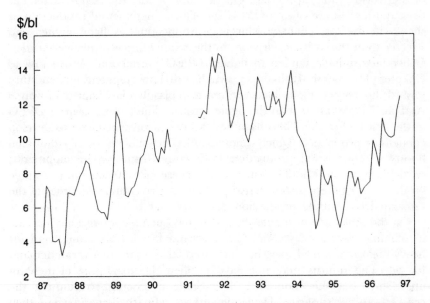

Figure 13.4: Differential Between 0.5% Sulphur Gasoil and 180 Centistokes Fuel Oil Prices in Singapore Market. Monthly Average. January 1987 to November 1996. $ Per Barrel.

markets. With high levels of backwardation reducing the possibilities for arbitrage, and in addition a stronger European market, the relative profitability of gasoil production reached its highest level for four years. Complex processes producing higher gasoil yields then performed considerably better over the course of 1996.

The divergence of gasoil and gasoline prices over the 1990s has been a strong feature of the market. Underlying Figures 13.3 and 13.4 is the other main feature, the strength of fuel oil prices. There have been some factors specific to the Asian market which have helped to maintain a higher relative price of fuel oil. First, there has been strong demand from the ship bunker market, consequent on the increasing volumes of freight movements as trade has expanded. Secondly, as we saw in Chapter 5 in the case of Korea, strong demand for electricity and lags in bringing new (non-oil) power stations on stream, have lead to the swing use of fuel oil. However, the strength of fuel oil after the Gulf War has been a global phenomenon, and at this level can be largely attributed to changes in the crude oil market.

The sharp narrowing of the light to heavy crude differential was shown in Figure 11.2 (for Saudi Arabian crudes) and in Figure 11.7 (for the differential between the Dubai/Oman average and dated Brent). This has been caused by a series of factors, on both the supply and demand side of the world crude oil market. On the supply side, the major factor has been the high share of non-OPEC supplies in incremental production as shown in the last chapter. Marginal production has been lighter and sweeter than the average, driving up the relative price of heavier crudes. Other factors have tended to lighten OPEC production. As we saw in Chapter 11, Saudi Arabia has switched its development expenditure towards lighter crudes, and has attempted to bias its spare capacity towards Arabian Heavy rather than the lighter grades. The hemming in of OPEC in the first half of the 1990s has also lead to a strong incentive to develop condensate production (which is not included in OPEC quotas). Additional boosts to condensate production have come from gas developments, particularly in the Middle East. Many of these projects are not economic on the basis of the gas, but derive profitability from the economics of the associated condensate production.

On the other side of the market, there has been a considerable increase in refining cracking capacities. The demand for heavier crudes and for heavy fuel oil as a feedstock, has increased faster than the overall demand for oil. The returns may not have justified the fixed cost element of upgrading, but the economics of variable costs implied running the crackers at full capacity. Developments in refinery demand were then showing the reverse asymmetry to developments in crude oil supply, and the relative price of heavy oil was forced up.

The behaviour of the heavy to light differential in the 1990s marks a notable failure of conventional wisdom, and, in terms of refinery planning, a failure to consider counterfactual arguments. This was more than just having been surprised by the narrowing of the differential; the conventional wisdom was that the differential would increase. The logic behind the argument ran as follows. With the Middle East producing one-third of world output, but having two-thirds of world reserves, over time incremental production would become biased towards heavy high sulphur Middle East oil. Not only would the Middle East have to meet incremental demand, it would also have to make up for falls in the supply of lighter non-OPEC oil. Further, incremental oil demand was biased towards lighter oil products, and environmental regulation was pushing towards ever lower sulphur limits. Put together, these arguments implied that the heavy high sulphur crudes would be priced at an ever greater discount to the increasingly scarce light low sulphur supplies.

For many this logic remains to a large extent the conventional wisdom, with a view that the early 1990s have been a short-term aberration. We would argue however that the logic is faulty, for two main reasons. First, the mapping from the composition of the incremental final demand barrel of oil products, to crude oil price differentials is not a direct one. The conventional wisdom only holds if refinery structures are taken as constant. In reality, the impact on crude oil differentials is determined by asymmetries in the development of demand and refinery structure. In the 1990s, refinery structure changes have been more than sufficient to meet demand composition changes. Indeed the major product surpluses (in Asia in particular) have emerged at the top of the demand barrel, namely in gasoline.

The second major fault with the logic of the conventional wisdom is that, as we have seen in previous chapters, the short- to medium-term dynamics of relative crude production changes bear little or no relationship to estimates of the relative resource base. That mapping may become stronger over time; however the discounted impact of these longer-term implications is small, and certainly too small to use in any contemporary planning process.

4.   The Arabian Light Netback in Singapore vs Rotterdam

In Chapter 11 we noted the persistence since 1992 of a higher realized price for sales of Arabian Light for eastbound loadings compared to those to the West. In Figure 13.5 we consider the impact of this disparity from the demand rather than the supply side. In the figure we have shown, by month and as a twelve-month rolling average, the difference between the

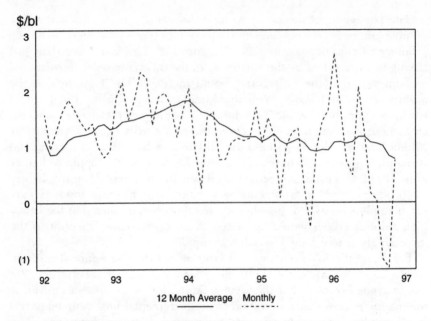

Figure 13.5: Differential Between Netback Value of Arabian Light from Singapore and Rotterdam. Monthly and Twelve-Month Rolling Average. January 1992 to November 1996. $ Per Barrel.

netback of Arabian Light from Singapore and the netback from Rotterdam (i.e. the difference between the GPWs net of freight costs).[1]

The netback value from Singapore is seen to have been higher than that from Rotterdam in all but three of the 59 months shown in Figure 13.5. The average value of the gap increased steadily to nearly $2 per barrel at the end of 1993, before declining to a level of around $1 per barrel through 1995 and into 1996. While refiners in many Asian countries may have been fairly price insensitive in their behaviour, Figure 13.5 does provide a further explanation as to their willingness to pay more than European refiners, i.e. their netback value of Arabian Light has been significantly higher. The figure of $1 per barrel is significant as it marks the extent of the higher realization from eastbound loadings found in Chapter 11. This raises the question of whether the differential pricing between regions has been an exercise in rent capture, with the effect of equalizing refinery profit margins across regions. A view of this is provided by Figure 13.6, which shows the difference between the simple refining margins in Singapore and Rotterdam (net of freight costs), and the difference between fob realizations for eastbound and westbound sales found in Chapter 11. Both series are shown as twelve-month rolling averages.

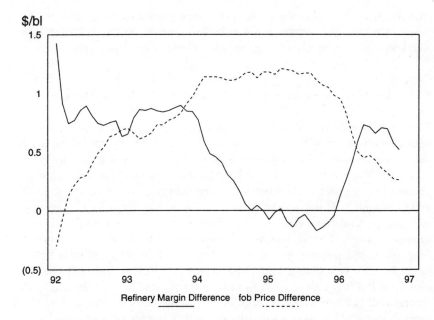

Figure 13.6: Differentials Between Refining Margins for Arabian Light in Singapore and Rotterdam, and Between Realized fob Arabian Light Prices for Eastbound and Westbound Sales. Twelve-Month Rolling Averages. January 1992 to November 1996. $ Per Barrel.

At first sight, Figure 13.6 does seem to indicate some link between the pricing of crude and a process of refinery margin equalization. By the end of 1994 (i.e. when the average represented calendar year 1994 as a whole), the twelve-month average Singapore refinery profits had fallen below those in Rotterdam, remaining there through 1995. The upturn in relative Singapore profitability at the start of 1996 is to a large extent a function of the distortion noted in Chapter 11, i.e. that high levels of market backwardation drove dated Brent prices sharply upwards against Dubai.

The apparent perfection of the link between crude oil export prices to Europe and Asia and relative refinery profits is misleading. An increase in relative crude acquisition costs will only lead to an equal (and opposite) change in relative profits if a series of extreme conditions is met. There would have to be no pass through of crude oil costs in product prices, no changes in relative demand conditions, arbitrage conditions, freight rates, and in the relative regional degree of excess refinery capacity. Closer inspection of Figure 13.6 also reveals how weak the link is. From the start of 1994, the movement to the parity of refining profits involved a fall of 80 cents in Singapore profits relative to Rotterdam. Over that period the relative crude acquisition price only increased by about 20 cents. Given that much of this 20 cents would have been passed through into product prices, we can not attribute a significant portion of the fall in relative Singapore profits to changes in relative crude acquisition costs.

5.   The Time Structure of Prices and the Refining Margin

The time structure of crude oil and product prices is an important determinant of refinery profitability. In particular, Asian refinery margins are generally boosted by conditions of backwardation (i.e. prompt prices above those for future distant delivery), and depressed by contangos (i.e. prompt prices lower than those for future delivery). To illustrate, consider a refiner who sells out product priced on monthly averages of Singapore spot prices, and buys crude oil at prices related to the monthly Dubai/Oman average. The refiner is assumed not to trade in paper markets. Singapore spot quotations represent cargoes for loading within 15 to 30 days. By contrast, Dubai and Oman quotes represent the next calendar month in the first half of each month, and the next but one month in the second half. The monthly average then contains a forwardness of 45 days until loading, 22.5 days more than product quotes. Assuming, for the moment, that the time structure of crude oil and product markets is the same, the refinery margin would then include 22.5 days of that time structure. Margins are then improved by backwardations, and depressed by contangos.

The above refiner gains in a backwardation, because the average time to loading represented by their output price linkage is less than that of their input price linkage. They also benefit from facing input prices based on a long haul crude with output prices based on shorter haul products. With our current assumption of identical time curves for crude and products, the longer sailing time of the reference crude compared to products obtained in Singapore adds a further two weeks or so of the time structure into the profit margin. The refiner then gains in a backwardation not just because Dubai quotes have a greater forwardness to loading, but also because, after loading, the journey time to the user is longer than average Singapore product exports.

Now let us allow crude and products to have a different time structure, while keeping the structure the same across products, and assume that the refiner faces spot crude and product prices, but also trades in paper markets. The refiner's general principle will then be to buy crude prompt when its prices are in contango, and buy further forward in paper markets during backwardations. Likewise, when product markets are in backwardation, products are sold prompt, while in contango paper sales are made further down the time curve.

As an example, Figure 13.7 shows a time curve for crude oil that is in backwardation, and a time curve for oil products in contango. Evaluating the refinery margin at spot prices (i.e. at time t_0), results in a loss of p_1–p_2 per unit. However, following the principles above, the refiners aim to lock in a margin by trading on the paper markets further along both time curves. They should trade as far out as there is sufficient liquidity before buy-sell spreads become too wide. If that depth is trading t_1 forward in crude markets, and t_2 forward in product markets, then (with further spread trading to lock in differentials), the refiners can achieve a profit of p_3–p_4 per unit, instead of the loss implied at t_0 prices.

The time structure of oil prices is perhaps the least understood aspect of oil price behaviour. Two main errors are prevalent. The first, and most common, is to see the time structure as being in some way an indication of the predicted level of prices in the future. As an example, consider the following explanation of why crude oil prices tend to be in backwardation a majority of the time: 'correct and unsurprising, considering that oil prices have been falling for the majority of the time since NYMEX introduced its WTI futures contract in 1983'.[2] The link apparently being made is that falling markets and expectations of lower prices in the future generate a backwardation.

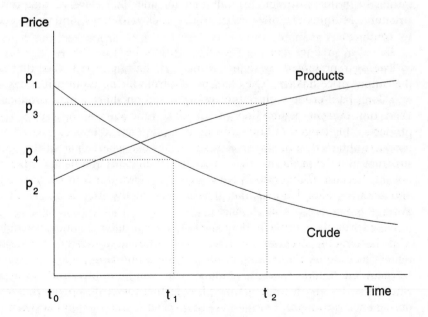

Figure 13.7: Illustration of the Effect of Trading Along Time Curves on Refinery Margins.

The fallacy of the time structure as an indicator of future price movements argument is best illustrated by considering the economic definition of an efficient market. A market is efficient if its prices contain all the information that is currently known, i.e. the only way in which they can change is when the information set changes. Given this definition, the rational expectation of the price that will prevail in the next time period, is, in fact, that which prevails today. Information that leads to an expectation of a future fall in prices will also impact on current prices. There would then be an element of prediction in the current time structure of prices. Empirically, futures prices are indeed an extremely poor predictor of future price movements.

If the time structure of prices was determined by predictions of future prices, it would be completely flat in an efficient market. The reason time structures are not flat, is the difference between the prediction made today of prices in the next period, and the price at which an agent is willing to trade today for quantities today as opposed to the next period. The difference is the balance of two effects. The first of these is given by the cost of carry. To transfer oil from the current period to the next, involves paying storage charges plus the opportunity cost (given by interest forgone) of the capital tied up in the oil. Should prices quoted today for the next period rise above current prices by more than this cost of carry, agents could then buy oil today, sell it immediately for next period delivery, store the oil for one period, and still emerge with a profit. As long as spare storage is available and the two prices can be arbitraged, the degree of contango in a market must be limited by the cost of carry.

The second effect is the convenience yield of holding physical oil, i.e. the additional valuation placed on having oil to use today, rather than receiving it in one period's time. Again, note that there is no element of prediction involved, an agent can have flat price expectations, while still placing a higher value for their own operations in having oil in the current rather than in the next period. For an agent with not enough stock cover of their own, there need be no limit to this value. Hence, while contangos are, subject to the provisos noted above, limited by the cost of carry, there is no theoretical limit on the extent of a backwardation. Further, far from backwardations being associated with falling prices, as implied in the quote above, they are normally associated with a tightness in the market. Backwardated markets are in fact more likely to be rising markets, since they tend to emerge as a result of short-term excess demand. In a similar fashion, periods of short-term excess supply tend to create contangos, i.e. the time structure provides increased incentives to carry over the excess supply into the next period. However, again note that the existence of a contango does not mean the prediction of higher prices in the next period; the prediction is already contained in the current price.

The impact of the time structure on inventory behaviour creates a further asymmetry between contangos and backwardations. The convenience yield is reduced by the level of current stocks. Thus, if a contango is large enough to provide incentives to build stocks, it will also have the effect of reducing the convenience yield, and thus reducing the possibility of a backwardation occurring. Contangos tend to be fairly stable, since they are caused by a short-term supply overhang which is then carried, *via* an inventory build, into the future. By contrast, a backwardation provides an incentive to release stocks and increase current production. The higher the level of backwardation, the greater are those incentives. As result, the level of a backwardation is normally volatile, and it becomes increasingly volatile the larger its absolute degree. In summary, contangos are bounded and stable, backwardations are unbounded and less stable.

Beyond the fallacious view of the time structure as a predictor of future prices, the other way of viewing it which we also believe is unhelpful in the case of oil, is the theory of normal backwardation. This was first propounded by Keynes (1930)[3] and Hicks (1946),[4] and still has some latter day adherents in the context of oil markets.[5] The premise of the theory (which was designed for the context of agricultural rather than oil markets), is that consumers have more flexibility in their buying decisions than producers have in their output decisions (i.e. in agricultural commodities the timing of output is dictated by the seasons). Thus, the argument runs, producers seek to sell further out in time than consumers wish to buy. At any one time, there would be in total a greater volume hedged by producers than by consumers. Along the time curve there is excess supply further out. To balance this, speculators need to be able to buy profitably further out and sell further in, receiving compensation in a risk premium, i.e. prices need to be lower for more distant delivery times. Normal backwardation is then a phenomenon necessary to induce speculators to balance the differing time preferences of consumers and producers, and to assume the risk involved.

The above argument, based on commodities where the seasonality is in production, does not carry into the oil market, where the seasonality is in the downstream rather than the upstream. To use the normal backwardation argument for crude and products, would be to hold the untenable proposition that crude producers necessarily have less flexibility than refiners, who in turn have less flexibility than distributors and retailers. Further, the neat division implied between hedgers (all involved in physical oil operations), and speculators (neither producers nor consumers), is considerably more blurred in reality.

To suggest that all trading activity by oil companies is hedging, and that all oil market trading by other firms is speculation, would be stretching the truth. In particular, there is no evidence on the basis of trading strategy or profitability, to suggest that the function of pure trading companies and Wall Street subsidiaries is to collect up the risk premia associated with the oil industry's hedging need.

The monthly average time structure of Dubai crude oil (represented by the differential between the first two forward months), is shown in Figure 13.8 for the period from 1986 to 1996. Figure 13.9 shows the monthly average time structure for Singapore gasoil prices (represented by the difference between spot quotes and the nearest swap market quotation), for the period after the Kuwait crisis.

We noted above that there are two asymmetries between the behaviour of price backwardations and contangos. Backwardations have no defined upper bound, whereas contangos are limited, and, while contangos tend to be fairly stable, backwardations become more volatile the greater their degree. The first of these features is shown clearly in Figures 13.8 and 13.9, with the potential for high levels of backwardation during tight markets when the convenience yield is large. In particular, note the high level of backwardation in Dubai during the Kuwait crisis, and the high level of gasoil backwardation in the first quarter of 1996. The second feature is illustrated by Figure 13.10, which shows, as a daily series on the same scale, the Singapore market time structure (expressed as the

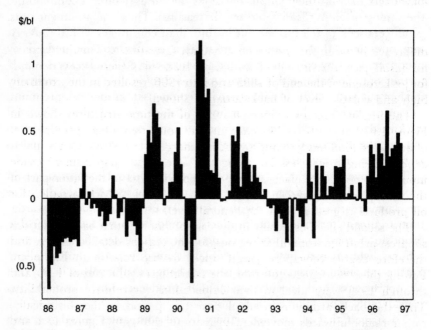

Figure 13.8: Time Structure of Dubai Crude Oil Prices. Monthly Averages. January 1986 to November 1996. $ Per Barrel.

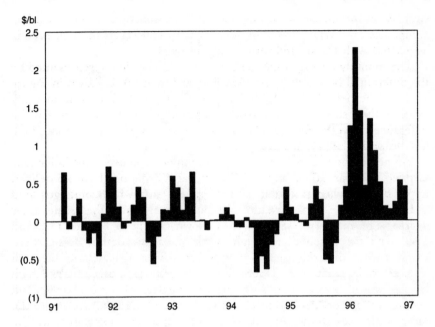

Figure 13.9: Time Structure of Singapore Gasoil Prices. Monthly Averages. March 1991 to November 1996. $ Per Barrel.

difference between first and second month forward swap market quotes) for gasoil, jet kerosene, fuel oil and Tapis crude oil. Severe excess demand for jet kerosene at the end of 1995 and into 1996, resulted in the extremely high and volatile level of backwardation shown.

The implications for refinery margins of the time structures shown in Figure 13.10 are twofold. We have already noted that a high general level of backwardation tends to improve Asian margins, as outputs are normally, *de facto*, being priced less far along the time curve than refinery crude input. The positive impact is compounded when, as in the latter part of the time period covered in Figure 13.10, the level of backwardation for oil products is greater than the general level.

The second effect, especially in the case of gasoil, is that backwardations suppress arbitrage from other regions, for the reasons detailed earlier, and therefore enable Singapore gasoil prices to rise relative to European. Putting these two effects together, the conditions at the end of 1995 and through 1996 were ideal in their implications for refinery profitability. That the backwardations resulted over this period in the (historically) rather minor upwards movement in margins shown in Figures 13.1 and 13.2, is an indication of the severity of the overall depression in profitability.

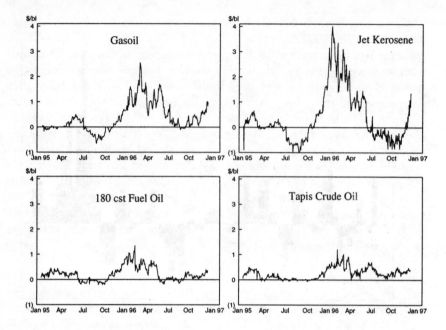

Figure 13.10: Time Structure of Prices for Gasoil, Jet Kerosene, 180 Centistokes Fuel Oil, and Tapis Crude Oil in Singapore Market. Daily. January 1995 to November 1996. $ Per Barrel.

6.   Downstream Profitability

For refiners with retailing outlets, or those in regulated markets, the changes in refinery profit margins discussed above have either been just one element of changes in their overall profitability, or have been totally unconnected to it. For these refiners, developments in the 1990s have in some cases represented even more downwards pressure on their profitability. Retailing margins in Asia have typically been very high relative to US and European levels. With the absence of the competitive effects that have produced wafer thin margins elsewhere, retailing has been able to compensate for weak refinery margins. A new refinery venture may on paper appear to offer a very low return to capital based on refinery margins alone. However, if building the refinery creates an otherwise impossible ingress into the domestic market, its profitability can derive from the returns to the integrated operation. Hence the reason why lack of access into the Chinese market has created the block on refinery joint ventures that we saw in Chapter 3. As mainly export orientated refineries they are unprofitable; as part of an integrated domestic operation they could be viable.

Given the importance of retailing margins to the Asian downstream, deregulation has in many cases created a further set of pressures. Where retailing has been opened to full competition, for instance in Thailand, margins have been severely impacted on. Deregulation may give access to the market, what it can not give is the same level of profitability that obtained pre-liberalization. We saw in Chapter 4 that import liberalization in Japan has had the side effect of beginning to create the conditions for the entry of hypermarkets into retailing. Elsewhere retailers are being affected by rising costs, particularly in the explicit or imputed land costs of their operations. In addition, poor refining margins have created an incentive to attempt to recoup profits by building up retail market share, creating some competitive pressures even in markets that have hitherto been uncompetitive. In total, while the retail sector is still the mainstay of downstream profitability, there are some strong downwards pressures bearing on the margins involved.

The greatest potential changes in refinery profitability occur where refiners have either faced administered input or output prices, or prices that are affected by existing regulation. For example, the greatest impact on Japanese margins has not come from the factors underlying Figure 13.1, but from the sharp fall in wholesale gasoline prices prior to, but caused by, the relaxation of the import regime in 1996. Profitability in this case has been primarily determined by the regulatory stance of government. In addition, product specification tightening, requiring additional investment in upgrading and desulphurization capacity, represents another source of fixed cost.

We have seen in this chapter that a variety of factors have combined to cause a notable weakening of Asian refinery profitability over the first half of the 1990s, with a particularly dramatic collapse in the return to upgraded capacity. In looking towards the future path of profitability, it is the behaviour of the same factors that will be crucial. On the specific question of the returns to upgraded capacity, the key factor driving the process is the balance between incremental OPEC and non-OPEC output, as addressed in the previous chapter. Within that picture there are two further slants. The first is the eventual return of full-scale Iraqi production, bringing a volume of medium to relatively heavy oil onto the market. The second is the tendency for incremental non-OPEC production to become slightly heavier – for example the development of the North Sea, and in particular the West of Shetland province, does involve developing some fields of heavier oil. Offsetting that is the tendency of a constrained Saudi Arabia to direct development towards lighter oil, and the sharp increases in world condensate production. Additional factors bearing on the upgrading margin include the extent to which gasoline demand is reined in by tax policy. In the determination of the level of simple refining margins, we would highlight four main factors that could occasion higher profitability. First, if Asian refinery expansions begin to slow relative to incremental demand, or alternatively if demand growth accelerates. Secondly, if closures and restructuring in Europe are sufficient in restoring some strength to that market, and thus reduce the strong dampening effect on Asia that Europe has exerted throughout the 1990s. Thirdly, if crude oil markets become weaker, for instance if OPEC attempts to maintain market share in the face of non-OPEC production increases.

The fourth key determinant is government policy. In most circumstances, any retrenchment or slowdown of policy can lead to less incremental capacity coming on stream, or at least arriving on a slower timetable. In others, some policy retrenchment can exacerbate the situation elsewhere. For instance, allowing access into a country's downstream, but deciding instead of full liberalization to leave some protection from imports, say through a differential tariff structure for crude and products, can make refinery expansion even more attractive. In China liberalization of the import regime and reduction of demand suppression would be beneficial to refinery margins generally as it boosts demand. On the other hand, liberalization of the access of foreign capital into retailing removes an obstacle which has prevented the building of a large tranche of new refining capacity. Likewise in India, fast liberalization involving the freeing of markets and the lifting of subsidies helps not only to damp demand increases, but also to bring on further tranches of capacity.

Notes

1. As noted in Chapter 11, Saudi Arabian formula prices to Europe and the USA include a freight adjustment factor, while those to Asia do not. This effect has been included in Figure 13.5.
2. Robert Weiner (1995), 'Brent's Hedging Efficiency Affirmed', *Argus Energy Trader*, no. 17.
3. J.M. Keynes (1930), *A Treatise on Money*, vol. 2.
4. J.R. Hicks (1946), *Value and Capital*, Oxford University Press, Oxford.
5. See P. Verleger (1996), *Paper Oil: The Link between Financial and Physical Markets*, Charles River Associates. reported in Argus Energy Trader, 19 April 1996.

PART VI

Conclusions

CHAPTER 14

ASIAN LIBERALIZATION AND SINGAPORE TRADING

The diversity of Asia does not lend itself to generalizations. One can, for instance, make observations on the thrust of European policy. Europe is on a global scale small, has a series of supranational institutions, and there is a broad similarity in national structures, religion, political orientation and policy concerns. Likewise, Latin America has some integrity as a concept, comprised of a series of nations of similar historical circumstances and levels of development, with a common religion and a small number of languages. By contrast, Asia is primarily defined as being made up of a series of countries that happen to be broadly contiguous. Geographical proximity has created interactions and a force towards uniformity in many characteristics within Europe. Such effects are absent in Asia, where, for example, the distance from Singapore to Tokyo is about the same as that between London and New York. The countries of Asia range from some of the very poorest in the world to some of the very richest. They span the full range of political orientation and political institutions. There is no uniformity in language (often not even within national borders), or in historical circumstances. Supranational institutions are rare. Even defining a geographical subset, for instance the countries of ASEAN, does little to reduce the diversity. The are no natural forces towards a convergence, other than, as argued by Yamazaki, the notion and implementation of modernity.[1]

While generalizations are difficult, there are some broad common threads in terms of oil market policy. The most obvious is simply that significant structural changes are occurring, often after decades of ossification. Within those changes, there is greater recourse to the use of market prices and competitive processes. However, the nature of oil market liberalization needs to be clearly stated. It is a process whereby government actions bear on the oil industry in a different fashion and produce different outcomes, not one whereby governments do not intervene or where government policy does not have an impact. Some have made the argument that liberalization will not occur because governments do not wish to lose control. We would argue that there is in fact no contradiction between governments exercising influence, and a more competitive, or market orientated, oil industry structure.

The lack of correspondence between allowing market processes to function and an inert government policy, is most clearly shown in the US situation. The market is highly liberal in terms of its operations, but the actions of government, at both a federal and a state level, are very far from being neutral. Prices are a political issue, to the extent that the federal government has used strategic reserves for their short-term price impact even outside of crisis situations. Government tax policy has a bearing on operations upstream and downstream, as do environmental legislation, product specifications, general competition law in both its domestic and supranational implications, commodities trading laws, export laws, tariffs, and energy specific regulatory bodies. Oil has an impact on US foreign policy targets and their implementation, which in turn affect oil markets and their perceptions. Sanctions and even secondary sanctions affect international capital flows and market operations, and constrain the actions of both US and non-US oil companies. In total, market liberalization and competitive processes should never be confused with a necessary reduction in government intervention. In terms of its impact on the oil market, not just domestically but worldwide, the US government is highly interventionist notwithstanding the competitive and liberal nature of its market structures.

Asian oil market liberalization does not imply then that government merely retreats into the shadows leaving market forces to be the only active consideration. Governments will remain as a dominant feature of the Asian oil market. What liberalization is creating is a greater sensitivity of oil markets to international market conditions, and a redefinition of the role of state enterprise, competition and foreign capital. The reverse side to this transition is less reliance on overt *dirigiste* policies motivated by non-economic strategic and supply security arguments. Outside China, current processes are to a large extent a rolling back of the sharp increase in state control that occurred as a result of the 1970s oil shocks. In many cases the industrial organization of oil has been treated differently to other sectors, due to a combination of the politicization of oil prices and awareness of their income distributional impact, and non-economic strategic and supply security grounds.

The attitude towards direct state involvement in the oil industry has varied across Asia. In some countries there has been no significant direct involvement in the domestic market (most notably Korea and Japan) and liberalization has involved reducing the degree of direct regulation over industry operations. In the 1990s, there is only one case (namely Cambodia) of a national oil company being set up where none previously existed, primarily to give the state involvement in any oil discoveries.[2] In China, the approach to national oil companies has been to create more of them, and (depending on which phase of the regulation–deregulation cycle policy is in), to facilitate competition between them. In Indonesia a monolithic state oil company survives, to date virtually unreconstructed.

At the other extreme there is Taiwan, no privatization yet but the aim of complete privatization, and the Philippines with a realized privatization of 60 per cent of the major part of the state oil holding.[3]

However, privatization can not be the sole determinant of what makes an internationalized market orientated company. The two Asian state companies that would most generally be accepted in this category are Petronas (limited privatization to date), and PTT (none to date). If privatization is not the prerequisite for becoming an international commercial company, what is?

We would argue that internationalization is most likely to arise either when the relationship between state and company is extremely well defined, or when it is extremely badly defined. The key features of this relationship are the degree of control the company has over capital, its freedom in raising capital, and the extent to which a government commitment not to appropriate the company's capital is credible. An excellent example of where the implicit contract was badly defined is provided by Boué[4] in the case of Venezuela, where a motive for the internationalization of the state oil company, PdVSA, was given by the government's appropriation of readily accessible domestically held PdVSA capital for other purposes.

If, and only if, that contract is well defined, and the relationship between government and company can credibly be signalled as a commercial and arms' length one, then the company can operate as a proxy to a private company. The absence of stock market effects on efficiency can to some extent be substituted for by benchmarking, for instance arising out of joint ventures with private companies. If the government acts in effect as a single shareholder out of commercial considerations, then the definition and crediblity of the contract produces what is, in terms of performance, a private company. We would suggest that to a large but not complete extent, the above represents what Statoil has been moving towards following the *dirigiste* inspired original form of its relationship with government.

In the above conditions, the motive for internationalization becomes the same as it would be for any private sector company. The corollary of this implies that if the major motivation for the internationalization of a state oil company is simply to diversify away from a once protected domestic market subject to deregulation, then the relationship between the company and the government is not defined in a way that is conducive to commerciality. The relationship between Petronas and PTT[5] and their respective governments appears to be moving in the direction that would facilitate their ultimate emergence in the status of effective international oil companies.[6] In other cases, we see little possibility of a number of Asian national oil companies becoming effective in international competition, without a dramatic shift in thinking with regard to their relationship with government and the constraints in the use of capital.

While change is taking different forms in different countries, and the speed varies considerably, overall there is a general tendency towards greater reliance on market solutions and freer competition. This change is likely to find a reflection in the operation of the Asian oil market, and also in the attitude to and response to the price signals generated by trading activity.

Consider first the impact on the downstream industry. Progress towards liberalization has been associated with a rush of new developments in some countries where capacity expansions were previously regulated (notably Korea and Thailand), while in others a perception that liberalization might proceed has led to a raft of proposed projects (notably India and China). The rush of capital, at least in its planning stage, seems to be unaffected by considerations of domestic energy balances in a single country, and is apparently focused purely on the idea of a voracious Asian appetite for oil products. At the level of a single domestic market, the implicit assumption that the demand from the rest of Asia for any exportable surplus is perfectly elastic, would mean that any capacity overshooting would have benign consequences.

When this reasoning is aggregated across countries, the logic breaks down, since demand for the sum total of exportable surpluses is of course not elastic. The result is simply a large number of countries all trying to sell excess production into the same markets. In the mid-1980s, Singapore was the only Asian country with a substantial exportable surplus of oil products. We have noted that Korea, the Philippines, Thailand and Taiwan already have joined or are about to join Singapore in this category, albeit it to varying degrees and in some cases perhaps only temporarily. We have discounted the idea that India will provide a significantly expanded vent for the surplus of others, at least in the immediate future and even if there are only a few new refinery projects there. Add to this picture new refinery projects in Malaysia and Indonesia, and overall the result is a large number of countries all looking primarily to the Chinese market to absorb their surplus.

At the start of the 1990s many would have considered the idea that there might be a significant overshoot in Asian refining capacity as one of the more outrageous of the 'what ifs' we referred to in Chapter 1. Within a few years much of that conventional wisdom has proved to be untenable. In Chapter 13 we considered the path and determination of profit refinery margins in greater detail. It is clear that expansions are exerting downward pressure on profits. Increasingly, the refiners in the best position appear to be those who can achieve cost savings while still in a large and regulated domestic market. If this becomes a common lobbying position (and we have already noted its use in Taiwan), it should be noted that the burden of the capacity overhang may not be absorbed totally by refinery profit margins. Part of the adjustment might occur by a slowdown in the pace of deregulation, or, through policy retrenchment, by following a path of less than full liberalization. As we have noted already, government policy is not an exogenous process with regard to developments in the industry.

Deregulation is leading to a greater exposure of Asian oil companies to international (i.e. Singapore) market prices, either full price liberalization and potential import competition or through domestic pricing formulae explicitly linked to Singapore prices. In the detailing of the operation of the Singapore oil market complex, it has been seen that to date participation by most of these companies has been limited. This raises the question of whether exposure to the market will lead to an infusion of new participants and liquidity in that market. The amount of risk management carried out through market operations by Asian companies has to date been fairly limited. While both Japanese and Chinese companies have been extremely active, it could be argued that in the past they have been primarily conducting pure trading activity rather than risk management. The domestic regulation of both economies meant that the risk exposure to fluctuating Singapore prices had been limited.

It does not immediately follow that exposure to price risk automatically generates trading activity. Management structures do have to be sufficiently developed in order to facilitate such trading. At first sight the conditions for this appear contradictory, for they involve both a decentralization of control and the maintenance of tight managerial control and account-ability. Decentralization of management control is necessary in that a climate has to prevail in which what appear as mistakes *ex post*, are tolerated. For example, consider the case of a trader who hedges a purchase of, say, gasoil, and the price of gasoil subsequently falls. Certainly, *ex post* the cost would have been lower if the hedge had not been placed. A management structure that holds the trader to account is not beneficial to trading, indeed at each point it encourages the course of taking no action that would expose the trader to any risk. In the above example the hedge removes the price risk from the company, but exposes the trader to the risk of being held responsible for its *ex post* outcome. The managerial structure must then include the delegation of responsibility in such a way that traders are free to trade. In many Asian companies (and indeed in most national oil companies throughout the world) this structure is not in place.

While traders need the conditions to trade, they must also not be allowed to do so in an uncontrolled and opaque fashion. In other words the risk profile of the company has to be constantly and correctly monitored. In reality this does not work against the operation of the first condition, as it is essentially a matter of information flows and the laying down of boundaries for the traders' operations. It should be noted that the most highly publicized trading failures had information failures at their root. In the case of Metalgesellschaft, upper management did not fully understand the nature of the risk the trading operations had assumed. When Showa Shell lost heavily in foreign exchange trading, the situation was compounded by the ability of the traders to roll over positions, and thus keep the paper losses from being realized and apparent in management accounts. Likewise in the Barings collapse a series of information and monitoring failures appeared to have facilitated the catastrophe, as they did with Sumitomo's copper trading. Simple rules on the nature and extent of the risks that can be assumed, the complete separation of trading and accounting (i.e. 'back room' and 'front room' operations), and managerial information systems that reflect the full nature of the risk profile at any one point in time need to be put in place.

However, despite the possibility of managerial constraints on trading, refiners and other downstream players that once had little or no risk to market prices, now face risk. International companies who already trade in Singapore find there has been an increase in the volume of their physical operations in Asia that are exposed to risk. Likewise, non-participants find that markets which they had no interest in, now directly impact on their operations. In some cases, most notably PTT in Thailand and CPC in Taiwan, trading and risk management activity have been greatly increased. In others the costs of not carrying out such activity will become more obvious. In total, liberalization in Asia is an important force behind the future development of markets.

The increasing importance of market prices in Asia is a unifying force for market evolution. There is however also an important separating and regionalizing force at work. We identified in previous sections the movement to less regulation of industry structure and pricing, simultaneous with a move to greater regulation of product quality. While the liberalization of the former generates new trading activity, so does the tighter regulation of the latter. Throughout Asia oil products are becoming less fungible across countries. Product specifications are tightening at different rates and for different final objectives across the region. While liberalization tends to lead to a unification of the market, the different product specifications are leading to greater dispersion.

A whole series of niche markets in different specifications is opening up. Some refiners will have the capacity to supply to most markets (indeed this is a source of considerable comparative advantage for the industry in Singapore), while others will be extremely limited. Blending facilities become more important as traders seek to arbitrage between the different national specifications. Logistical problems and bottlenecks become more important, and different specifications for the same product may experience price spikes at different times. Further, the total volume of trade can become far greater than the addition of the absolute size of national product imbalances. For example, we noted in Chapter 5 that the configurations of Korean refineries are becoming increasingly out of line with the tightening of the domestic product specifications. The result will be large exports of high sulphur products, but simultaneously significant imports of low sulphur versions of the same products. In summary, liberalization brings far greater importance to the price setting markets we now describe, but environmental factors mean that the degree of friction in the operation of those markets may become greater.

Supply security is perhaps the largest red herring involved in considerations of oil policy. In the Asian context, this is most manifest in the issue of dependency on the Middle East for crude oil imports. This is frequently portrayed as either a major strategic weakness of the Asian economies, or as creating a geopolitical tinderbox with, for example, scenarios of a blue water Chinese navy attempting to police the Straits of Hormuz. Supply security was also the pretext for much of the advance in government intervention in oil in the 1970s, and still used in policy today. There is a notion that Middle East oil is in some way different in a strategic sense from other oil, and that a country which can obtain oil from other regions is in some way less vulnerable. Supply security is still an obsession of governments, and is an issue at which they are apparently prepared to throw money. For instance the Korean Ministry of Industry, Trade and Energy (MOTIE), as of 1996, is still building up the funds available to refiners to buy oil from non-Middle East sources.

There are several reasons why we would contend that supply security is not a valid issue. First, with the development of markets supply disruptions do not result in physical shortfalls, but in a price impact. The price impact is common to all oil, regardless of its origin. A country with a low reliance on Middle East supplies in its overall oil imports is thus as exposed as one with a high reliance. Likewise, any use of market power by Middle East producers affects the markets as a whole, not simply those who import from them. Secondly, the disruption of an existing import flow can be covered rapidly by purchases elsewhere, given the high degree of fungibility of waterborne oil cargoes. Thirdly, oil does tend to flow whatever the regime. The only case of a long-term withdrawal of Middle East oil from the market has been occasioned by consumer country boycott, not by producer country policy. Finally, we would note that the combined impacts of dollar oil price weakness and domestic currency appreciation mean that oil is extremely cheap in most Asian economies. That Japan in the mid 1990s only uses as large a proportion of its GDP to pay for its entire oil import bill as it did in the early 1950s and less than one-eighth of the proportion it used in 1980, is a stark illustration of how circumstances have changed since the 1970s oil shocks. In total, it is extremely wasteful to create distortionary policies that carry implicit or explicit costs, purely on the basis of supply security arguments for a cheap, plentiful and fungible commodity. Compared to receiving gas through a transnational pipeline or from a LNG plant, the supply security aspects of oil rather pale into insignificance.

Raising the issue of Middle East dependency in terms of a spectre at the banquet of fast economic growth is rather too common. Asian governments have no real need to fear any such development. While we believe that dependency on Middle East supplies is not an issue that should motivate policy in the way it often has, we also question the degree to which it will actually happen. The premise that Asia will swiftly become almost totally dependent on the Middle East for oil imports is often made, but it should not be taken as a given.

There are two main factors that detract from the supposed inevitability of swift and dramatic increases in dependency. The first, and lesser, is the path of regional production. We have argued that the main determinant of this path is not oil prices or the size of estimates of the existing recoverable reserve base, but government policy. Asian fiscal regimes are hostile, are not effective in leading to capital accumulation, and are extremely bad in providing incentives for any capital that is attracted. To re-use the example of Vietnam, the reason why it swiftly turned from being a major upstream prospect to a large disappointment was in the first instance geology. However, disappointment was then turned into a state of disillusionment by the lack of flexibility in fiscal conditions. Likewise, we do not see any movement of Indonesia into the ranks of oil importers as an inevitable process. Sharp falls in production are more than likely while conventional fiscal terms rank 222nd in the world, but would be avoidable if the fiscal regime were to react to conditions. In China, where the fiscal regime is slightly more liberal, the problem is the role given to foreign capital. In particular, confining foreign interest to the less attractive parts of the Tarim is not the optimal strategy if the aim is to reduce the potential scale of future import dependency. China is not providing the conditions that would facilitate large-scale capital inflow in both the upstream and the downstream. Combining high risk with low return in the upstream, and large capital commitment without sufficient domestic market access in the downstream, produces an industry with low appeal to international capital.

The second, and more important, factor that weighs on Asia's future Middle East dependence is the balance at the global level between OPEC and non-OPEC production. The swiftly increasing dependency argument rests on the implicit assumption that some process is about to halt permanently the decades long and near linear increase in non-OPEC production. The Asian dependency argument was also made in the mid 1970s, when the six Middle East members of OPEC exported 5.1 mb/d to Asia and Australasia. Twenty years later the volume had only grown to 6.6 mb/d. Those twenty years may have been special, but oil markets rarely develop in any linear fashion, and the experience is enough to dismiss the idea of the inevitability of sharp and prolonged increases in dependency.

Liberalization is creating some significant changes in market structure. Some Asian oil companies have expanded and marketized in such a way as to become significant internationally. Domestic markets have also moved to greater competition. In some cases, such as Japan, this has created the conditions for a major reorganization of the industry, and some degree of capital outflow. In other countries it has facilitated capital inflow, or a significant diversion of private domestic capital into the oil sector. Overall these changes have resulted in three main outcomes. First, the Asian downstream has become less profitable. Secondly, there has been a tendency towards refinery capacity expansions overshooting the level of demand increases. Thirdly, the role of market signals has been expanded. The latter aspect led us to consider the mechanisms whereby these signals are generated. In the case of oil products, this involved detailing the development of trading in Singapore, a local market entering the transition to becoming a regional market. In the case of crude oil, we found a complete vacuum in terms of price formation within Asia, and highly imperfect mechanisms in the Middle East. We return to the implications of this below.

Singapore developed into the centre of Asian oil trading through a combination of natural advantages, the historical development of the physical oil infrastructure, and government policy. Among its natural advantages are geographical location as a crossroads in trade. Among its induced advantages are market orientation, infrastructure, and a government which has been accommodatory and provided incentives, without producing a heavy layer of market distortions and constraints. The Singapore oil industry has developed through cooperation between companies and government, with the government providing incentives and creating conditions conducive to each development. Trading has been a further addition to a complex that began in the 1870s with distribution, then went on to storage, blending, bunkering and the provision of upstream supply services. After independence came export refining, trading, petrochemicals, independent storage and paper trading. Trading did not develop in isolation, nor did it expand spontaneously.

Oil trading in Asia has some special characteristics. The small size of the Singapore market relative to London and the USA, has contributed to a noticeable reluctance for traders in Singapore to make a market. We can think of this as the 'wallflower complex', i.e. never take to the dance floor before others, if ever at all. This has manifested itself in several ways. First, there is the tendency not to trade during the day until the International Petroleum Exchange opens in London. The Singapore market takes its cue from London, in such a way that very little trade is carried out in the Singapore morning and early afternoon. This could be looked at in two ways. It might be seen as a natural aversion to the risk of making a price in a small market, only to see the larger market move in another way. However, it could also be seen as a market that is still in the early stages of transition to a regional rather than a local market.

The second aspect to the wallflower complex has been the reluctance, even among major multinational companies, to act as a market maker in the physical markets. We have seen how the gasoil market is dominated by one trading company, and in particular how that company is involved in the majority of deals that trade absolute price levels. We note that this dominance also extends to the psychology of the market, i.e. few traders are prepared to trade in a direction in which they believe the dominant company does not want the market to go. In a room of wallflowers, the solitary dancer is the centre of attention, particularly when they seem to be influencing the choice of music. While such a market is not particularly healthy held up against an ideal, we would emphasize that we do not believe there are any abuse of market power issues involved. While the dominant firm does exert considerable influence on prices, it does not do so by the maintenance of any barriers to entry. Indeed, without a refinery or retail network, compared to many firms it is especially unsuited to a role of achieving full entry deterrence to the price setting mechanism. The market is fully contestable, and the ultimate source of any market failure is not the actions of the dominant firm, but the reluctance of others to contest.

A further form of the wallflower complex, particularly in the past, has been the tendency to think and act locally rather than regionally and globally. Singapore's potential role is as the regional trading and supply centre, strongly linked into a world system by the binding threads of constant arbitrage right along the time curve of prices. At points, however, we have noted a certain residual, but diminishing, parochialism in trading strategy and the concerns of the market.

Trading in other regions began with spot markets, before further layers of derivative markets evolved. Informal forward markets developed, followed by futures, options and swaps, resulting in a complex of interrelated markets. The transition in Singapore has taken a different course, primarily because the logistics of a large cargo size export market differ from the high liquidity spot markets which arose from trading along

US pipelines or in the Rotterdam barge trade. The growth in Singapore has occurred in informal swaps markets, while the formal futures market was slow to develop. The swaps markets have added the dimension of time to trading, and in particular facilitated the growth of efficient arbitrage between Singapore and the European market.

The exploitation of arbitrage opportunities has meant that there are no convenient ring fences to be put around the Asian market. It is not isolated from other centres, and indeed we have argued that one of the most important reasons for the extremely weak profit margins in Asian refining over the mid 1990s, has been the weakness of the European market. From a physical trading perspective, the Asian market only offers two (not mutually exclusive) forms of organization. Trading can be at the global level, using inter-regional flows and requiring full participation in paper markets. It can also be focused on the expanding logistically driven niche markets, for example supplying Indian subcontinent tenders or the trade of oil products into southern China. The scope for trading at any level between the macroscopic global level and the niche markets is now extremely limited.

The demand for risk management in Asia outside Singapore, has been very limited due to the regulation of the industry and the presence of national oil companies which have often not been exposed to risk. To some extent the early development of the paper markets overshot the need for them – there was perhaps an excess supply of traders in effect, producing a market with wafer thin margins. In addition, further development was hampered by the frailties of, and lack of definition in, the underlying physical market, most noticeably in the case of the rise and fall of Singapore naphtha swaps.

However, market liberalization has increased the exposure to floating prices, and the demand for risk management. We are here concentrating primarily on short-term commercial risk, and exposure to timing, quality and location differentials, rather than any form of routine hedging. The latter can sometimes be justified on cash flow optimization grounds. On the other hand, many companies are unwilling to pay the in/out costs of such hedging, and consider that some of the equity valuation of oil companies arises because they do carry absolute price risk. We see the major role of paper markets as being in the area of relative price risk in all its forms. There is no contradiction in not wishing to reduce absolute price exposure, while wishing to remove relative price exposure, and indeed the companies who do not are disadvantaged compared to those that do in what is an increasingly global market.

Crude oil price formation in Asia has primarily relied on the use of monthly averages derived from the Dubai market. Dubai is now a weak market, with little liquidity and a declining physical base. Under current

Middle East marketing policies there is no other grade in the region that serves as a workable alternative. More importantly, Dubai no longer serves as an adequate marker crude oil for Asia, because Dubai related oil is no longer at the margin of the market. That margin is now made up of Atlantic basin crude oils linked to the price of dated Brent. Those flows are marginal in that the volumes moving east swing sharply according to market differentials. The development of the balance between OPEC and non-OPEC production leads to the conclusion that areas of strongest incremental crude production (i.e. the western hemisphere) are not those of strongest incremental demand (i.e. Asia). As a result, West African crude oil is being forced out of the US market, and increasingly displaced into Asia. The consequence of this marginal role is that the only logical price marker for Asia is North Sea Brent.

The change of the margin of the Asian market is primarily a function of the changing geography of world crude oil trade. Where deregulation and the attendant squeeze on profitability bear on this is in the enhanced importance of that margin. In regulated markets where refiners act with little effective sensitivity to prices, any deficiencies in the marker crude oil are of little import. In such circumstances, when using adjustment terms from monthly averages of an index, over time the nature and functioning of that index matters little. If buyers are acting with little sensitivity to short-term price movements, at the extreme it could be argued that the index need not even be oil, if its short-term price dynamics and approximation to changes in opportunity costs are unimportant.

In less regulated markets price sensitivity increases, especially when there are alternative sources of oil, and when the buffer of regulation induced profits is reduced. The role of the index as a price signal and approximation to marginality now becomes more of an issue, as does, in some cases, the ability to hedge that index and its differentials from output prices. We suspect that the maintenance of Dubai related pricing has much to do with the desire to be able to influence relative regional crude oil prices. The ability to do so has been reduced greatly by both the breakdown of the compartmentalization of the world crude oil market, and by the changes in buyer behaviour induced by liberalization. Given this, together with the frailty of the Dubai market itself, a change to Brent would now appear to be the optimal course of action.

In terms of the use of markets, the Asian oil industry is in the main still at the start of the process. There is some correspondence between Asia in the mid 1990s and Europe in the early and mid 1980s. At that time the use of markets in Europe was very unsophisticated, and there was a deep scepticism about the merits of market participation. The nascent futures markets were slow to develop, and attracted considerable cynicism. It took three attempts to launch a successful crude oil futures contract on the IPE. Informal forward markets had begun to develop, but the motivations of traders (beyond fiscal concerns) were highly speculative, and trading strategies were rather primitive bludgeons. Trading in the morning was extremely limited, as participants normally waited for the all powerful New York market to open to provide the lead. There are strong enough parallels to lead us to be optimistic about the longer-term development of the Singapore market complex, in both its informal and formal incarnations.

Notes

1. Masakazu Yamazaki (1996), 'Asia, A Civilisation in the Making', *Foreign Affairs*, July/August 1996.
2. However, downstream the Cambodian government has moved to privatize the state owned distributor, Compagnie de Kampuchea des Carburants.
3. Admittedly, the largest part of this was to the state oil company of another country. However, we would claim that in terms of its relations with Petron, and in terms of measuring the degree of privatization achieved, the ownership status of Aramco is largely irrelevant.
4. Juan Carlos Boué (1993), *Venezuela: The Political Economy of Oil*, Oxford University Press, Oxford.
5. PTT has a far lesser degree of vertical integration than Petronas (or indeed Statoil) which raises a series of subsidiary questions. However, we do not believe that this should necessarily affect its ability to compete internationally as a predominantly downstream and trading company.
6. It could be argued that the change in the relationship between companies and governments in these two cases has primarily arisen out of pre-privatization restructuring and commitments. Whether, like Statoil, that relationship could have arisen in any case, and they could have emerged as commercialized international but wholly state owned firms within the implicit privatization agenda, is not a testable proposition.

APPENDICES

APPENDIX 1

OIL PRICE ASSESSMENT

1. Introduction

Most physical oil that is traded, whether through tenders, term contracts or spot markets, is priced through a formula in relation to market prices. For example, a typical deal for gasoil might relate the price of cargo to a, say, five-day average of Singapore gasoil prices, plus or minus an agreed premium. The exact value of the cargo is unknown at the time of the deal, it is left to the market to price it at a specified time in the future. Likewise, many crude oil producers tend to link their sales to market prices of given grades of crude oil, with pricing again being determined by the market at some point in the future.

This raises the question of how the market price is defined. The Singapore markets we consider in this study are, with one exception, informal. Trades are carried out by telephone, telex and fax, and there is no central organization in authority collating information. Unlike a futures market, there is no electronic display on a trading floor describing the exact alignment of prices in the market at that moment. No automatic process within an informal market generates observable price levels. Filling in this gap is the domain of price assessment, and the task for price assessment agencies is to seek to report the level of prices in such informal market situations. The market prices on which most trade is based are then assessed prices, estimates of what the (unobservable) market price is. It is important therefore to give some explanation of where these prices come from, and the methodologies behind their compilation.

As well as providing an explanation of the process of price assessment, we use this appendix to provide an overview of the complex of markets that together make up the Singapore market, the grades traded and their locations. Further analysis of the markets themselves rather than the methodology and ambit of the price assessments is given in the main text of the study. To avoid lengthy explanations elsewhere, we also detail some of the technical specifications of oil products that play a role in trading. Section 2 concerns the general methodology of price assessment. We consider the four main price assessment agencies in the Asia-Pacific region in this appendix.[1] Within the general category of price assessment, we draw a distinction between price reporting and panel prices. Sections 3 and 4 cover respectively price reporting and panel price assessments of oil product prices, with crude oil price assessments being covered in Section 5. Section 6 presents some statistical analysis of the beliefs held in the market concerning price levels at any given time, and a final section offers some conclusions.

2.   Principles of Assessment and Standardization

Any form of price assessment consists of the production of a vector of prices on the basis of a sample of the total amount of information generated by the market. The information from a physical market will tend to have multiple dimensions. Deals done are likely to be for different specifications within each broad oil product category, for different volumes, for loading at different times in the future, for loading or delivery in different locations, and will be transacted at different times of the day. In addition, valuable information is generated by market participants signalling the prices at which they would deal (particularly at times when few, if any, actual deals are made), and by their interpretation of the impact of new information (in general terms 'news') as it reaches the market.

The major methodological difference between price assessment agencies lies in how the sample of information is generated, and through what mechanism this information generates the price. Price reporting agencies produce the sample through daily telephone contact with market particip-ants and other involved parties such as brokers. The prices are then generated by the price reporters' evaluation of their sample of information. Panel pricing also relies on a transfer of information from the market to the assessment agency via companies. The difference arises in the form that the information takes. In a panel it is transmitted to the agency as the companies' own assessment of the price, and then these assessments are turned into the published price by a defined formula.

Whether the transformation of information into prices is undertaken by price reporters or by market participants themselves, the fundamental problem remains how to interpret the multidimensionality of the informa-tion. The approach that is taken in solution to this problem is to provide a benchmark against which the information in all deals done or in market talk can be gauged. This benchmark consists of the standardization of five main parameters – quality, location, quantity, the timing of the deal itself, and the timing of the physical transference of oil.

Standardization of quality involves the definition of the full vector of characteristics that constitute the putative benchmark oil cargo. An assessment of the value caused by a deviation from this specification can then be made, so that deals for off specification oil still carry some usable information. Oil products in particular have many quality characteristics that can vary cargo by cargo, and the specification of the benchmark needs to contain a definition of the acceptable levels for any characteristic that could cause a difference in value. For example, Table A1.1 shows some characteristics drawn from a typical gasoil specification, together with the standard industry test for each as defined by the assessment agency.[2]

Table A1.1:   Selected Specifications for a Typical Singapore Gasoil Assessment.

Characteristic	Minimum	Maximum	Test Method
Ash % of mass	-	0.01	ASTM D 482
Cetane Index	48	-	ASTM D 976
Colour	-	2	ASTM D 1500
Copper Corrosion	-	No. 1	ASTM D 130
Density at 15°C	0.82 kg/l	0.87 kg/l	ASTM D 1298
Distillation 90 % recovery	-	370°C	ASTM D 86
Flash Point	60°C	-	ASTM D 93
Pour Point	-	9°C	ASTM D 97
Sediment % of mass	-	0.01	ASTM D 473
Sulphur % of mass	-	0.5	ASTM D 129
Viscosity at 40°C	1.5 cst	5.6 cst	ASTM D 445
Water % by volume	-	0.05	ASTM D 95

As shown in Table A1.1, the specification lays down a series of quality criteria. Refined oil will not be 100 per cent pure hydrocarbons, so some of the criteria involve the proportion of other material that will be found, i.e. water, ash, sediment and sulphur. Other criteria involve the physical characteristics of the gasoil. For instance, the flash point is the temperature at which the oil will ignite on application of a flame, and the pour point is the temperature 3 degrees centigrade above that at which the oil ceases to flow in a standard test tube. Viscosity, i.e. the rate of flow or 'thickness', is measured in Table A1.1 in centistokes, with lower centistoke values relating to freer flowing oil. A measure such as the cetane index seeks to summarize the combustion properties of the oil. In particular, the cetane index is a measure of the ability of the oil to ignite spontaneously under compression, i.e. exactly the property that is needed in a diesel engine (and exactly the reverse of what is needed in a gasoline engine). The higher the cetane index, the better the combustion properties in a diesel engine, and an index of 48 as shown in Table A1.1 implies that the performance of the gasoil as diesel would be in most circumstances satisfactory.

The numerous dimensions of oil quality raises the question of how a buyer knows the specifications of a cargo at point of purchase. Cargoes are examined by companies that provide testing services, so that it can be ascertained whether the specification stated in the deal has been met by the supplier. In Singapore such survey companies currently charge about 10 Singapore cents per tonne for products, 7 cents for crude oil and 75 cents for chemicals, and obtain revenue of around $20 million per year for these testing services.[3]

The second parameter in the standardization that produces the benchmark assessments is location. Oil is traded as exportable cargoes in production areas, or as delivered cargoes in consumption areas. In the case of delivered cargoes, prices will then include a freight element, i.e. pricing is for cargo and freight (c+f). At point of production the value reflects oil as transferred onto a tanker, i.e. free on board (fob). Price reporters are then likely to receive information on both fob and c+f trades from various locations, and standardization (involving adjustments for freight differences) produces a benchmark for the standardized locations. Hence, the reporting of a c+f trade for oil to be delivered into, say, Taiwan, contains information that is useful in compiling a fob Singapore quote, and the multidimensionality of the information is thus further reduced.

The value per unit of oil may vary according to the volume traded in a deal. More important, however, is the definition of the minimum quantity that can be taken to represent a 'genuine' trade. For example, through trading small parcels of oil, either as tank transfer or a small cargo 'topping-up' an already partially loaded tanker, a large number of deals can potentially be reported. With the suspicion that these deals may be made to generate information to influence the price assessment, standardization normally involves the definition of a minimum cargo size. The information from the trading of smaller parcels can then be discounted, in some cases completely. Trading of cargoes larger than the maximum used in the standardization does not of course present the same potential problem (trading very large cargoes to influence quotes is not exactly an efficient *modus operandi*) and the information from these deals is usable with an adjustment for any quantity discounts.

Timing factors represent the fourth and fifth parameters for standardization. Price levels may vary considerably over a day's trading, and the information set thus consists of pieces of data with different timings. In producing a quote for the day, one option would be to simply produce a range reflecting the minimum and maximum prices during the day. However, given the often high levels of intra-day price volatility, the range produced would often be large. Averaging the two numbers does not produce any meaningful quote, and in particular does not produce an assessment of either the time or volume of trade weighted daily average.

Trying to produce either of these averages directly both increases the information demands for the assessment, as well as meaning that by the end of the day the market could be trading at prices far away from the quote. While the daily trading range methodology has been used in the past, the universal practice today is to 'time stamp' the assessment. In other words the assessment represents a standardized putative deal taking place at a defined moment in the day.

Given that, due to logistics, 'spot' oil trading does involve a delay between a deal and the physical transfer of oil, the length of this delay is also a factor. A deal where the transfer is very prompt may involve some element of a distressed cargo, and pricing will reflect this. If, say, there is short-run tightness in the market, oil to be transferred in two weeks will be more valuable than an identical cargo where the physical transfer is six weeks away. In other words, the time structure of prices matters, and standardization involves the specification of the loading time (for fob trades) or the delivery time (for c+f trades) of the cargoes.

The definition of a benchmark specification then provides a way of reducing the multidimensionality of the information being generated by the market. To a large extent, the difficulty of producing an unbiased assessment (be it by price reporters or by a panel) does vary according to what is being assessed. The relationship between ease of assessment and the liquidity of a market is non-linear. A market with a large number of deals is easy to assess. Likewise, in a highly illiquid market (in terms of physical deals done), assessment relies solely on market talk about standardized cargoes. The difficulty of assessment arises in a thin market, where a small number of reported deals done coexist with market talk. It is these circumstances where methodologies and the relative treatment given to talk and deals can become significant.

It should be noted that methodology is not exogenously imposed. Given that there is a constant interface between assessment agencies and their subscribers, in the longer term methodology will be able to evolve, and is thus, at least to a certain extent, endogenous to the market. Although the nature of that interface and the speed of response in reaction to it may often be a matter for debate between market participants, the point remains that the methodologies we describe are not set in stone; they are the result of a more fluid process.

3. Oil Products Markets (1) – Price Reporting

The two major price reporting agencies are the UK company Petroleum Argus, and the US company Platt's, both of whom maintain offices in Singapore. The quotations for oil products in Asia and the Middle East made on a daily basis by the two companies are shown in Table A1.2.[4] These are the quotes made as of 1997, and the range offered tends to gradually change over time. Markets will change, and therefore there will be the need for new quotes or the removal of existing ones; for example, as trade patterns or national product specifications change. Gasoline specifications, for instance, may change by lead content or octane requirement, and the maximum sulphur levels mandated in other products are also in a state of flux.

Table A1.2 lists the physical quotations, and in addition both agencies produce quotes for forward months in the swaps markets considered in Chapter 8. In the physical market, we consider first those products that are assessed free on board (fob) Singapore, that is the price of Singapore's exportable surplus exclusive of transport costs to other locations. While, as noted above, the specifications for any oil product involve a large

---

Table A1.2:   Physical Markets Assessed by Petroleum Argus and Platt's.

Petroleum Argus Asia Pacific Products

	fob Singapore	c+f Japan	fob Middle East	fob Indonesia
Gasolines	Unleaded 95 RON Unleaded 97 RON		Unleaded	
Distillates	Jet/Kerosene Gasoil 0.5%S HP Gasoil 0.5%S LP	Jet/Kerosene Gasoil 0.5%S LP	Kerosene Gasoil	
Fuel Oil	HSFO 180 cst HSFO 380 cst 2%S Fuel Oil Bunker 180 cst Bunker 380 cst	HSFO 180cst	HSFO 180cst HSFO 380 cst Bunker 180cst Bunker 380 cst	LSWR
Other	Naphtha	Naphtha	Naphtha	

Platt's Oilgram Price Report

	fob Singapore	c+f Japan	fob Middle East	c+f South China	c+f Hong Kong
Gasolines	Unleaded 92 RON Unleaded 95 RON Unleaded 97 RON	Unleaded		Leaded 70 RON Leaded 83 RON Leaded 90 RON	
Distillates	Kerosene Gasoil 0.5%S HP Gasoil 0.5%S LP Gasoil 1%S HP	Kerosene Gasoil Cracked Gasoil Pure	Kerosene Gasoil	Jet Kerosene Gasoil 0.5%S LP	Marine Diesel
Fuel Oil	HSFO 180 cst HSFO 380 cst 2%S Fuel Oil LSWR Mixed/Cracked	HSFO 180cst	HSFO 180cst	HSFO 180 cst HSFO 380 cst	HSFO 180 cst HSFO 380 cst
Other	Naphtha MTBE (c+f)	Naphtha	Naphtha		

number of parameters, there are two main potential parameters in gasoline trading.

The first factor is lead content. However, with the phase-out of leaded gasoline in many countries, the bulk of volume of trade in gasoline has switched to unleaded grades. As a result, both Petroleum Argus and Platt's have removed their fob Singapore leaded gasoline quotes, and now focus solely on unleaded. The major distinguishing factor for unleaded gasoline is now its octane content, expressed as a Research Octane Number (RON). The higher the RON, the lower the tendency for engine knocking, i.e. premature ignition due to compression, and hence the higher the quality of the gasoline.

Distillate assessments comprise of kerosene and gasoil. Kerosene is normally assessed for use as jet fuel in commercial aviation, for instance Singapore quotes do not reflect the use of kerosene in illumination or heating. As the product specifications for gasoil vary across countries and uses, a range of gasoil quotes is produced. As Table A1.1 showed, a gasoil specification has many parameters. However, the major distinguishing features are sulphur content and pour point. The most actively traded grade is that whose specifications were shown in Table A1.1, i.e. high pour with 0.5 per cent sulphur content, with quotations being produced for low pour material (defined usually as a pour point of 0 centigrade), and for material of 1 per cent sulphur content. For fuel oil the major distinguishing features are sulphur content and viscosity. The major cargo market is in high sulphur fuel oil (HSFO), with a sulphur content of 3.5 per cent and a viscosity of 180 centistokes, with separate quotations being produced for HSFO of 380 centistokes. There are also assessments for fuel oil with 2 per cent sulphur content, recording the relatively thin market for cargoes that meet the specifications for import into Thailand. As well as its use in power generation, fuel oil has a market as bunker fuel for ships. Petroleum Argus reports bunker fuel quotations separately, representing a higher sulphur level than HSFO of 4 per cent, in both 180 and 380 centistokes viscosity, with the latter being the more active market. In fact the bunker market in Singapore is larger than that for HSFO, with Singapore being the largest bunkering port in the world. There are over 150 companies supplying a market of some 300 thousand b/d.

Low sulphur waxy residue (LSWR) is a form of fuel oil specific to the Asia Pacific market. LSWR is the residue from distillation of crudes with the high wax content which is characteristic of many regional crude oils, and Indonesian in particular. It has two main markets, the major one being its use as a direct burning fuel for power generation, particularly in Japan where it represents a swing fuel at the margin. It can also be traded into the USA, sometimes spiked with gasoil, or used in the region as a vacuum distillation feed for use in cracking. As the major source of LSWR is from Indonesian refineries, Petroleum Argus assesses it fob Indonesia, while Platt's assesses it in Singapore in cracked form. Naphtha, for petrochemical use or as an input to reformers to produce gasoline, is rarely traded fob Singapore, but assessments are produced following the methodology detailed later in this section. Finally (as a c+f Singapore quotation), Platt's assesses the market for the gasoline additive methyl tertiary butyl ether (MTBE).

Beyond assessments for cargoes loading at Singapore, prices are also assessed for other locations. As shown in Table A1.2 there are quotations for delivered cargoes into the Japanese market, where the price reflects not only the value of the cargo, but also the freight charge necessary to transport it to Japan (hence c+f). Platt's also assesses the import markets into South China and Hong Kong. The grades assessed in c+f markets reflect local product requirements, and hence the use of leaded gasoline quotes including extremely low quality 70 RON material. The final market location assessed in the region reflects the export trade in oil products from refineries in the Middle East, i.e. fob prices from the Gulf. Given the lack of a viable spot market in this area, prices tend to be assessed on the basis of the Singapore price minus a freight adjustment using rates for specified tanker types.

Having established the specifications and locations, as noted in the last section there are still three parameters that oil price reporting agencies have to standardize for their quotes for physical markets. The first is the time stamp, i.e. the time of day for which assessments attempt to reflect market values. Petroleum Argus uses a time stamp of 7 p.m. Singapore time, while Platt's uses 6.30 p.m. Singapore time. The second feature is the size of cargo, normally expressed as a range between minimum and maximum sizes, to be assessed for each grade. For instance, typically for fob Singapore quotations, benchmarks are normally for 25 to 100 thousand barrels of gasoline, 100 to 200 thousands of gasoil and kerosene, and 20 to 50 thousand tonnes. The final aspect is the time of delivery of the cargo, which is normally taken as being for loading three to five weeks in the future.

The major difference between Platt's and Petroleum Argus probably lies in the issue highlighted at the end of the previous section, i.e. how to treat the relative information contained in done deals and in market talk. Platt's has tended to give priority to deals, while Petroleum Argus tends to place the emphasis on market talk. As noted above, the importance of this depends non-linearly on the liquidity of the market. On a day when many deals have been reported for a grade, or a day when none have been reported, assessment can rely exclusively on deals and market talk in the respective cases.

The difference may arise when there are, say, just one or two deals reported. The problem can occur because of the potentially high leverage that an individual reported deal might have. For instance, leverage can be high in the market for gasoil, where traders selling small cargoes can also be active in the market for large arbitrage cargoes from Europe (each cargo up to fifteen times larger than those in the Singapore market). In such circumstances, i.e. when the number of deals reported is low, while providing an accurate measure of the (truncated) physically transacted market, the information contained in the reported deal should perhaps be tempered with the consideration that there could be other leveraging factors at work. As noted in the last section, in such cases over time there should be an element of endogeneity in methodology to reduce the degree of any possible distortion that might arise.

A further example of the need for consistency in assessment is shown in the Sinagpore naphtha market, as described in Chapter 8. A methodology that was based on deals when available, and otherwise a netback, resulted in the distortion that was shown in Figure 8.1. Other agencies did not rely on the reported deals, and a large differential opened up between rival naphtha quotes. The endogeneity arose when, following strong consumer resistance, the methodology was swiftly changed to a netback only basis. There is an obvious potential discontinuity involved in assessing on the basis of deals when those deals are extremely rare.

4. Oil Products Markets (2) – Panel Pricing

The basic premise behind panel price reporting is to take a sample of price assessments made by market participants, and then to statistically arrive at the panel index for that day on the basis of the distribution of panellists' responses. While methodologies can differ, the common facet is to use criteria to reject outliers in the distribution, and then to average the remaining inputs. We consider the methodology of the two panel oil pricing systems in Asia in this and the next section.

The Far East Oil Price (FEOP) index, based in Singapore, provides daily assessments for a series of oil products and one crude oil, these grades being shown in Table A1.3, together with the location and standard cargo size used for the quote. Assessments represent physical delivery of cargoes between three and five weeks forward, except for c+f Far East Naphtha. This quote relates to the open specification naphtha contract, and provides a 'near' and a 'far' price, the former being for delivery four half months in advance, and the latter for five half months.

The methodology of the FEOP index is as follows. The time stamp for all assessments is 5.45 p.m. Singapore, and panellists have from this time until 8 p.m. to submit their inputs (aiming to reflect the market at the

Table A1.3: Assessments Made by Far East Oil Price Index.

Product	Location	Cargo Size	Comments
Naphtha	c+f Far East	25 thousand tonnes +/-10%	Open specification format
Naphtha	fob Singapore	70-100 thousand barrels+/-10%	Open specification format
Gasoline	fob Singapore	25-30 thousand barrels	Malaysian leaded specification
Gasoline	c+f Taiwan	250 thousand barrels +/-5%	Taiwan unleaded specification
Jet	fob Singapore	25 thousand tonnes +/-10%	Standard industry specification
Jet	c+f Japan/Korea	25 thousand tonnes +/- 10%	Standard industry specification
Gasoil	fob Singapore	150 thousand barrels +/-10%	0.5% Sulphur high pour
Gasoil	c+f Shanghai	150 thousand barrels +/-10%	0.5% Sulphur low pour
LSWR	fob Indonesia	200 thousand barrels +/-10%	Mixed/cracked quality
HSFO	fob Singapore	20-30 thousand tonnes	180 centistokes, 3.5% Sulphur
HSFO	fob Singapore	20-30 thousand tonnes	380 centistokes, 3.5% Sulphur
Tapis	fob Terengganu	450 thousand barrels +/-10%	Tapis crude oil export quality

time stamp regardless of the actual time of submission). Inputs are made by entering assessments into a computer spreadsheet, and then transmitting this by modem directly to the processor, the entity responsible for compiling and distributing FEOP prices. The processor for the system is Reuters Singapore, with overall coordination and administration for the system provided by Oil Trade Associates, a Singapore based oil consultancy.

After the deadline for submission of inputs, the processor proceeds as follows for each grade assessed. The rejection criterion used is to reject the lowest and highest 25 per cent of inputs (rounded up to the nearest integer), with the remaining inputs being averaged to calculate the FEOP index. Thus, for example, if 18 inputs are received for a given grade, the lowest and highest five are discarded, and the remaining eight averaged. The assessments for the day are then distributed to subscribers by the processor. In addition each panellist receives (confidentially), a record of how their assessment differed from the index, and how it has differed on average over the week, month and year to date. Panellists therefore have feedback on their relative position in the distribution of responses.

The identity of the panellists is in the public domain (although their individual inputs are of course confidential). Table A1.4 provides a cross-tabulation of the 38 FEOP panellists as of 1996 by the grades they assess.

5. Crude Oil Markets

The second panel pricing agency is the Asian Petroleum Price Index (APPI), whose prices, as noted in Chapter 10, have been central in the determination of the Indonesian tax reference price, the ICP. APPI

Table A1.4: Panellists for Far East Oil Price Index by Grades Assessed.

Company	(1)	(2)	(3)	(4)	(5)	(6)	(7)	(8)	(9)	(10)	(11)	(12)
Astra Oil					X	X	X	X		X	X	
BHP Petroleum					X					X		X
BP Singapore		X	X		X		X		X	X	X	
Caltex Trading	X	X	X	X	X	X	X		X	X	X	X
Chevron			X	X	X	X	X		X	X	X	X
Coastal	X	X	X	X	X	X	X	X	X			X
Conoco					X		X			X		X
Daxin Petroleum		X	X				X	X		X		
Elf Trading Asia	X	X	X	X	X	X	X	X		X	X	X
Glencore	X	X	X	X	X		X	X	X	X	X	X
Idemitsu		X			X		X			X		
Indoil		X			X		X		X			X
Itochu Petroleum		X	X	X	X		X			X		
JP Morgan		X			X		X			X	X	X
Libra Petroleum	X	X										
Marubeni		X	X	X	X		X			X	X	X
Mitsui	X		X	X			X	X	X	X	X	
Morgan Stanley			X	X	X	X	X			X	X	X
Neste Oy	X	X	X	X			X	X		X	X	X
Nicor Petroleum		X	X	X			X	X				
Nippon Oil	X	X			X	X	X			X	X	X	X
Petrochemical Corp of Singapore	X	X					X			X	X	
Perta Oil					X		X		X			
Petrodiamond		X			X		X		X	X	X	
Petroleum Authority of Thailand		X					X			X		
Petronas												X
Shell Australia		X	X	X			X		X	X		X
Shell Singapore	X	X	X	X	X	X	X	X	X	X	X	X
Shogun Oil		X	X	X	X	X	X	X		X	X	
Sinochem							X	X		X	X	
Singapore Petroleum Company		X	X	X	X		X			X	X	X
Star Energy		X	X									
Starsupply		X	X	X	X	X	X	X	X	X	X	
Statoil					X		X			X		
Texaco					X		X			X	X	X
Tosco		X	X	X	X	X	X	X	X	X	X	X
Ultramar		X			X		X		X			
Wickland		X	X	X			X			X	X	X	X

KEY : (1) Naphtha (Japan/Korea) (7) Gasoil (Singapore)
   (2) Naphtha (Singapore)   (8) Gasoil (Shanghai)
   (3) Gasoline (Taiwan)   (9) LSWR (Indonesia)
   (4) Gasoline (Singapore) (10) HSFO 180 cst (Singapore)
   (5) Jet (Singapore)   (11) HSFO 380cst (Singapore)
   (6) Jet (Japan/Korea)   (12) Tapis crude oil

quotations began in April 1985 for oil products, but here we will concentrate on their crude oil assessments which started in January 1986.

Quotations are produced from Hong Kong, with the system being administered by the trading and marketing company Seapac Services. The role of processor equating to Reuters in the FEOP system, is taken by the Hong Kong offices of the accountants KPMG Peat Marwick. The index is produced weekly, with assessments being reported to the processor by noon on the Thursday of each week. The identity of the panellists (who currently number about thirty) is not in the public domain.

There are twenty-one grades of crude oil assessed, with the index for the week in each grade being derived by the following methodology. Panellists submit their assessment as a 10 cent spread, for example as $15.30–$15.40. The APPI index is then calculated by first computing the simple average of the low point of the spread and its sample standard deviation.

Panellists' inputs which fall outside of the range of the average plus or minus one standard deviation are discarded, and the remaining inputs are then averaged. To illustrate this process, imagine that there are twelve panellists, submitting inputs in dollars per barrel as follows (arranged in ascending order).

Panellist	Input
1	15.00–15.10
2	15.10–15.20
3	15.15–15.25
4	15.20–15.30
5	15.30–15.40
6	15.30–15.40
7	15.35–15.45
8	15.40–15.50
9	15.55–15.65
10	15.85–15.95
11	15.90–16.00
12	16.30–16.40

The simple average of the low points is $15.45, with a sample standard deviation of 38 cents. The new sample is set as being inputs whose low points lie within the range of one standard deviation either side of the average, i.e. the range from $15.07 to $15.83, and thus the first one and the last three inputs are discarded. The simple average of inputs 2 through to 9 is $15.29, and hence the APPI index will be given as $15.29–$15.39. The APPI midpoint is then given by $15.34.

The benchmark details for APPI crude oil quotations are for fob cargoes with a cargo size for LR1 tankers, assuming payment 30 days after the bill of lading. The timing of delivery for Asian crudes assumes same month delivery for assessments in the first half of a month, and next month for assessments in the second half. For Middle East crudes, the corresponding delivery periods are next month and next but one month for assessments in the first and second half of the month respectively. Thus, for instance, an APPI assessment for Dubai crude made on 20 May will reflect the value of oil for July loading.

Petroleum Argus and Platt's both produce quotes for Asian and Middle East crude oils as part of their overall coverage. Both produce Asian quotes once per trading day, with a timestamp of 6.30 p.m. Singapore time. Platt's methodology uses cargoes loading 20 to 40 days ahead. Petroleum Argus produces quotes for a specified month, with the quotes rolling on the 16th of a month for the Indonesian grades, and at the beginning of the month for the other grades. Standard cargo sizes, which vary by producing country, are used.

For Middle East crudes, Petroleum Argus timestamps at 6.30 p.m. London time, while Platt's uses a window between 3.30 p.m. and 4 p.m. New York time. In addition, Petroleum Argus produces further quotes for the Dubai forward market at noon London time and 5 p.m. Houston time. For both agencies, the months represented in the Middle East quotes roll over on the 16th of the month (or first trading day after). A summary, as of 1996, of the Asian, Australasian and Middle East crude oil grades assessed by the three agencies we have considered in this section is shown in Table A1.5.

In addition to the above absolute price quotes, both Platt's and Petroleum Argus produce prices for the Indonesian crudes as differentials from the Indonesian Crude Price (in line with the usual pricing method used in trading in the spot market). Petroleum Argus also reports the non-Indonesian Asian crudes as a differential from the Tapis APPI price. For the paper Tapis market, Platt's produces quotes for the next two months, and Petroleum Argus for three months plus a quote for swaps in both of the next two quarters. Among Middle East crudes, Petroleum Argus produces four months of Dubai quotes, and also expresses spot prices for the other crudes in terms of differentials from the respective (retroactive) official prices. Platt's reports three months of Dubai prices, and also assesses the Oman price as a differential from the retroactive official price.

6.   The Statistics of Oil Price Perceptions

In the example used in the previous section to illustrate the methodology of APPI prices, using FEOP methodology produces a different outcome.

Table A1.5:   Crude Oil Assessments Made by APPI, Petroleum Argus and Platt's.

		APPI	Argus	Platt's
Australia	Cossack		x	x
	Gippsland	x	x	x
	Griffin			x
	Jabiru	x	x	x
	NW Shelf		x	x
	Thevenard			x
China	Daqing	x		x
	Nanhai			x
	Shengli			x
Indonesia	Arun	x		x
	Ardjuna	x		x
	Attaka	x		x
	Belida	x		x
	Cinta	x	x	x
	Duri	x	x	x
	Handil	x		x
	Lalang	x		x
	Minas	x	x	x
	Widuri	x	x	x
Malaysia	Labuan			x
	Miri			x
	Tapis	x	x	x
	Paper Tapis		x	x
PNG	Kutubu	x	x	x
Vietnam	Bach Ho		x	
Oman	Oman	x	x	x
Qatar	Qatar Land		x	
	Qatar Marine		x	
Saudi Arabia	Arab Light	x		
	Arab Heavy	x		
UAE	Dubai	x	x	x
	Lower Zakum		x	
	Murban	x	x	x

The APPI midpoint in the example came to $15.34, while following the FEOP methodology of discarding the upper and lower 25 per cent produces an average for the remaining inputs of $15.40. The numbers used in the illustration were artificial (and indeed they were very deliberately skewed), but the divergence raises the question as to whether these methodological differences are significant. More generally, if we take inputs also to be representative of the distribution of perceptions that are expressed to a price reporter sampling market talk, analysis of the distribution has a more general interest as a method of market description.

To investigate this we have taken a sample consisting of all inputs for

FEOP during 1994 that assessed the price of gasoil fob Singapore. This represents a total of some 3500 observations. We have first expressed all inputs in terms of their deviation from the average of all inputs for the same day, and then aggregated across all days in 1994. The first property we wish to investigate is the symmetry of the distribution. The reason APPI and FEOP methodology produced different results for the example above was due to a lack of symmetry in the inputs we used, which had a longer tail to the low side. Thus APPI methodology rejected three low inputs and only one high input, while FEOP methodology rejected three of each. The closer the distribution of inputs is to symmetry, the less difference would be expected between the two methodologies.

Our aggregate distribution of oil price perceptions around the average is indeed symmetric. Its median is shown not to be significantly different from zero (using a Mann-Whitney U test), and the distribution is symmetric around this median. Looking at the inputs on a daily basis rather than in aggregate, finds that the hypothesis of symmetry can only be rejected on less than 5 per cent of trading days. However, oil price perceptions are not normally distributed. The distribution has a skewness of -0.54, and kurtosis of 5.33. The Kolmogorov-Smirnoff statistic is 0.076, which means that the hypothesis of normality can be rejected.

It is rare to find an oil price statistic that follows a normal distribution. Oil price changes tend to show leptokurtosis, and in particular they tend to have too many observations in the tails to be normally distributed. If oil prices themselves rarely follow the normal distribution, it is perhaps not too surprising that point perceptions of oil prices also do not. However, in this case the departure from normality is not that the tails are too 'fat', as with oil price changes, for price perceptions the tails are too thin. For example, under a normal distribution 90 per cent of observations will lie within 1.65 standard deviations of the mean. Our gasoil input sample is more concentrated around the mean, with 90 per cent of observations falling within 1.06 standard deviations of the mean. Inputs are more clustered around the daily averages than a normal distribution would suggest.

The above has implications for the number of observations likely to be rejected by the APPI methodology. Using normal distributions the rejection rate would be 31.7 per cent, under the distribution we have estimated for FEOP gasoil it would be expected to be lower at around 27 per cent. There is no *a priori* reason for believing that crude oil price perceptions reported to APPI should follow the same statistical distribution as gasoil price perceptions. However, our evidence tends to suggest that the two distributions are more like one another than they are like the normal distribution. We have information for two full quarters of the total rejection rate from in excess of 5000 inputs across all APPI crude assessments over the course of each quarter. These rates are 27.3 per cent and 29 per cent, less than that which would be produced by a normal distribution, and consistent with a similar to the thin tailed distribution we have found for FEOP gasoil.

As one further check on this process, we have calculated an assessment by using APPI methodology on FEOP gasoil inputs, which produces a total rejection rate of 26.7 per cent. Given that the distribution of perceptions is symmetric, then for gasoil the APPI methodology is akin to using a rejection rate of 13.3 per cent in each tail under the FEOP system.

To summarize, market perceptions of the oil price are symmetric around their mean, and follow a non-normal distribution. In particular, the symmetry property implies that assessments will not be highly sensitive to changes in the proportion of inputs in each tail of the distribution rejected. However, what would happen if there were to be differential reporting rates among panellists? For example assume that each panellist occupies the same relative position on the distribution each day, i.e. some of the panellists always provide lower than average assessments, some average, and some always produce higher than average.

Imagine there are eight panellists in each category, and on a given day they all produce an input. The FEOP methodology will then reject the highest and lowest six, leaving the assessment to be made as the mean of two low, two high and eight average assessments. On the next day, assume that none of those who normally provide higher than average assessments provide an input, leaving the assessment as the average of four low and four average assessments. While an extreme case of differential reporting rates, would it produce a major bias?

With all twenty-four panellists submitting, the FEOP price is the average of the inputs between 25 per cent and 75 per cent of the distribution. On the next day, if panellists always kept the same order along the distribution, it would effectively take the average from 20.8 per cent to 50 per cent (i.e. with sixteen inputs the average from input 5 to input 12 is taken, representing 5/24 and 12/24 along the whole distribution). In fact, such is the concentration of inputs around the average, the difference between these two averages using our calculated distribution of inputs is minor, amounting to just 4.8 cents.

In the above example, the failure of all panellists who usually submit high quotations to provide any input only depressed prices, relative to a case where they had, by less than 5 cents. In other words, large relative shifts in the balance between buyers and sellers reporting, perhaps counter intuitively, do not produce any large swings in the level of the assessed quotation. However, even this limited movement in the example relied on the assumption that panellists do occupy the same position in the relative ordering of inputs. If there was no consistent pattern, i.e. if the fact that a given panellist was higher than average today carried no information about his relative position tomorrow, then there would be no bias at all. Each day the inputs to the panel, (or by analogy the companies a price reporter speaks to) would represent a complete distribution.

We have analysed the inputs to see if there is any systematic bias according to category of panellist. Again perhaps counter intuitively, there would appear to be no significant biases involved. Taking averages across 1994 and categorizing the companies making inputs for gasoil, there are no categories more than 3 cents away from the average. Oil broking firms tended to be higher than the average by 2.6 cents, Singapore refiners higher by 1.1 cents, trading companies were lower by 1.1 cents, and Japanese traders and refiners together with Asian national oil companies were 2.2 cents lower. These differences by categories from the average then amount at maximum to less than 0.2 per cent.

Given this extremely close clustering of assessments around the average, changes in the composition of the panel day by day are likely to have a negligible effect. The further implication is that there would be necessary bias if similar daily shifts occurred in the composition of companies that a price reporter contacted.

7. The Role of Price Assessment

The role of price assessment has received little attention in most of the literature on oil markets. One exception is Roeber,[5] whose views merit some consideration as being representative of a more cynical view of how price assessment works. Roeber sees price reporting, in a definition in which he includes panel pricing, as a statistically invalid anomaly. He sees several problems. The first is that sample base is not random, and it varies day by day. We would contend that this is not a problem; a price reporter will seek to cover the whole market in a balanced way, just as panels seek to have a balanced panel. There is no reason to believe that the changes day by day in the companies a price reporter talks to or in the companies that submit inputs to a panel are anything but random. Roeber puts the variation down to holidays, illness and (rather unfairly and unnecessarily in our opinion) drunkenness, which can certainly be taken to be random processes. Further, as was demonstrated in the previous section, even highly asymmetric changes in the companies providing information need not have anything but a minor impact on the outcome.

The second problem that Roeber sees is the selective revelation and invention of information. A price reporter should not use unconfirmed deals, which takes care of invention of information, and as fabrication of information is almost impossible to conceal for any length of time, liars merely signal their unreliability as a source. For selective revelation to be a problem there has to be some suggestion that it naturally biases prices up or down, and in a market with two sides there can be no such assumption. Further, imagine a net buyer who wishes to keep prices down, and adopts the method of only reporting the lowest priced deals that they are involved in. The price reporter will however be hearing of the higher priced deals from the other parties involved, which will then be unconfirmed. However, they still carry information to the reporter. A succession of unconfirmed deals all involving the same party, and all at prices higher than that party reported, will tend to expose the stratagem fairly swiftly. For panel pricing the equivalent would be consistently submitting off market assessments, which is of course observable and also will merely input being discarded according to the rejection criteria employed. In total, neither selective revelation nor invention are likely to be serious problems.

The final problem Roeber sees is that information is highly variable and takes many forms, and thus requires a subjective decision. This does not apply to panel pricing, and for price reporters it constitutes the job description rather than the problem. Roeber sees the solution to the perceived problem as the increased use of oil futures market prices. This fits in with an evolutionary view of the industry that sees the development of futures as sweeping away the need for both informal markets and for price assessment. It is a strange view, and seems to imply that there are two price discovery methods in the world – futures prices, and the informal markets assessed by price reporters or panels. In reality, neither moves independently of the other. The price discovery function of futures markets is fully reflected in informal market prices, and what is left is the correct placement of prices between products, intra-product specifications and regional locations.

There is no possibility of the futures complex enlarging in such a way as to render the informal markets, and thus price assessment, unnecessary. Nearly twenty years after the launch of the first oil futures in the modern era, there are currently only five successful oil futures contracts in the world.[6] There is no suggestion that there is any process at work leading to an increasing multiplicity of such contracts, the current contracts being launched in 1978, 1983, 1984 (two), and 1987. Given the multiplicity of crude oil and oil product types, specification differences within types and multiple geographical trading and delivery locations, no oil futures complex can capture the heterogeneity of the physical oil trade. The idea that futures prices will in Roeber's evolutionary view be able to remove the need for price assessment is wrong in any oil trading region, but it is most clearly inappropriate within the context of the Singapore market.

We suggest instead that, far from being transient, for a series of reasons demand for price assessment in the Asia-Pacific region is likely to increase very sharply. The liberalization of oil industries across the region brings an increased need to trade. Product specifications, such as sulphur content, are being tightened across the region but at a differential rate. As a result the quality differences within markets are increasing, producing a need for extensions of price assessment. The growth of paper markets and in particular price swaps that are moving further out in time of coverage, increases the need for assessed numbers to settle those swaps against. If futures markets in the region did increase in liquidity, far from reducing the need for price assessment the development would increase it. The presence of extra hedging mechanisms increases the use of risk management, which involves hedging differentials between the heterogenous physical market and homogenous futures contracts. Finally, this study has lead to the conclusion that (despite the play of some strong globalizing factors) the Asia-Pacific market is becoming increasingly regionalized, with several distinct markets arising. This will further increase the desire for price transparency, and the complexity of price assessment.

Notes

1. We aim to give a balanced description of all four services. No endorsement of any single, or group of services over others is intended or should be inferred. Choice of assessment agency is a commercial decision, and, (just as it would be inappropriate in a study of this kind to, say, commend the gasoline of one Singapore refiner over another) the author has no wish to affect this decision. Interviews were made with all four agencies, and every attempt has been made to represent each one in a fair manner.
2. Table A1.1 is drawn from the specifications used by the Far East Oil Price index.
3. *Singapore Oil Report*, July 1995. The leading survey companies in Singapore include Caleb Brett (UK), Det Norske Veritas (Norway), Nippon Kaiji Kentei Kyokai (Japan), Société Générale de Surveillance (Switzerland) and Saybolt (Netherlands).
4. These are the quotes shown in the respective agencies' main Asian product price listings. Both agencies produce other services which often contain quotes relevant to specific grades in the region. For reasons of parsimony we have not attempted to include these.
5. Joe Roeber (1993), *The Evolution of Oil Markets: Trading Instruments and the Role in Oil Price Formation*, Royal Institute for International Affairs, London.
6. The five are light sweet crude oil, heating oil and unleaded gasoline in New York, gasoil in London, and Brent blend crude oil in London and Singapore.

APPENDIX 2

REFINERY TECHNOLOGY

They must meet this wizard; so they went over to the laboratory, which was on a little hill-top away off by itself, so that the inmate might be free to blow himself up as many times as he wanted....There was a chance here to effect the biggest saving in refining history, but the trouble was, the maximum percent of olefins demanded by the simple general equation – and here the chemist began to write on the blackboard $RCH_2 - CH_2 - CH_2R_1 \rightarrow RCH_3 + CH_2 = CH.R_1$ was seldom attained owing to polymerization of the olefins and the formation of naphthenes. After learning which, they went back to the ranch house for a supper of fried chicken.
Upton Sinclair (1927), **Oil! A Novel.**

Figure A2.1 presents a highly abstracted flow diagram for a complex refinery. Crude distillation produces straight run products such as gases, naphtha and distillates. The gases can be produced as liquid petroleum gases (LPGs), i.e. butane and propane, or sent through other processes

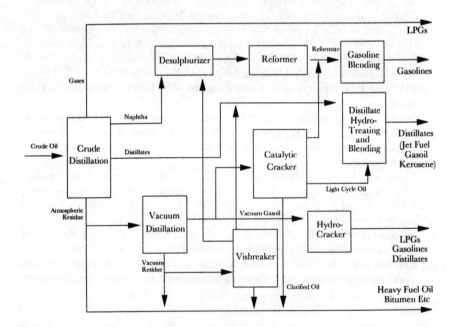

Figure A2.1: Simplified Refinery Flow Diagram.

such as isomerization and alkylation to provide components to improve the quality of finished gasolines.

Naphtha goes through a desulphurizer and then to a reformer producing reformate, which can then go through a blending unit to produce gasolines. Distillates flow through hydrotreating and then blending to produce grades such as jet fuel, kerosene, and gasoil (diesel). Distillation is a process that causes the separation of the hydrocarbon molecules according to their size. The qualities of the straight run products produced can then be improved by desulphurization, hydrotreating and blending. Reforming is a more complex process, which increases the octane of gasoline production through the chemical reconfiguration of naphtha molecules. The more severe processes break down residual fuel oil. In Singapore, for example, there are three such processes currently operative – visbreaking, residue fluid catalytic cracking and hydrocracking.

The first step in all of the processes is to run the residual fuel oil produced by straight distillation through a *vacuum distillation unit*. The use of a vacuum allows higher temperature distillation than that which is possible in a normal distillation unit, and results in two products. The first is vacuum gas oil (VGO). This would produce very low quality final distillate products, but as a feedstock to cracking units, as shown in Figure 6.4. The other product is vacuum residual, a highly viscous form of heavy fuel oil, which can be used as feedstock to thermal processes.

Thermal processes include such units as thermal crackers and cokers. However, the thermal process employed by Singapore refiners is *visbreaking*. A visbreaker helps to make the vacuum residual less viscous. Its primary output is fuel oil, but quantities of low quality gasolines and distillates are also produced which can mixed with the flows from straight run distillation.

While visbreaking is a very mild cracking procedure, *fluid catalytic cracking* (FCC) is a more severe process. This is the mainstay of the US refining industry, common throughout Asia, but is not employed in the Singapore refineries. While the process can be set up to produce high yields of distillates, it is normally seen as a process for increasing the yield of gasoline. The gasoline produced is of high quality, i.e. it has a high octane number, and can be blended without the need for further processing. Light cycle oil is also produced, which can be added to the distillate stream, as well as small quantities of clarified oil which is blended with heavy fuel oil. In the FCC process, heated VGO from vacuum distillation is brought into contact with a chemical catalyst. As a result, large molecules are broken down into smaller, less dense molecules. *Hydrocracking* in essence adds hydrogen to the molecular composition of the feedstock. It can be used to produce gasoline, but the result is poor quality output. Instead, hydrocracking is normally used to produce high quality middle distillates.

Approximate yields as a percentage of the feedstock for the four processes considered above, are shown in Table A2.1, assuming a FCC has been set for maximum gasoline production and a hydrocracker for maximum middle distillates, and that the original crude oil input is Arabian Light.[1]

Table A2.1:   Yields from Refinery Processes. Per Cent.

Process	Feedstock	Output	Yields	
Vacuum Distillation	Residual Fuel Oil	Vacuum Gas Oil	65.5	
		Vacuum Bottoms	34.5	
Visbreaking	Vacuum Bottoms	Gases	2.5	
		Gasoline	5.9	(Octane 40–45)
		Distillates	28.2	
		Fuel Oil	63.4	
Fluid Catalytic Cracking	Vacuum Gas Oil	Gases	17.2	
		Gasoline	48.9	(Octane 91.5)
		Light Cycle Oil	15.7	
		Clarified Oil	8	
		Coke	5.5	
Hydrocracking	Vacuum Gas Oil	LPG	3.6	
		Gasoline	11.5	(Octane 65)
		Distillates	84.6	
		Sulphur and other	2.7	

A more advanced variant of catalytic cracking is *residue fluid catalytic cracking* (RFCC), such as the units in the Shell and Singapore Refining Company (SRC) refineries in Singapore. A RFCC unit can process atmospheric residue from some crude oils, and thus also serves as a fuel oil yield reduction unit as well as a producer of high quality products. Figure A2.2 shows the structure of the RFCC complex at the SRC refinery. The elements below the line (together with one of the crude distillation units) were added in the major expansion and upgrading of the refinery that came on stream over the course of 1995 and 1996.

Note
1. The yields are taken from Lakdasa Wijetilleke and Anthony J. Ody (1984), *World Refinery Industry: Need for Restructuring*, World Bank, Washington.

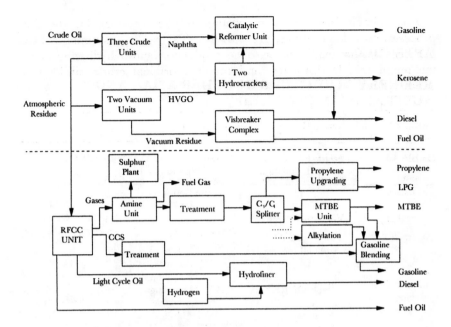

Source: SRC reported in *Oil and Gas Journal*, 14 August 1995.

Figure A2.2: SRC Refinery Structure After Addition of RFCC Complex.

APPENDIX 3

REFINERY STRUCTURE IN OTHER ASIAN AND AUSTRALASIAN COUNTRIES

Table A3.1:   Refineries in Selected Asian and Australasian Countries. Refiner, Location and Crude Distillation Capacity. Thousand b/d.

Country	Refiner	Location	Crude Capacity
Australia	Ampol (NSW)	Sydney	111
	Ampol (Queensland)	Lytton	95
	BP	Bulwer Island	54
	BP	Kwinana	111
	Inland Oil Refiners	Eromanga	1.5
	Mobil	Adelaide	67
	Mobil	Altona	103
	Shell	Clyde	80
	Shell	Geelong	110
Bangladesh	Eastern Refinery	Chittagong	31
Brunei	Shell	Seria	9
Burma	Myanmar Petrochemical	Chauk	6
	Myanmar Petrochemical	Malun	25
	Myanmar Petrochemical	Thanlyin	26
Indonesia	Pertamina	Balikpapan	214
	Pertamina	Cepu	3
	Pertamina	Cilacap	285
	Pertamina	Dumai	115
	Pertamina	Musi	109
	Pertamina	Pangakalan Brandan	5
	Pertamina	Sungai Pakning	48
Malaysia	Esso Malaysia	Port Dickson	71
	Petronas/Conoco	Melaka	100
	Sarawak Shell	Lutong	45
	Shell	Port Dickson	105
New Zealand	New Zealand Refining	Whangarei	95
Pakistan	Attock Refinery	Rawalpindi	30
	National Refinery	Karachi	62
	Pakistan Refinery	Karachi	45
Sri Lanka	CPC	Sapugaskanda	50

Sources: *Oil and Gas Journal* with modifications from various sources.

APPENDIX 4

CRUDE OIL STREAMS IN ASIA AND AUSTRALASIA

Table A4.1: Gravity and Sulphur Content of Major Asian and Australasian Crude Oil Streams.

		API Gravity	*% Sulphur*
Australia	Barrow Island	36.7	0.04
	Challis	39.5	0.07
	Cooper Basin	48.1	0.03
	Cossack	47.3	0.03
	Gippsland	48.0	0.10
	Griffin	54.6	0.01
	Jabiru	41.6	0.04
	Jackson	42.1	0.03
	NW Shelf Condensate	55.0	0.01
	Skua	41.5	0.04
	Talisman	41.1	0.05
	Thevenard	37.6	0.03
	Wandoo	19.3	0.14
Brunei	Brunei Condensate	66.5	0.01
	Brunei Light	40.3	0.06
	Champion	23.7	0.13
	Seria Light	34.6	0.08
China	Daqing	32.7	0.10
	Shengli	24.0	0.90
	Nanhai	39.5	0.05
India	Bombay High	39.4	0.17
Indonesia	Anoa	45.9	0.01
	Ardjuna	35.3	0.11
	Arimbi	35.5	0.17
	Arun	54.3	0.01
	Attaka	42.9	0.04
	Bekapai	39.8	0.07
	Belida	45.1	0.02
	Bontang	50.8	0.03
	Bunyu	33.4	0.07
	Camar	36.0	0.13
	Cinta	31.1	0.09
	Duri	21.5	0.20
	Handil	33.8	0.07
	Kakap	51.8	0.01
	Lalang	39.6	0.05
	Minas	35.2	0.08

Table A4.1: Gravity and Sulphur Content of Major Asian and Australasian Crude Oil Streams. (continued)

		API Gravity	*% Sulphur*
Indonesia (cont.)	Udang	37.6	0.02
	Walio	35.0	0.66
	Widuri	33.2	0.07
Malaysia	Bintulu	26.5	0.13
	Dulang	38.6	0.04
	Labuan	31.5	0.08
	Miri	31.9	0.08
	Tapis	43.0	0.03
New Zealand	Maui Condensate	57.3	0.01
	McKee	38.8	0.07
Pakistan	Badin	44.6	0.04
	Mazari	42.9	0.05
Papua New Guinea	Kutubu	44.0	0.04
Thailand	Erawan Condensate	54.3	0.01
Vietnam	Bach Ho	40.9	0.03
	Dai Hung	30.7	0.09
	Rong	22.1	0.10

Sources: Company Assays, Petroleum Argus, *Petroleum Intelligence Weekly, Oil and Gas Journal* and Platt's.

APPENDIX 5

DESTINATION OF OIL PRODUCT EXPORTS FROM SINGAPORE

Table A5.1: Destination of Gasoil Exports from Singapore. 1986–95. Thousand b/d.

	1986	1987	1988	1989	1990	1991	1992	1993	1994	1995
Hong Kong	20.3	19.0	23.0	36.9	36.3	40.2	39.4	65.1	86.7	103.6
Thailand	31.7	45.8	72.7	77.1	95.5	78.3	61.7	63.4	59.4	47.5
Vietnam	5.0	6.2	7.6	0.9	3.3	8.1	14.4	13.8	27.2	38.8
Malaysia	26.6	22.6	23.1	26.2	28.1	38.1	50.2	44.8	40.8	30.4
Bangladesh	2.5	3.1	3.6	2.4	8.0	10.3	11.4	13.1	18.5	22.4
China	21.4	21.1	23.7	33.0	13.7	11.0	10.5	37.4	41.0	21.1
India	2.5	0.5	1.2	3.5	4.9	13.7	23.2	19.8	23.5	20.3
Australia	12.3	11.4	9.6	10.6	9.1	6.4	6.2	7.5	11.1	13.3
Guam	1.6	3.4	3.1	3.6	5.4	5.0	5.6	5.8	5.9	8.9
Philippines	0.1	3.9	2.3	1.8	3.6	2.8	3.4	17.1	20.7	5.8
Cambodia	0.0	0.0	0.0	0.0	0.0	0.0	0.2	0.4	2.9	4.5
Papua New Guinea	3.1	1.9	1.1	2.4	2.6	1.2	2.5	3.2	4.3	3.6
Burma	0.0	0.0	0.1	0.1	0.9	0.6	1.0	3.0	2.4	1.8
Japan	1.0	6.3	4.3	0.3	1.0	0.6	1.5	0.5	2.3	1.4
USA	4.3	3.5	1.6	0.4	0.5	2.4	2.4	1.9	0.7	0.6
Iran	15.7	0.5	1.9	0.0	0.0	0.6	0.0	0.0	0.0	0.0
Others	15.4	10.6	8.2	8.3	10.1	9.8	11.7	6.4	10.6	6.9
TOTAL	164	159.8	187.0	207.5	223.0	229.2	245.4	303.2	358.0	330.8

Source : Calculations from Singapore Trade Statistics, various years.

Table A5.2:   Destination of Fuel Oil Exports from Singapore. 1986–95. Thousand b/d.

	1986	1987	1988	1989	1990	1991	1992	1993	1994	1995
Hong Kong	34.4	27.0	36.8	34.2	26.5	22.8	30.8	39.4	44.0	32.4
Malaysia	29.0	28.9	31.4	30.9	31.6	41.0	39.5	33.2	33.0	31.1
Thailand	1.9	5.8	5.7	4.3	26.5	22.1	20.6	24.9	14.4	29.2
China	5.6	1.0	2.4	6.2	2.5	5.9	8.4	17.8	29.1	20.6
Vietnam	0.7	2.5	1.2	2.4	5.8	8.1	7.8	3.5	5.1	15.6
USA	23.6	23.5	18.3	16.9	18.4	12.4	17.7	16.8	17.7	12.7
Taiwan	1.4	0.6	2.7	6.9	3.8	0.2	0.7	3.0	4.7	12.6
Australia	9.2	5.3	10.5	7.6	8.1	13.6	19.5	11.1	11.0	10.3
New Caledonia	0.1	0.2	0.0	0.0	0.0	1.0	0.1	2.4	3.8	5.6
Philippines	4.7	6.3	3.6	7.0	0.0	3.9	7.4	1.7	4.8	3.1
Japan	44.4	42.4	35.3	39.7	34.8	17.0	9.8	10.2	9.3	2.1
India	0.0	0.0	0.0	0.0	0.0	0.0	0.0	0.2	1.1	1.9
Guam	1.1	0.0	1.8	3.9	5.9	0.4	0.2	0.6	0.8	1.5
Korea	3.1	2.6	3.7	4.3	4.7	7.8	3.5	2.4	1.4	0.9
Others	6.8	2.6	2.8	2.7	1.6	2.4	3.4	1.6	1.0	4.2
TOTAL	166.0	148.7	156.2	167.0	170.1	158.4	169.3	168.9	181.2	183.8

Source :   Calculations from Singapore Trade Statistics, various years.

Table A5.3:   Destination of Jet Fuel Exports from Singapore. 1986–95. Thousand b/d.

	1986	1987	1988	1989	1990	1991	1992	1993	1994	1995
Hong Kong	14.2	18.2	18.9	21.0	27.8	26.3	32.0	44.3	50.7	53.0
Japan	26.7	45.4	53.2	41.9	53.7	42.9	32.6	26.5	25.7	27.7
Vietnam	0.1	0.0	0.0	1.4	1.4	3.0	3.4	3.9	6.2	10.3
China	0.0	0.0	0.0	0.0	0.0	0.4	3.0	8.9	10.3	9.5
Guam	4.9	5.7	5.7	5.1	7.7	8.3	5.4	5.1	8.8	6.8
Australia	3.1	2.1	2.2	3.6	2.6	2.1	1.3	2.2	3.8	5.3
Taiwan	0.0	0.4	0.2	0.3	0.7	0.0	0.0	0.0	4.6	4.7
Thailand	3.5	6.2	8.7	8.2	8.8	10.5	11.1	10.7	7.4	2.9
Bangladesh	1.1	0.7	0.1	0.6	2.0	1.7	1.5	1.7	2.3	2.6
USA	7.8	9.6	5.8	8.4	3.3	0.9	0.6	3.7	4.4	2.4
Malaysia	3.4	3.8	4.4	3.9	5.5	5.8	8.7	8.9	6.1	1.9
Others	10.1	8.2	6.1	6.7	11.9	20.5	9.6	5.0	2.7	5.4
TOTAL	74.7	100.2	105.4	101.1	125.6	122.5	109.1	121.0	133.0	132.6

Source :   Calculations from Singapore Trade Statistics, various years.

Table A5.4: Destination of Gasoline Exports from Singapore. 1986–95.
Thousand b/d.

	1986	1987	1988	1989	1990	1991	1992	1993	1994	1995
Malaysia	25.5	25.9	27.3	26.5	31.1	32.0	39.9	36.3	51.5	52.7
Taiwan	0.2	0.8	6.2	8.9	24.0	24.5	18.9	9.8	13.6	22.7
Japan	0.2	3.5	5.8	10.8	12.2	14.9	18.4	17.0	12.7	10.7
Thailand	2.9	3.2	4.4	7.1	8.6	7.6	9.5	13.9	12.2	9.3
Hong Kong	0.2	0.1	0.3	1.8	0.7	1.7	3.9	9.5	11.7	9.3
Vietnam	0.0	0.0	0.5	2.8	0.5	0.1	1.3	22.5	10.5	8.4
Korea	8.3	10.1	5.1	10.1	7.5	3.2	4.0	3.2	7.5	8.3
Australia	3.5	2.1	8.5	9.5	3.9	6.1	2.6	1.8	5.4	4.5
China	2.0	2.9	3.1	3.2	3.5	3.6	3.6	3.8	4.3	4.5
Guam	0.0	0.0	0.6	0.8	1.3	1.2	0.9	1.0	3.7	4.3
Philippines	11.1	4.5	3.1	5.2	7.8	13.7	7.7	5.8	5.0	3.1
New Zealand	53.9	53.1	64.9	86.7	101.0	109.0	111.0	125.0	138.0	2.5
India	0.0	0.0	0.0	0.0	0.0	0.0	0.0	0.0	0.0	1.2
Others	11.0	3.7	2.6	4.8	6.0	12.2	7.8	4.1	2.7	4.4
TOTAL	53.9	53.1	64.9	86.7	101.0	108.6	110.5	124.7	138.0	145.9

Source : Calculations from Singapore Trade Statistics, various years.

Table A5.5: Destination of Naphtha Exports from Singapore. 1986–95.
Thousand b/d.

	1986	1987	1988	1989	1990	1991	1992	1993	1994	1995
Japan	38.9	30.1	42.6	46.0	50.5	57.6	55.7	44.5	51.6	60.4
Korea	4.8	7.6	1.6	1.2	6.5	15.2	27.5	49.4	33.5	19.1
Hong Kong	7.3	7.1	6.9	8.8	8.3	9.0	11.2	12.4	12.3	12.7
Malaysia	0.2	0.3	0.4	0.3	0.4	0.3	0.2	0.4	7.8	7.1
Thailand	0.2	0.2	1.0	0.3	0.7	0.5	0.3	0.4	0.8	1.7
Taiwan	7.0	5.7	2.0	0.4	0.4	0.6	0.6	0.6	1.7	1.6
China	0.0	0.2	0.0	3.0	0.5	0.0	0.0	4.0	1.9	0.6
Others	3.0	3.3	1.1	1.7	0.7	1.2	0.3	1.3	1.1	2.3
TOTAL	61.5	54.5	55.6	61.8	68.1	84.4	95.8	113.1	110.7	105.5

Source : Calculations from Singapore Trade Statistics, various years.

INDEX